Exploring
Anatomy&Physiology
in the Laboratory
Core Concepts
SECOND EDITION

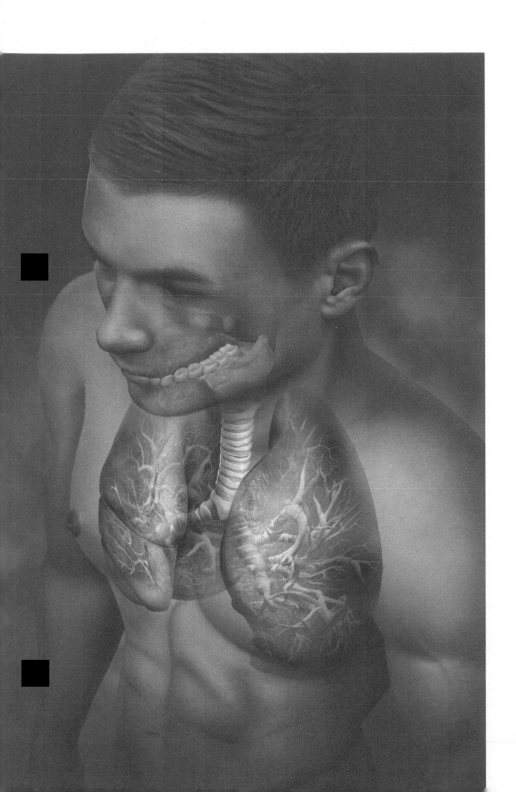

Erin C. Amerman

MORTON
PUBLISHING

925 W. Kenyon Avenue, Unit 12
Englewood, CO 80110

morton-pub.com

Book Team

President and CEO	David M. Ferguson
Senior Acquisitions Editor	Marta R. Pentecost
Developmental Editor	Sarah D. Thomas
Editorial Project Managers	Trina Lambert, Rayna S. Bailey
Production Manager	Will Kelley
Production Assistants	Joanne Saliger, Amy Heeter
Cover and Illustrations	Imagineering Media Services, Inc.

Dedication

For Doug Morton, whose support of my vision, commitment to this book, and belief in the value of higher education to transform lives will not be forgotten.

For Elise, who performs amazing feats with the human body.

Printed in the United States of America

10 9 8 7 6 5 4 3 2

ISBN-10: 1-61731-780-2

ISBN-13: 978-1-61731-780-4

Library of Congress Control Number: 2017962086

Preface

I first started writing lab procedures for my students 15 years ago in response to frustration my students and I were feeling with the anatomy and physiology lab and the accompanying lab manual. My students wanted—and needed—focused activities, clear objectives and explanations, and exercises that enabled them to appreciate the often-missed "big picture" of A&P. None of the available lab manuals provided this, so I was left with no choice but to write my own lab procedures. The results of my efforts were well worth it, as my students were engaged and active the entire lab period. Their lab and class grades improved.

In 2003, I met with David Ferguson, now the president of Morton Publishing Company, and he offered me a dream opportunity: to share my exercises with students and instructors throughout the country. The result of that meeting was the text *Exercises for the Anatomy and Physiology Laboratory*, a simple black-and-white manual with focused activities. My goal with this book was to solve the teaching problems in the A&P lab and to enhance the experience for my colleagues and our students.

Exercises was enthusiastically received, suggesting that we had indeed provided instructors and students something that they had been needing. We were so encouraged that we set out to produce an expanded, full-color version of the exercises that included more explanations, new activities, a complete art program, and new pedagogy. This book became *Exploring Anatomy and Physiology in the Laboratory*, or *EAPL*. Like its predecessor, it was warmly received.

EAPL was originally intended for both one- and two-semester anatomy and physiology courses. However, we received feedback that there was simply too much content in *EAPL* for the one-semester course. I have taught one-semester anatomy and physiology, and I agreed—the book was too big. So, we put together *EAPL: Core Concepts* (*EAPL CC*), which retained all the enhanced features of *EAPL*, but was streamlined for use in one-semester courses.

With the first edition of *EAPL CC*, we consolidated the anatomy and physiology chapters into one chapter on each organ system. In addition, the number of key terms for students to learn was reduced significantly, and we focused more on the foundational principles of anatomy and physiology instead of details. This proved to be a good recipe for a one-semester A&P lab manual.

However, like all first editions, there was room for improvement. With your feedback, we aimed to make those improvements with this second edition, which we present to you now. As you peruse this book, please notice the following changes, which were a direct result of your helpful suggestions:

- **Added text narrative to make *EAPL CC* 2e a self-contained lab manual.** One of the most frequent requests we received was to make *EAPL CC* more self-contained so that students didn't need an additional textbook to complete the activities. We heard your requests, and responded. Every effort was made in the revisions for this edition to ensure that the text narrative defines and explains all key terms.

- **Expanded and improved art program.** In line with our goal of making *EAPL CC* a self-contained lab manual, we further expanded the art program with nearly 240 new and improved figures. This ensures that all key structures are clearly shown, often from multiple views. We are also excited to be able to add select photos of anatomical models for the first time.

- **New quiz questions**. Even the best quiz questions can grow stale after a while. With that in mind, the end-of-unit quiz questions have been updated, and approximately 50 to 70 percent of the questions have been altered or replaced in each unit.

- **Fine-tuned activities**. Many of the procedures or activities were altered to make them more time-efficient. In addition, certain exercises that just weren't working were cut. These were replaced with other, better exercises. For example, we added drawing activities in many units, as research has shown that students retain information better when they draw the structures they are studying.

- **Improved student accessibility**. We want to maximize this book's accessibility for our students so they have the best possible lab experience. To that end, we reorganized the model inventories so they match the order in which the terms are presented in the text. In addition, we ensured that nearly all terms labeled on figures are discussed in the text (any time extra structures are labeled, students are alerted to these extra reference labels). Both changes are to save students the frustration of digging through the unit to find structures and terms. Finally, we also added four new Hints & Tips boxes to help students better navigate the lab.

We hope that you enjoy the second edition of *EAPL: Core Concepts*. Please share your thoughts with us about the book so that we can continue to improve it for our students.

—Erin C. Amerman

About the Author

Erin C. Amerman has been involved in anatomy and physiology education for more than 17 years as an author and professor, currently at Florida State College at Jacksonville in Jacksonville, Florida. She received a B.S. in Cellular and Molecular Biology from the University of West Florida and a Doctorate in Podiatric Medicine from Des Moines University. She is the author of six textbooks on the subject of anatomy and physiology, four of which are with Morton Publishing Company.

Acknowledgments

Textbooks are an enormous undertaking. Many people were integral to the production and development of this edition, and I would like to take this brief opportunity to express my gratitude.

First and foremost I would like to thank my family and friends, particularly my daughter Elise, my mother Cathy, and my husband Chris. Without your unwavering support and patience, none of my work would be possible. I'd also like to thank Dr. Lourdes Norman-McKay, whose advice, wisdom, and friendship helps to keep me (mostly) sane. Lastly, I can't forget my animals: my dogs, who drop toys and bones in the middle of my laptop, and my cats, who always manage to do exactly the least helpful thing possible.

Next I would like to extend my gratitude to the talented book team with whom I was fortunate enough to work: Joanne Saliger, who expertly designed the book as she always does; Will Kelley, who typeset the text; Trina Lambert, who skillfully copyedited the manuscript; Sarah Thomas, who coordinated everything as the developmental editor; Marta Pentecost, who oversaw the project as the acquisitions editor in Portuguese, French, and English; Carolyn Acheson, who provided the index; the team at Imagineering, who provided the beautiful illustrations; and John Crawley, Michael Leboffe, and Justin Moore, who allowed me to use several of their excellent photos and photomicrographs. I truly appreciate all of your hard work and generosity.

I would also like to thank the following reviewers for their invaluable suggestions for this and other titles in the *EAPL* series that helped to improve this edition:

- Diana M. Coffman, Lincoln Land Community College
- Angela Corbin, Nicholls State University
- Dr. Cassy Cozine, University of Saint Mary
- Molli Crenshaw, Texas Christian University
- Kathryn A. Durham, R.N., Ph.D., Lorain County Community College
- Jill E. Feinstein, Richland Community College
- Nancy E. Fitzgerald, M.D., Alvin Community College
- Carrie L. Geisbauer, Moorpark College
- Carol Haspel, Ph.D., LaGuardia Community College
- Stephanie Ann Havemann, Ph.D., Alvin Community College
- Elizabeth Hodgson, York College of Pennsylvania
- Steven Leadon, Durham Technical Community College
- Eddie Lunsford, Southwestern Community College
- Dr. Shawn Macauley, Muskegon Community College
- Darren Mattone, Muskegon Community College
- John David Matula, Alvin Community College
- Justin Moore, American River College
- Tommy D. Morgan, Alvin Community College
- Dr. Anita Naravane, St. Petersburg College
- Michele Robichaux, Nicholls State University
- Deanne Roopnarine, Nova Southeastern University
- Amy Fenech Sandy, Columbus Technical College in Columbus, GA
- Lori Smith, American River College
- Valory Thatcher, Mt. Hood Community College
- Dr. Donald R. Tredway, Tulsa Community College
- Cathy Whiting, Gainesville State College

The acknowledgments would be incomplete without thanking Doug Morton, to whom I will be eternally grateful for adding me to the Morton family. And finally, I extend a special thank you to President David Ferguson for his support, patience, friendship, Broncos games, and willingness to go hiking with me to look for snakes even if he is unwilling to actually touch a snake himself.

Be Prepared

Objectives set learning goals to prepare students for what they are expected to know after completing the lab. The numbered objectives also aid in the review of material.

Pre-Lab Exercises encourage students to actively prepare for the lab by defining key terms, doing labeling and coloring exercises to learn anatomical structures, and reviewing vital material from previous units, saving instructors from having to spend extra time reviewing material from the lecture. These exercises have been updated with new terms and new figures and can be completed using information available in the lab manual. By asking students to draw their own leader lines and write out definitions, the pre-lab exercises are designed to help students retain information and build a deeper understanding of the content.

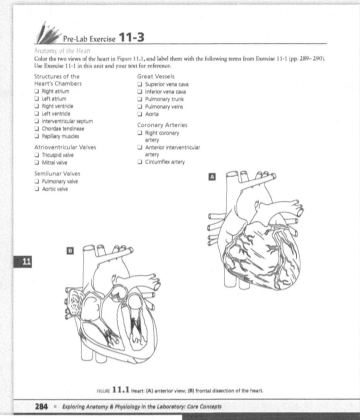

Pre-Lab Exercise **11-3**

Anatomy of the Heart

Color the two views of the heart in Figure 11.1, and label them with the following terms from Exercise 11-1 (pp. 289– 290). Use Exercise 11-1 in this unit and your text for reference.

Structures of the Heart's Chambers
- ❑ Right atrium
- ❑ Left atrium
- ❑ Right ventricle
- ❑ Left ventricle
- ❑ Interventricular septum
- ❑ Chordae tendineae
- ❑ Papillary muscles

Atrioventricular Valves
- ❑ Tricuspid valve
- ❑ Mitral valve

Semilunar Valves
- ❑ Pulmonary valve
- ❑ Aortic valve

Great Vessels
- ❑ Superior vena cava
- ❑ Inferior vena cava
- ❑ Pulmonary trunk
- ❑ Pulmonary veins
- ❑ Aorta

Coronary Arteries
- ❑ Right coronary artery
- ❑ Anterior interventricular artery
- ❑ Circumflex artery

11

FIGURE **11.1** Heart: (A) anterior view; (B) frontal dissection of the heart.

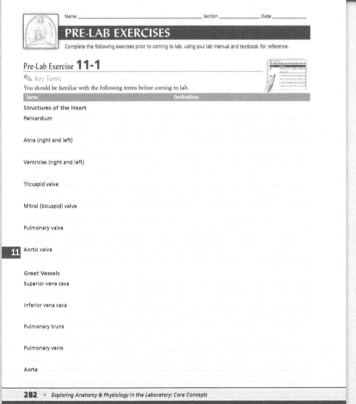

Name _____ Section _____ Date _____

PRE-LAB EXERCISES

Complete the following exercises prior to coming to lab, using your lab manual and textbook for reference.

Pre-Lab Exercise **11-1**

Key Terms

You should be familiar with the following terms before coming to lab.

Term	Definition
Structures of the Heart	
Pericardium	
Atria (right and left)	
Ventricles (right and left)	
Tricuspid valve	
Mitral (bicuspid) valve	
Pulmonary valve	
Aortic valve	
Great Vessels	
Superior vena cava	
Inferior vena cava	
Pulmonary trunk	
Pulmonary veins	
Aorta	

11

Be Organized

Model Inventories provide organized and easily referenced lists of anatomical structures students are responsible for identifying. These lists help students catalog the specimens they see in the lab. The emphasis on examination, description, pronunciation, and writing the names of anatomical structures encourages students to be actively involved in the learning process and allows them to better retain the material.

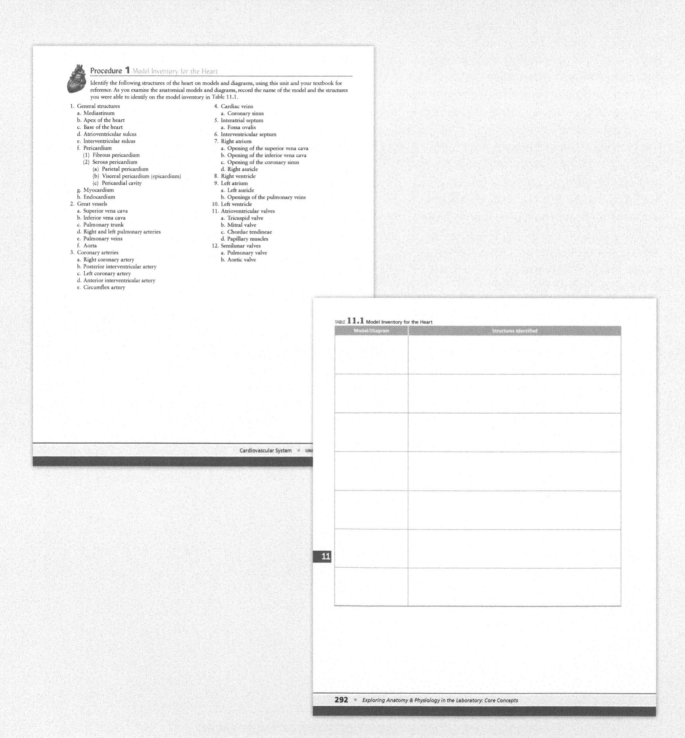

Be Focused

Illustrations and **Photographs** in *Exploring Anatomy & Physiology in the Laboratory: Core Concepts*, 2e, were specifically designed to improve student understanding of important concepts and procedural instructions in the laboratory setting. (Many lab manuals simply reproduce artwork and exposition from the related textbook, unnecessarily increasing the redundancy, size, and price of the manual.)

With more than 200 new and revised illustrations and photographs, this edition offers a more detailed and realistic view of human anatomy. Many of the histology images are new, taken from the best sources of commercially available slides. Images of models have been added in this edition.

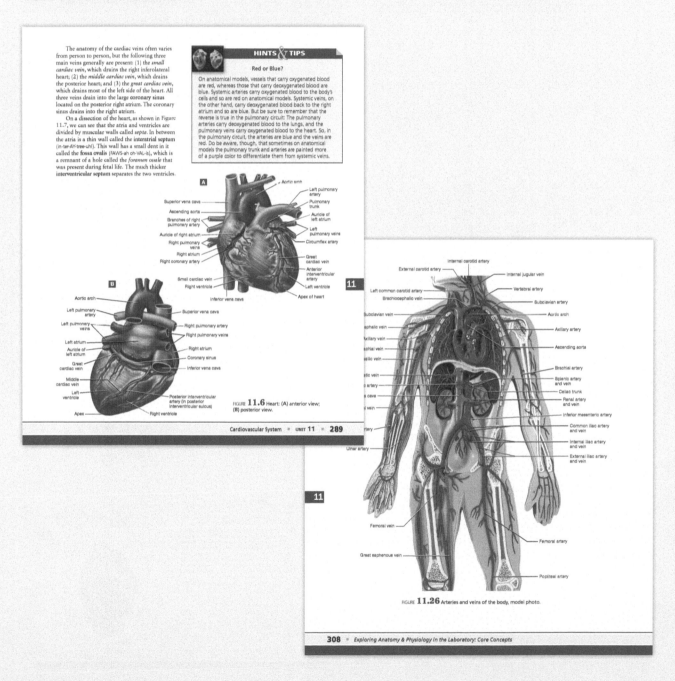

Be Active

Focused Activities are the guiding philosophy of this lab manual. Students learn best when they are actively engaged in the laboratory. In this manual, students are asked to be active by describing, labeling, writing, coloring, and drawing. 13 new activities have been added to this edition, including a starch solubility exercise, special senses activities, and many more.

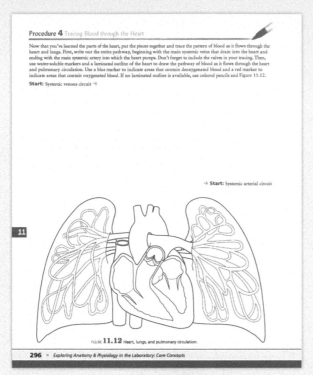

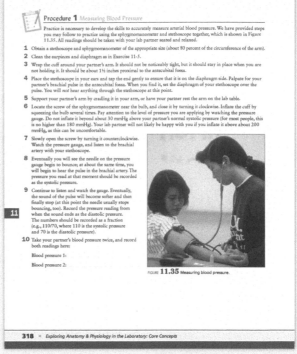

Tracing Exercises ask students to write step-by-step, turn-by-turn directions to follow substances (blood cells, food molecules, waste by-products, electrical events) through the human body, then trace the substances' path on a "map" of the body. These exercises allow students to see the big picture of how the body systems interact and to understand the relationship between structure and function.

Hints & Tips sidebars appear throughout the book to help students navigate some of the more difficult topics in A&P.

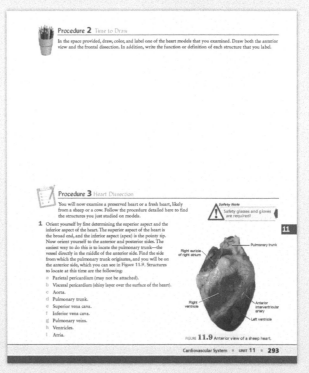

Be Sure

Unit Quiz sections completing each unit ensure that students understand key concepts and achieve the learning objectives, signaling that they are ready to move on. This new edition features new assessment questions throughout. These quizzes consist of labeling, fill-in-the-blank, multiple choice, and sequencing questions that test the students' ability to retain the material they completed in the lab. Critical thinking questions, frequently related to clinical scenarios, check students' deeper understanding by challenging them to provide answers they cannot find verbatim in either the lab manual or the textbook. These sheets can be used as graded lab quizzes and/or to check attendance in the lab.

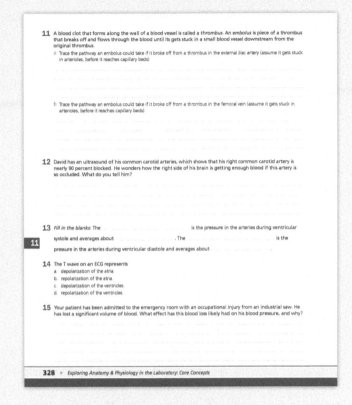

11 A blood clot that forms along the wall of a blood vessel is called a *thrombus*. An *embolus* is piece of a thrombus that breaks off and flows through the blood until its gets stuck in a small blood vessel downstream from the original thrombus.

a Trace the pathway an embolus could take if it broke off from a thrombus in the external iliac artery (assume it gets stuck in arterioles, before it reaches capillary beds):

b Trace the pathway an embolus could take if it broke off from a thrombus in the femoral vein (assume it gets stuck in arterioles, before it reaches capillary beds):

12 David has an ultrasound of his common carotid arteries, which shows that his right common carotid artery is nearly 90 percent blocked. He wonders how the right side of his brain is getting enough blood if this artery is so occluded. What do you tell him?

13 *Fill in the blanks:* The _____ is the pressure in the arteries during ventricular systole and averages about _____ . The _____ is the pressure in the arteries during ventricular diastole and averages about _____

14 The T wave on an ECG represents
a. depolarization of the atria.
b. repolarization of the atria.
c. depolarization of the ventricles.
d. repolarization of the ventricles.

15 Your patient has been admitted to the emergency room with an occupational injury from an industrial saw. He has lost a significant volume of blood. What effect has this blood loss likely had on his blood pressure, and why?

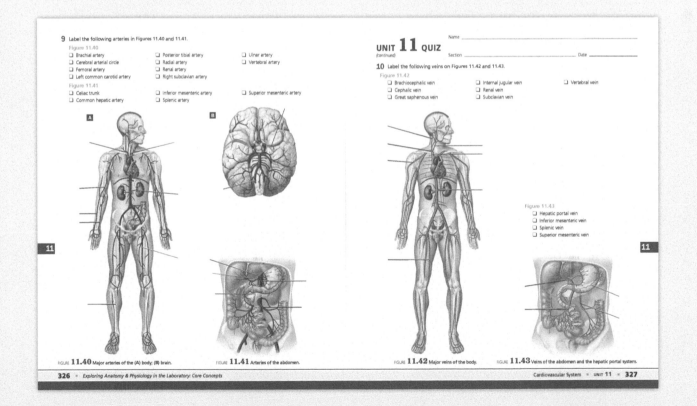

9 Label the following arteries in Figures 11.40 and 11.41.

Figure 11.40
❏ Brachial artery
❏ Cerebral arterial circle
❏ Femoral artery
❏ Left common carotid artery

❏ Posterior tibial artery
❏ Radial artery
❏ Renal artery
❏ Right subclavian artery

❏ Ulnar artery
❏ Vertebral artery

Figure 11.41
❏ Celiac trunk
❏ Common hepatic artery

❏ Inferior mesenteric artery
❏ Splenic artery

❏ Superior mesenteric artery

FIGURE **11.40** Major arteries of the (A) body; (B) brain.

FIGURE **11.41** Arteries of the abdomen.

Name _____

UNIT 11 QUIZ
(continued)

Section _____   Date _____

10 Label the following veins on Figures 11.42 and 11.43.

Figure 11.42
❏ Brachiocephalic vein
❏ Cephalic vein
❏ Great saphenous vein

❏ Internal jugular vein
❏ Renal vein
❏ Subclavian vein

❏ Vertebral vein

Figure 11.43
❏ Hepatic portal vein
❏ Inferior mesenteric vein
❏ Splenic vein
❏ Superior mesenteric vein

FIGURE **11.42** Major veins of the body.

FIGURE **11.43** Veins of the abdomen and the hepatic portal system.

Be Aware

Textbooks are expensive, and the last thing students need is to spend too much for a lab manual. Morton Publishing is committed to providing high-quality products at reasonable prices.

It is our sincere hope that *Exploring Anatomy & Physiology in the Laboratory: Core Concepts*, 2e, will provide you the tools necessary for a productive and interesting laboratory experience. We welcome all comments and suggestions for future editions of this book. Please feel free to contact us at eapl@morton-pub.com or visit us at www.morton-pub.com.

Be Choosy

MortyPak Options

Bundle *Exploring Anatomy & Physiology in the Laboratory: Core Concepts*, 2e, with one or more of the following supplemental titles:

- *A Visual Analogy Guide to Human Anatomy*
- *A Visual Analogy Guide to Human Physiology*
- *A Visual Analogy Guide to Human Anatomy and Physiology*
- *A Visual Analogy Guide to Chemistry*
- *A Photographic Atlas of Histology*
- *A Photographic Atlas for the A&P Laboratory*
- *A Dissection Guide and Atlas to the Rat*
- *A Dissection Guide and Atlas to the Fetal Pig*
- *A Dissection Guide and Atlas to the Mink*
- *Mammalian Anatomy: The Cat*
- *An Illustrated Atlas of the Skeletal Muscles*

Morton CustomLab

In an effort to lower the prices of books, and to provide instructors and students books tailored to their needs, we offer the enhanced Morton CustomLab program. With CustomLab, instructors can remove or combine material from our existing lab manuals or photographic atlases, or may use photographs and illustrations from our extensive online library, to create their own, personalized lab manual. Enrollment minimums apply.

E-books

As higher education continues to evolve and incorporate technology, we are pleased to offer our titles as e-books for students who prefer an e-format.

Pronunciation Guide

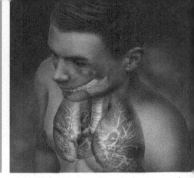

Note: Accented syllables are capitalized; for example, "a-NAT-oh-mee and fiz-ee-AHL-oh-jee."

abdominal (ab-DAH-mih-nuhl)

abdominopelvic (ab-dom-ih-noh-PEL-vik)

abducens (ab-DOO-senz)

acetabulum (aeh-seh-TAB-yoo-lum)

acromial (ah-KROH-mee-uhl)

acromioclavicular (ah-KROH-mee-oh-klah-VIK-yoo-lur)

acromion (ah-KROH-mee-ahn)

acrosomal (ak-roh-ZOH-muhl)

adenohypophysis (ad-in-oh-hy-PAWF-ih-sis)

adipocytes (AD-ih-poh-syt'z)

adrenal (uh-DREE-nuhl)

adrenocorticotropic (ah-dree-noh-kohr-tih-koh-TROH-pik)

agglutination (uh-gloo-tin-AY-shun)

agranulocytes (AY-gran-yoo-loh-syt'z)

ala (AY-luh)

aldosterone (al-DAHS-tur-ohn)

alveolar (al-vee-OH-lahr)

alveoli (al-vee-OH-lye)

amnion (AM-nee-ahn)

amylase (AM-uh-layz)

antebrachial (an-tee-BRAY-kee-uhl)

antecubital (an-tee-KYOO-bih-tuhl)

anterior (an-TEER-ee-ur)

antidiuretic (an-tee-dy-yoo-RET-ik)

aorta (ay-OHR-tah)

aponeurosis (ap-oh-noo-ROH-sis)

arachnoid mater (ah-RAK-noyd MAH-tur)

arcuate (ARK-yoo-it)

areola (aehr-ee-OH-lah)

arrector pili (ah-REK-tohr PIL-aye)

arteriosus (ahr-TEER-ee-oh-suss)

atrioventricular (ay-tree-oh-ven-TRIK-yoo-lur)

atrium (AY-tree-um)

auricle (OHR-ih-kuhl)

auscultation (aws-kuhl-TAY-shun)

axillary (AX-il-ehr-ee)

axolemma (aks-oh-LEM-uh)

axon (AX-ahn)

azygos (ay-ZY-gus)

baroreceptor (BEHR-oh-reh-sep-ter)

basal lamina (BAY-zul LAM-in-uh)

basale (bay-SAY-lee)

basilar (BAY-zih-lur)

basophils (BAY-zoh-filz)

biceps brachii (BY-seps BRAY-kee-aye)

blastocyst (BLAST-oh-sist)

brachial (BRAY-kee-uhl)

brachialis (bray-kee-AL-iss)

brachiocephalic (bray-kee-oh-seh-FAL-ik)

brachioradialis (bray-kee-oh-ray-dee-AL-iss)

bradycardia (bray-dih-KAR-dee-uh)

bronchi (BRONG-kye)

bronchial (BRONG-kee-uhl)

bronchioles (BRONG-kee-ohlz)

bronchomediastinal (brongk-oh-mee-dee-ah-STYN-uhl)

buccal (BYOO-kuhl)

buccinator (BUK-sin-ay-tur)

bulbourethral (bul-boh-yoo-REETH-ruhl)

bursae (BURR-see)

calcaneal (kal-KAY-nee-uhl)

calcaneus (kal-KAYN-ee-us)

calcitonin (kal-sih-TOH-nin)

calvaria (kal-VEHR-ee-uh)

calyces (KAY-lih-seez)

canaliculi (kan-ah-LIK-yoo-lee)

canines (KAY-nynz)

capitulum (kah-PIT-yoo-lum)

carina (kuh-RY-nuh)

carotid (kuh-RAWT-id)

carpal (KAR-puhl)

cauda equina (KOW-dah eh-KWY-nah)

cecum (SEE-kum)

celiac (SEE-lee-ak)

centrioles (SEN-tree-ohlz)

centromere (SIN-troh-meer)

centrosome (SEN-troh-sohm)

cephalic (sef-AL-ik)

cerebellum (sehr-eh-BELL-um)

cerebrospinal (seh-ree-broh-SPY-nuhl)

cerebrum (seh-REE-brum)

cervical (SUR-vih-kuhl)

chiasma (ky-AZ-mah)

chiasmata (ky-az-MAH-tah)

chondrocytes (KAHN-droh-syt'z)

chordae tendineae (KOHR-dee tin-din-EE-ee)

chorion (KOHR-ee-ahn)

chorionic villi (KOHR-ee-ahn-ik VILL-aye)

choroid (KOHR-oyd)

chromatin (KROH-mah-tin)

chromosomes (KROH-moh-sohmz)

chylomicrons (ky-loh-MY-krahnz)

chyme (KY'M)

cilia (SIL-ee-uh)

ciliary body (SILL-ee-ehr-ee)

circumflex (SIR-kum-flex)

cisterna chyli (sis-TUR-nuh KY-lee)

cisternae (sis-TER-nee)

clitoris (KLIT-uhr-iss)

coccyx (KAHX-iks)

cochlea (KOHK-lee-ah)

colloid (KAWL-oyd)

conceptus (kun-SEPT-uhs)

conchae (KAHN-kee)

conjunctiva (kon-junk-TY-vah)

conus medullaris (KOHN-us med-yoo-LEHR-us)

coracoid (KOHR-ah-koyd)

cornea (KOHR-nee-ah)

coronal (koh-ROH-nuhl)

coronary (KOHR-oh-nehr-ee)

corpora cavernosa (kohr-POHR-uh kah-ver-NOH-suh)

corpus callosum (KOHR-pus kal-OH-sum)

corpus luteum (KOHR-pus LOO-tee-um)

corpus spongiosum (KOHR-pus spun-jee-OH-sum)

cranial (KRAY-nee-uhl)

cremaster (kreh-MASS-ter)

cricoid (KRY-koyd)

cricothyroid (kry-koh-THY-royd)

crista galli (KRIS-tah GAL-ee)

cruciate (KROO-shee-iht)

crural (KROO-ruhl)

cubital (KYOO-bit-uhl)

cystic (SIS-tik)

cytokinesis (sy-toh-kin-EE-sis)

cytoplasm (SY-toh-plaz-m)

cytosol (SY-toh-sahl)

dendrites (DEN-dryt'z)

detrusor (dee-TROO-sur)

diaphragm (DY-uh-fram)

diaphysis (dy-AEH-fih-sis)

diastole (dy-AEH-stoh-lee)

diencephalon (dy-en-SEF-ah-lahn)

digital (DIJ-it-uhl)

diploid (DIH-ployd)

dorsalis pedis (dohr-SAL-iss PEE-diss)

ductus deferens (DUK-tuss DEF-er-ahnz)

duodenum (doo-AW-den-um)

dura mater (DOO-rah MAH-tur)

embryogenesis (em-bree-oh-JEN-ih-sis)

endolymph (EN-doh-limf)

endometrium (en-doh-MEE-tree-um)

endomysium (en-doh-MY-see-um)

endoplasmic reticulum (en-doh-PLAZ-mik reh-TIK-yoo-lum)

endosteum (en-DAH-stee-um)

eosinophils (ee-oh-SIN-oh-filz)

ependymal (eh-PEN-dih-muhl)

epicardium (ep-ih-KAR-dee-um)

epicranius (eh-pih-KRAY-nee-uhs)

epididymis (ep-ih-DID-ih-miss)

epidural (ep-ih-DOO-ruhl)

epiglottis (ep-ih-GLAW-tiss)

epimysium (ep-ih-MY-see-um)

epiphyseal (eh-PIF-ih-seel)

epiphysis (eh-PIF-ih-seez)

epithalamus (ep-ih-THAL-ih-mus)

epithelial (ep-ih-THEE-lee-uhl)

eponychium (ep-oh-NIK-ee-um)

erector spinae (eh-REK-tohr SPY-nee)

erythrocytes (eh-RITH-roh-syt'z)

esophagus (eh-SOF-ah-gus)

fascia (FASH-uh)

fascicles (FASS-ih-kullz)

femoral (FEM-oh-ruhl)

fibroblasts (FY-broh-blastz)

fibula (FIB-yoo-lah)

fibularis (fib-yoo-LEHR-iss)

fimbriae (FIM-bree-ay)

flagella (flah-JEL-uh)

fossa ovalis (FAWS-ah oh-VAL-is)

fovea centralis (FOH-vee-uh sin-TRAL-iss)

frontal (FRUHN-tuhl)

frontalis (frun-TAL-iss)

gametes (GAM-eetz)

gametogenesis (gah-meet-oh-JEN-uh-sis)

gastrocnemius (gas-trawk-NEE-mee-uhs)

gingivae (JIN-jih-vay)

glomerulus (gloh-MEHR-yoo-luhs)

glossopharyngeal (glah-soh-fehr-IN-jee-uhl)

glucagon (GLOO-kah-gawn)

gluteal (GLOO-tee-uhl)

Golgi (GOHL-jee)

gonads (GOH-nadz)

gracilis (gruh-SILL-iss)

granulocytes (GRAN-yoo-loh-syt'z)

granulosum (gran-yoo-LOH-sum)

gyri (JY-ree)

haploid (HAP-loyd)

hematocrit (heh-MAEH-toh-krit)

hemoglobin (HEE-moh-glohb-in)

hemolysis (heem-AW-lih-sis)

hepatopancreatic ampulla (heh-PAEH-toh-payn-kree-at-ik am-POOL-ah)

hilum (HY-lum)

humerus (HYOO-mur-us)

hyaline (HY-ah-lin)

hypophyseal (hy-PAW-fih-see-uhl)

hypothalamus (hy-poh-THAL-uh-muss)

ileocecal (ill-ee-oh-SEE-kuhl)

ileum (ILL-ee-um)

iliac (ILL-ee-ak)

iliacus (ill-ee-AK-uhs)

iliocostalis (ill-ee-oh-kawst-AL-iss)

iliopsoas (ill-ee-oh-SOH-uhs)

ilium (ILL-ee-um)

incus (ING-kus)

infundibulum (in-fun-DIB-yoo-lum)

inguinal (IN-gwin-uhl)

integument (in-TEG-yoo-ment)

integumentary (in-TEG-yoo-MEN-tuh-ree)

interatrial (in-ter-AY-tree-uhl)

intercalated (in-TUR-kuh-layt-ed)

intraperitoneal (in-trah-pehr-ih-toh-NEE-uhl)

ischium (ISS-kee-um)

jejunum (jeh-JOO-num)

jugular (JUG-yoo-lur)

keratin (KEHR-ah-tin)

keratinocyte (kehr-ah-TIN-oh-sy't)

Korotkoff (koh-RAWT-koff)

labia (LAY-bee-ah)

lacrimal (LAK-rih-muhl)

lacteal (lak-TEEL)

lacunae (lah-KOO-nee)

lambdoid (LAM-doyd)

lamellae (lah-MELL-ee)

laryngopharynx (lah-ring-oh-FEHR-inks)

larynx (LEHR-inks)

latissimus dorsi (lah-TISS-ih-muss DOHR-sye)

leukocytes (LOO-koh-syt'z)

lingual (LING-yoo-uhl)

lipase (LY-payz)

longissimus (lawn-JISS-ih-muss)

lucidum (LOO-sid-um)

lumbar (LUHM-bahr)

lunula (LOON-yoo-luh)

lymph (LIMF)

lymphatic (limf-AEH-tik)

lymphocytes (LIMF-oh-syt'z)

lysosomes (LY-soh-zohmz)

macrophages (MAK-roh-feyj-uhz)

macula lutea (MAK-yoo-lah LOO-tee-ah)

malleolus (mal-ee-OH-lus)

malleus (MAL-ee-us)

mammary (MAM-uh-ree)

manual (MAN-yoo-uhl)

manubrium (mah-NOO-bree-um)

masseter (MASS-uh-tur)

maxillae (mak-SILL-ee)

mediastinum (mee-dee-uh-STY-num)

medius (MEE-dee-uhs)

medulla (meh-DOOL-uh)

meiosis (MY-oh-sis)

melanin (MEL-uh-nin)

melanocytes (mel-AN-oh-syt'z)

melatonin (mel-uh-TOH-nin)

meninges (meh-NIN-jeez)

menisci (men-ISS-kee)

mental (MEN-tuhl)

mesenchyme (MEZ-en-ky'm)

mesenteric (mez-en-TEHR-ik)

mesentery (MEZ-en-tehr-ee)

metatarsals (met-uh-TAHR-sulz)

micelles (my-SELLZ)

microglial (my-kroh-GLEE-uhl)

microvilli (my-kroh-VIL-aye)

micturition (mik-choo-RISH-un)

mitochondria (my-toh-KAHN-dree-ah)

mitosis (my-TOH-sis)

mitral (MY-trul)

monocytes (MAHN-oh-syt'z)

mons pubis (MAHNS PYOO-biss)

morula (MOHR-yoo-luh)

musculocutaneous (musk-yoo-loh-kyoo-TAY-nee-us)

myelin (MY-lin)

myocardium (my-oh-KAR-dee-um)

myocytes (MY-oh-syt'z)

myofibrils (my-oh-FY-brillz)

myometrium (my-oh-MEE-tree-um)

myosin (MY-oh-sin)

nasal (NAY-zuhl)

nasopharynx (nayz-oh-FEHR-inks)

nephrons (NEF-rahnz)

neurilemma (noor-ih-LEM-uh)

neuroglial (noor-oh-GLEE-uhl)

neurohypophysis (noor-oh-hy-PAWF-ih-sis)

neurons (NOOR-ahnz)

neutrophils (NOO-troh-filz)

nuchal (NOO-kuhl)

nucleolus (noo-klee-OH-lus)

nucleus (NOO-klee-us)

obturator (AHB-too-ray-tur)

occipital (awk-SIP-ih-tuhl)

occipitalis (awk-sip-ih-TAL-iss)

ocular (AWK-yoo-lur)

oculomotor (awk-yoo-loh-MOH-tohr)

olecranon (oh-LEK-rah-nahn)

oligodendrocytes (oh-lig-oh-DEN-droh-syt'z)

omentum (oh-MEN-tum)

oocytes (OH-oh-syt'z)

oogenesis (oh-oh-JEN-ih-sis)

oogonia (oh-oh-GOHN-ee-uh)

oral (OH-ruhl)

orbicularis oculi (ohr-bik-yoo-LEHR-iss AWK-yoo-lye)

orbital (OHR-bit-uhl)

organelles (ohr-gan-ELLZ)

oropharynx (ohr-oh-FEHR-inks)

osmosis (oz-MOH-sis)

osseous (AHS-see-us)

ossicles (AW-sih-kullz)

osteoblasts (AH-stee-oh-blasts)

osteoclasts (AH-stee-oh-klasts)

osteocytes (AHS-tee-oh-syt'z)

osteons (AHS-tee-ahnz)

otic (OH-tik)

ovale (oh-VAL-ay)

oxytocin (awks-ee-TOH-sin)

palate (PAL-it)

palatine (PAL-uh-ty'n)

palmar (PAHL-mur)

palpebrae (pal-PEE-bree)

pancreas (PAYN-kree-us)

papillae (pah-PILL-ee)

parietal (puh-RY-ih-tuhl)

parotid (puh-RAWT-id)

patella (puh-TEL-uh)

pedal (PEE-duhl)

pelvic (PEL-vik)

pericardial (pehr-ee-KAR-dee-uhl)

pericardium (pehr-ee-KAR-dee-um)

perilymph (PEHR-ee-limf)

perimetrium (pehr-ee-MEE-tree-um)

perimysium (pehr-ih-MY-see-um)

periosteum (pehr-ee-AH-stee-um)

peristalsis (pehr-ih-STAHL-sis)

peritoneal (pehr-ih-toh-NEE-uhl)

peroxisomes (per-AWKS-ih-zohmz)

phalanges (fuh-LAN-jeez)

pharyngotympanic (fah-ring-oh-tim-PAN-ik)

pharynx (FEHR-inks)

phospholipid (FAHS-foh-lip-id)

phrenic (FREN-ik)

pia mater (PEE-ah MAH-tur)

pineal (pin-EE-uhl)

pituitary (pih-TOO-ih-tehr-ee)

placenta (plah-SIN-tuh)

plantar (PLAN-tahr)

platelets (PLAYT-letz)

platysma (plah-TIZ-muh)

pleural (PLOO-ruhl)

pneumothorax (noo-moh-THOHR-ax)

podocytes (POH-doh-syt'z)

popliteal (pahp-lih-TEE-uhl)

porta hepatis (POHR-tuh heh-PAEH-tis)

prostate (PRAW-stayt)

proximal (PRAWKS-ih-muhl)

pseudostratified (SOO-doh-strat-ih-fy'd)

psoas (SOH-us)

pterygoid (TEHR-ih-goyd)

pubic (PYOO-bik)

pubis (PYOO-bis)

pylorus (py-LOHR-us)

radius (RAY-dee-us)

rami (RAY-mee)

Ranvier (rahn-vee-ay)

renal (REE-nuhl)

rete testis (REE-tee TES-tis)

retina (RET-in-ah)

retroperitoneal (reh-troh-pehr-ih-toh-NEE-uhl)

ribosomes (RY-boh-zohmz)

Rinne (rinn-ay)

rugae (ROO-ghee)

sacroiliac (say-kroh-ILL-ee-ak)

sacrum (SAY-krum)

sagittal (SAJ-ih-tuhl)

saphenous (SAF-en-us)

sarcolemma (sar-koh-LEM-uh)

sarcomere (SAR-koh-meer)

sarcoplasm (SAR-koh-plazm)

sarcoplasmic reticulum (sar-koh-PLAZ-mik reh-TIK-yoo-lum)

sartorius (sar-TOHR-ee-uhs)

scapula (SKAP-yoo-lah)

sciatic (sy-AEH-tik)

sclera (SKLEHR-ah)

sebaceous (seh-BAY-shuhs)

sebum (SEE-bum)

sella turcica (SELL-uh TUR-sih-kuh)

semimembranosus (sem-aye-mem-brah-NOH-suhs)

seminiferous (sem-ih-NIF-er-us)

semitendinosus (sem-aye-ten-din-OH-suhs)

serous (SEER-us)

soleus (SOHL-ee-uhs)

spermatogenesis (sper-mat-oh-JEN-ih-sis)

spermatogonia (sper-mat-oh-GOH-nee-ah)

sphenoid (SFEE-noyd)

sphygmomanometer (sfig-moh-muh-NAH-muh-ter)

spinosum (spin-OH-sum)

spirometer (spih-RAH-meh-ter)

splenic (SPLEN-ik)

squamous (SKWAY-muss)

stapes (STAY-peez)

stenosis (sten-OH-sis)

sternal (STUR-nuhl)

sternocleidomastoid (stern-oh-kly-doh-MASS-toy'd)

stethoscope (STETH-oh-skohp)

stratum corneum (STRAT-um KOHR-nee-um)

striated (STRY-ayt-ed)

subclavian (sub-KLAY-vee-in)

sublingual (sub-LING-gwuhl)

sulci (SUL-kee)

sural (SOO-ruhl)

sutures (SOO-tchurz)

symphysis (SIM-fih-sis)

synovial (sih-NOH-vee-uhl)

systole (SIS-toh-lee)

tachycardia (tak-ih-KAR-dee-uh)

talus (TAY-luss)

tarsal (TAR-suhl)

telodendria (tee-loh-DEN-dree-uh)

telophase (TEL-oh-phayz)

temporalis (tem-pur-AL-iss)

thalamus (THAL-uh-muss)

thoracic (thoh-RAEH-sik)

tibia (TIB-ee-ah)

trabeculae (trah-BEK-yoo-lee)

trachea (TRAY-kee-uh)

trapezius (trah-PEE-zee-uhs)

triglycerides (try-GLISS-er-aye'dz)

trigone (TRY-gohn)

triiodothyronine (try-aye-oh-doh-THY-roh-neen)

trochanter (TROH-kan-tur)

trochlea (TROH-klee-uh)

trophoblast (TROHF-oh-blast)

tropomyosin (trohp-oh-MY-oh-sin)

troponin (TROH-poh-nin)

umbilical (um-BIL-ih-kuhl)

ureteral (yoo-REE-ter-uhl)

ureters (YOOR-eh-terz)

urethra (yoo-REETH-ruh)

uterus (YOO-ter-us)

uvea (YOO-vee-uh)

uvula (YOO-vyoo-luh)

vagus (VAY-gus)

vena cava (VEE-nah KAY-vah)

venosus (veh-NOH-suss)

vertebral (vur-TEE-bruhl)

vestibulocochlear (ves-tib-yoo-loh-KOHK-lee-ur)

visceral (VISS-er-uhl)

vitreous (VIT-ree-us)

vomer (VOH-muhr)

xiphoid (ZY-foyd)

zygote (ZY-goh't)

Contents

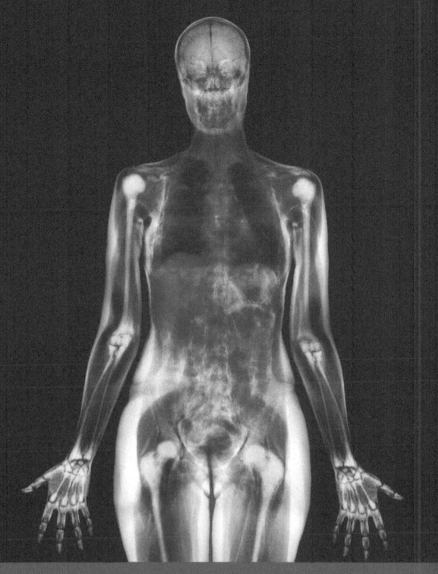

Introduction to Anatomy and Physiology

When you have completed this unit, you should be able to:

1 Demonstrate and describe anatomical position.

2 Apply directional terms to descriptions of anatomical parts.

3 Use regional terms to describe anatomical locations.

4 Locate and describe the divisions of the major body cavities and the membranes lining each cavity.

5 Demonstrate and describe anatomical planes of section.

6 Identify the organ systems, their functions, and the major organs in each system.

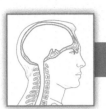

Name _____ Section _____ Date _____

1

PRE-LAB EXERCISES

Complete the following exercises prior to coming to lab, using your lab manual and textbook for reference.

Pre-Lab Exercise **1-1**

✎ Key Terms

You should be familiar with the following terms before coming to lab.

Term	Definition
Directional Terms	
Anterior	
Posterior	
Superior	
Inferior	
Proximal	
Distal	
Medial	
Lateral	
Superficial	
Deep	
Body Cavities and Membranes	
Posterior body cavity	
Anterior body cavity	
Serous membrane	
Planes of Section	
Sagittal plane	
Frontal (coronal) plane	
Transverse plane	

Pre-Lab Exercise **1-2**

Body Cavities

Color the body cavities in Figure 1.1, and label them with the terms from Exercise 1-2 (p. 12). Use Exercise 1-2 in this unit and your text for reference.

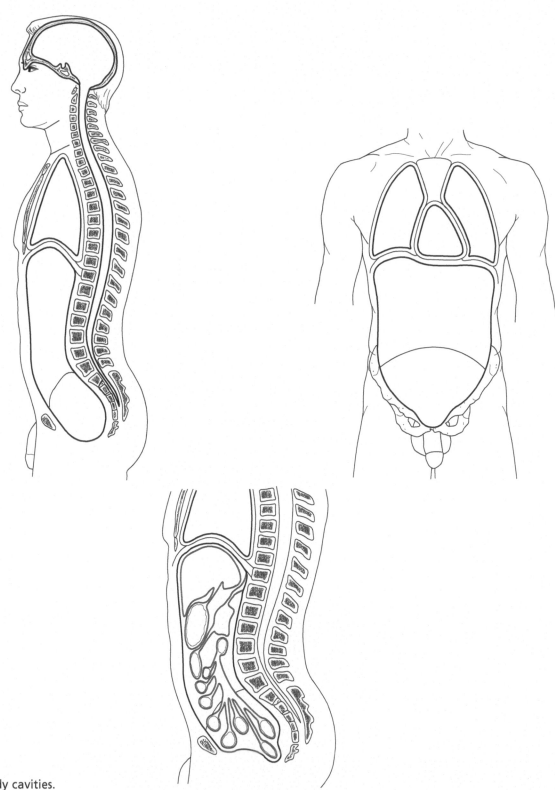

FIGURE **1.1** Body cavities.

The bullet entered the right posterior scapular region, 3 centimeters lateral to the vertebral region, 4 centimeters inferior to the cervical region, and penetrated deep to the muscle and bone, but superficial to the lung and parietal pleura . . .

Whole body scan using colored magnetic resonance imaging (MRI) of a woman in frontal section.

Would you believe that, by the end of this unit, you will be able to translate the above sentence and also locate the hypothetical wound? Unit 1 introduces you to the world of **anatomy** and **physiology**, or A&P, which is the study and science of the structure and function of the human body. We start this unit with the language of A&P. Like learning any new language, this may seem overwhelming at first. The key to success is repetition and application: The more you use the terms, the easier it will be for them to become part of your normal vocabulary.

From terminology, we move on to the organization of the internal body into spaces called **body cavities**, and to how we can examine the body's cavities and organs with specific cuts known as **anatomical sections**. Finally, we look at how the body's organs are combined into functional groups called **organ systems**. When you have completed this unit, return to the opening statement, and challenge yourself to use your new knowledge of anatomical terms, body cavities, and organs to locate the precise position of the bullet wound on an anatomical model.

Exercise **1-1**

Anatomical Terms

MATERIALS
- ❑ Laminated outline of the human body
- ❑ Water-soluble marking pens

Accurate communication among scientists in the fields of anatomy and physiology is critical, which means we must describe body parts, wounds, procedures, and more in a specific, standardized way. The first way in which we standardize communication is in the presentation of specimens in **anatomical position**. As you can see in Figure 1.2, in anatomical position the specimen is presented facing forward, with the toes pointing forward, the feet shoulder-width apart, and the palms facing forward.

Another method that makes communication easier and less prone to errors is to use **directional terms** to define the location of body parts and body markings. For example, when describing a wound on the chest, we could say either of the following:

❚ The wound is near the middle and top of the chest.

❚ The wound is on the right *anterior* thoracic region, 4 centimeters *lateral* to the sternum, and 3 centimeters *inferior* to the acromial region.

The second option is precise and allows the reader to pinpoint the wound's exact location. Remember that these descriptions are referring to a body in anatomical position.

Most directional terms are arranged into pairs of opposite directions. Some of the common pairs you will use in this course are illustrated in Figure 1.3 and include the following:

❚ **Anterior/Posterior.** The term **anterior** (an-TEER-ee-ur), also known as **ventral,** refers to the front of the body or of a body part. For example, we could say that the nose is on the anterior side of the body, or we may describe the surface of a bone that faces the front of the body as its *anterior surface*. However, we may also use the term anterior to describe a structure whose anatomical course takes it toward the front of the body. For example, the branches of spinal nerves called *anterior rami* (RAY-mee) are named this way because they travel toward the anterior side of the body. The opposite of

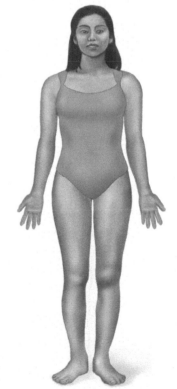

FIGURE **1.2** Anatomical position.

anterior/ventral is **posterior/dorsal**. Just as with the term anterior, the term posterior can have two meanings. Posterior can refer to the back side of the body or of a body part, or to a structure that travels toward the back side of the body (such as the *posterior rami*).

- **Superior/Inferior**. The directional term **superior** is used to describe structures that are toward or closer to the head. The opposite term of superior is **inferior**, which means away from the head or toward the tail. Both superior and inferior may be used to describe a structure's absolute position in the body or the position of a structure relative to another structure. For example:

  - The two largest veins in the body are the *superior vena cava* and the *inferior vena cava*. The superior vena cava is the vessel located nearer the head, and the inferior vena cava is the one located farther away from the head (or closer to the tail).

  - We can describe the head as being *superior* to the neck, and we can say that the abdomen is *inferior* to the chest.

  A very important rule to remember with superior and inferior is that these terms are used to describe structures only on the head, neck, and trunk, which are located in the part of the body known as the *axial region*. With very few exceptions, we do not use superior and inferior on the upper and lower limbs.

- **Proximal/Distal**. The upper and lower limbs are the part of the body known as the *appendicular region*. We use the terms proximal and distal on the appendicular region instead of using superior and inferior. The directional term **proximal** (PRAWKS-ih-muhl) refers to the closeness—or *proximity*—of a structure to its point of origin, which is the shoulder for the upper limb and the hip for the lower limb. You know this from the everyday term *approximate*, which refers to closeness. The opposite term, **distal**, refers to the farness—or *distance*—of a structure from its point of origin. As with superior and inferior, the terms can describe either a structure's absolute position or the position of a structure relative to another structure. Here are a couple of examples:

  - The part of the femur (thigh bone) that is closest to the hip is called the *proximal* end of the bone. The end of the femur that is farthest away from the hip is its *distal* end.

  - We can describe the knee as being *proximal* to the ankle, because it is closer to the hip than is the ankle. Similarly, the fingers are *distal* to the wrist because they are farther away from the shoulder than is the wrist.

HINTS & TIPS

Sorting Out Superior and Inferior versus Proximal and Distal

One of the most common problems students have with respect to directional terms is how to use the terms superior/inferior and proximal/distal. Superior and inferior are easy enough on the head, neck, and trunk, but many students forget the rules and want to use them on the upper and lower limbs. So to start, here's a mnemonic to help: "Use your **superior mind** to remember when to use **superior** and **inferior**." This will help remind you to use these terms only on the head, neck, and trunk, as these are the locations in the body that house structures associated with your "mind" (the brain and spinal cord).

Now let's think about proximal and distal for a minute. To start, why even have different terms on the upper and lower limbs? Well, stand with your hands by your side. If we were to use superior and inferior, we would say that your hand is inferior to your shoulder. But lift your arm in the air above your head—where is your hand now? It's superior to your shoulder, right? This is the problem with limbs: They can change position. Ideally, every specimen we describe is in anatomical position, but we can't guarantee that every patient we treat or body we find will be in anatomical position. So for the upper and lower limbs, we need a different set of terms.

Once you understand the "why" of proximal and distal, all that's left is to understand how to apply them. That part is easy, because both are words that you already know, even if you don't immediately recognize them. The word root for "proximal" is proxim-, which means "near." You use words with this word root all the time, such as "proximity" and "approximate." The word root for "distal" is dist-, which means "far." This is another very common word root—you use it in words like "distant," "distinct," "distance," and "distend."

Knowing the meanings of the word roots for the terms proximal and distal makes it very simple to come up with mnemonics for their use. Here are a couple of easy ones:

- ℹ️ The more **proximal** structure is the one in the closest **proximity** to the hip (or shoulder).

- ℹ️ The more **distal** structure is the one that is the most **distant** from the hip (or shoulder).

However, don't forget that these terms are only used on the upper and lower limbs, with very few exceptions. Just as you can't use superior and inferior on the upper and lower limbs, you also can't use proximal and distal on the head, neck, and trunk.

- **Medial/Lateral.** The terms medial and lateral reference an imaginary line running down the middle of the body called the *midline*. A structure is described as **medial** when its position is closer to this midline, and its position becomes more **lateral** as it moves farther away from the midline. Again, we can use these terms to describe a structure's absolute location or its location relative to another structure. For example:
  - The ulna, or the inner forearm bone, is the *medial* bone of the forearm because it is closest to the midline of the body. The radius, or the outer forearm bone, is the *lateral* bone of the forearm because it is farthest away from the midline of the body.
  - We can describe the ears as being *lateral* to the eyes because the ears are farther away from the midline than are the eyes. Conversely, we could say that the eyes are *medial* to the ears for the same reason.

- **Superficial/Deep.** The final pair of directional terms we use is one with which you are probably already familiar just from your everyday language: superficial and deep. As you might expect, the term **superficial** refers to a position that is closer to the surface of the body or closer to the skin. Structures that are **deep** are farther away from the skin's surface.

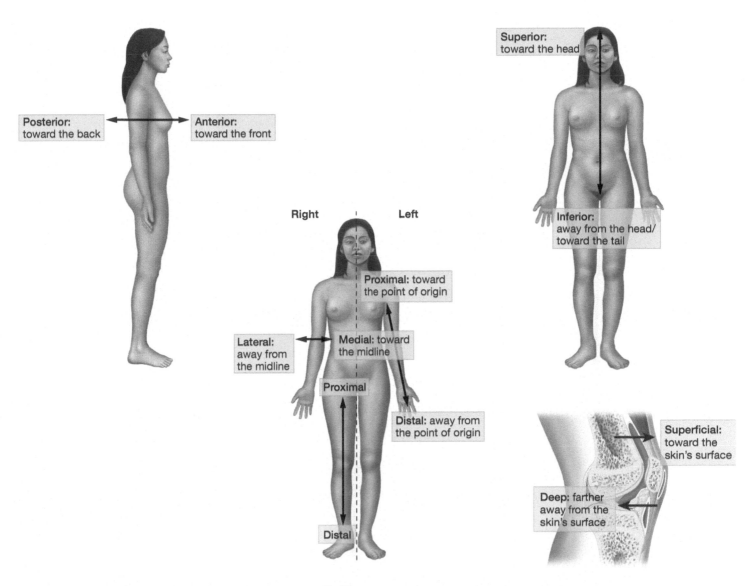

FIGURE **1.3** Directional terms.

A third way to ensure good communication among professionals is to use specific words known as **regional terms** to describe locations of parts of the body. The regional terms illustrated in Figure 1.4 and defined in Table 1.1 are among the more common terms you will encounter in your study of anatomy and physiology. Note that most of these terms are adjectives rather than nouns. This means that the term is not complete unless it is paired with the term "region." For example, we cannot say, "The wound is in the antebrachial." We instead must say, "The wound is in the antebrachial *region*."

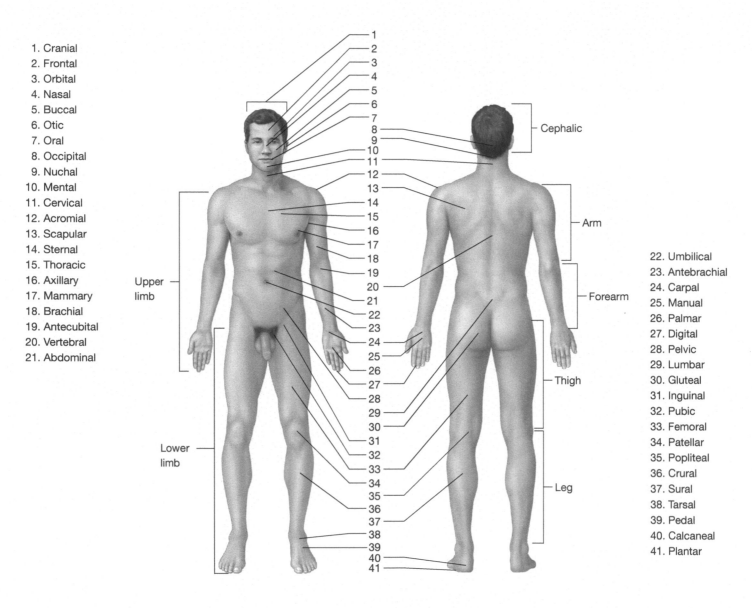

1. Cranial
2. Frontal
3. Orbital
4. Nasal
5. Buccal
6. Otic
7. Oral
8. Occipital
9. Nuchal
10. Mental
11. Cervical
12. Acromial
13. Scapular
14. Sternal
15. Thoracic
16. Axillary
17. Mammary
18. Brachial
19. Antecubital
20. Vertebral
21. Abdominal

22. Umbilical
23. Antebrachial
24. Carpal
25. Manual
26. Palmar
27. Digital
28. Pelvic
29. Lumbar
30. Gluteal
31. Inguinal
32. Pubic
33. Femoral
34. Patellar
35. Popliteal
36. Crural
37. Sural
38. Tarsal
39. Pedal
40. Calcaneal
41. Plantar

FIGURE **1.4** Regional terms.

TABLE **1.1** Regional Terms

Term	Definition
Adjectives	
Abdominal region	The area over the abdomen that is inferior to the diaphragm and superior to the bony pelvis
Acromial region	The area over the lateral part of the shoulder that contains the acromion of the scapula
Antebrachial region	The anterior forearm
Antecubital region	The anterior upper limb between the forearm and arm (over the elbow joint)
Axillary region	The area in and around the axilla (armpit)
Brachial region	The anterior and posterior arm (between the elbow and the shoulder)
Buccal region	The lateral portions of the face corresponding to the cheeks
Calcaneal region	The heel of the foot
Carpal region	The wrist
Cephalic region	The entire head from the chin to the top of the head
Cervical region	The neck
Cranial region	The top of the head, or the portion of the skull that encases the brain
Crural region	The anterior leg, or the shin
Digital region	The fingers or the toes
Femoral region	The thigh
Frontal region	The forehead
Gluteal region	The buttock
Inguinal region	The area along the inguinal ligament that divides the pelvis from the thigh
Lumbar region	The lower back
Mammary region	The area around the breast
Manual region	The general area of the hand
Mental region	The chin
Nasal region	The nose
Nuchal region	The ridge that runs along the back of the skull at the superior boundary of the occipital region
Occipital region	The general area of the back of the skull
Oral region	The mouth
Orbital region	The area around the eye
Otic region	The area around the ear
Palmar region	The anterior hand (the palm of the hand)
Patellar region	The anterior part of the knee over the patella (kneecap)
Pedal region	The foot
Pelvic region	The anterior pelvis
Plantar region	The bottom of the foot
Popliteal region	The posterior side of the knee joint
Pubic region	The area over the pubic bone
Scapular region	The area over the scapula in the superior back
Sternal region	The area in the middle of the chest over the sternum

(continues)

TABLE **1.1** Regional Terms (*cont.*)

Term	Definition
Adjectives	
Sural region	The posterior part of the leg (the calf)
Tarsal region	The ankle region
Thoracic region	The general chest area
Umbilical region	The area around the umbilicus (belly button)
Vertebral region	The area over the vertebral column (spine)
Nouns	
Arm	The portion of the upper limb from the elbow to the shoulder
Forearm	The portion of the upper limb from the elbow to the wrist
Leg	The portion of the lower limb from the knee to the ankle
Lower limb	The entire portion of the body from the hip to the digits of the foot
Thigh	The portion of the lower limb from the hip to the knee
Upper limb	The entire portion of the body from the shoulder to the digits of the hand

Procedure **1** Demonstrating Anatomical Position

Have your lab partner stand in a normal, relaxed way, and then adjust his or her position so it matches anatomical position.

Procedure **2** Directional Terms

Fill in the correct directional term for each of the following items. Note that in some cases more than one directional term may apply.

The elbow is _____ to the wrist.

The chin is _____ to the nose.

The shoulder is _____ to the clavicle (collarbone).

The forehead is _____ to the mouth.

The skin is _____ to the muscle.

The esophagus is _____ to the sternum (breastbone).

The nose is _____ to the cheek.

The spine is on the _____ side of the body.

The arm is _____ to the torso.

The knee is _____ to the hip.

Procedure 3 Labeling Body Regions

Use water-soluble markers to locate and label each of the following regions on laminated outlines of the human body. If outlines are unavailable, label the regions on Figure 1.5. The following list may look daunting, but you are probably familiar with several of the terms already. For example, you likely know the locations of the "oral," "nasal," and "abdominal" regions. Watch for other terms you may know.

Adjectives

- ❏ Abdominal (ab-DAH-mih-nuhl)
- ❏ Acromial (ah-KROH-mee-uhl)
- ❏ Antebrachial (an-tee-BRAY-kee-uhl)
- ❏ Antecubital (an-tee-KYOO-bih-tuhl)
- ❏ Axillary (AX-il-ehr-ee)
- ❏ Brachial (BRAY-kee-uhl)
- ❏ Buccal (BYOO-kuhl)
- ❏ Calcaneal (kal-KAY-nee-uhl)
- ❏ Carpal (KAR-puhl)

- ❏ Cephalic (sef-AL-ik)
- ❏ Cervical (SUR-vih-kuhl)
- ❏ Cranial (KRAY-nee-uhl)
- ❏ Crural (KROO-ruhl)
- ❏ Digital (DIJ-it-uhl)
- ❏ Femoral (FEM-oh-ruhl)
- ❏ Frontal (FRUHN-tuhl)
- ❏ Gluteal (GLOO-tee-uhl)
- ❏ Inguinal (IN-gwin-uhl)
- ❏ Lumbar (LUHM-bahr)
- ❏ Mammary (MAM-uh-ree)
- ❏ Manual (MAN-yoo-uhl)
- ❏ Mental (MEN-tuhl)
- ❏ Nasal (NAY-zuhl)

- ❏ Nuchal (NOO-kuhl)
- ❏ Occipital (awk-SIP-ih-tuhl)
- ❏ Oral (OH-ruhl)
- ❏ Orbital (OHR-bit-uhl)
- ❏ Otic (OH-tik)
- ❏ Palmar (PAHL-mur)
- ❏ Patellar (puh-TEL-ur)
- ❏ Pedal (PEE-duhl)
- ❏ Pelvic (PEL-vik)
- ❏ Plantar (PLAN-tahr)
- ❏ Popliteal (pahp-lih-TEE-uhl)
- ❏ Pubic (PYOO-bik)
- ❏ Scapular (SKAP-yoo-lur)
- ❏ Sternal (STUR-nuhl)

- ❏ Sural (SOO-ruhl)
- ❏ Tarsal (TAR-suhl)
- ❏ Thoracic (thoh-RAEH-sik)
- ❏ Umbilical (um-BIL-ih-kuhl)
- ❏ Vertebral (vur-TEE-bruhl)

Nouns

- ❏ Arm
- ❏ Forearm
- ❏ Leg
- ❏ Lower limb
- ❏ Thigh
- ❏ Upper limb

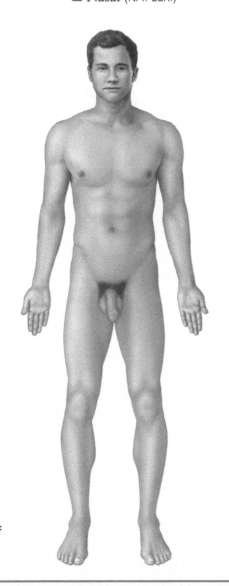
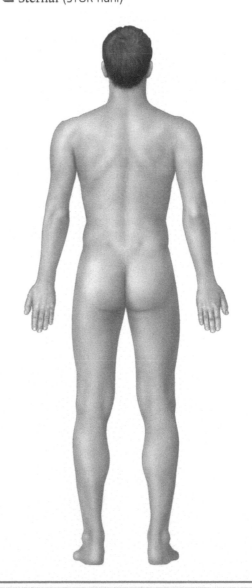

FIGURE **1.5** Anterior and posterior views of the human body in anatomical position.

Exercise 1-2

Body Cavities and Membranes

MATERIALS

❏ Human torso models
❏ Fetal pigs (or other preserved small mammals)
❏ Dissection kits/dissection trays
❏ Laminated outline of the human body
❏ Water-soluble marking pens

The body is divided into several cavities, many of which are fluid-filled and each of which contains specific organs. In this exercise, you will identify the body cavities and the organs contained within each cavity.

Several of the fluid-filled body cavities are formed by thin sheets of tissue called **serous membranes** (SEER-us). Cells of these membranes produce a thin, watery fluid called **serous fluid** that lubricates organs so they move within the cavity with little friction. Serous membranes are composed of two layers: an outer **parietal layer** (puh-RY-ih-tuhl) that is attached to the body wall and surrounding structures, and an inner **visceral layer** (VISS-er-uhl) that is attached to specific organs. Between the parietal and visceral layers is a thin potential space that contains serous fluid. This potential space is referred to as a cavity.

As you can see in Figure 1.6, there are two major body cavities, each of which is subdivided into smaller cavities, as follows:

1. **Posterior (dorsal) body cavity.** As implied by its name, the **posterior body cavity** is largely on the posterior side of the body (Fig. 1.6A). It is divided into two smaller cavities:

   a. **Cranial cavity.** The **cranial cavity** is the area encased by the skull. It contains the brain and a fluid called *cerebrospinal fluid.*

   b. **Vertebral** (or **spinal**) **cavity.** The **vertebral cavity** is the area encased by the vertebrae. It contains the spinal cord and is also filled with cerebrospinal fluid.

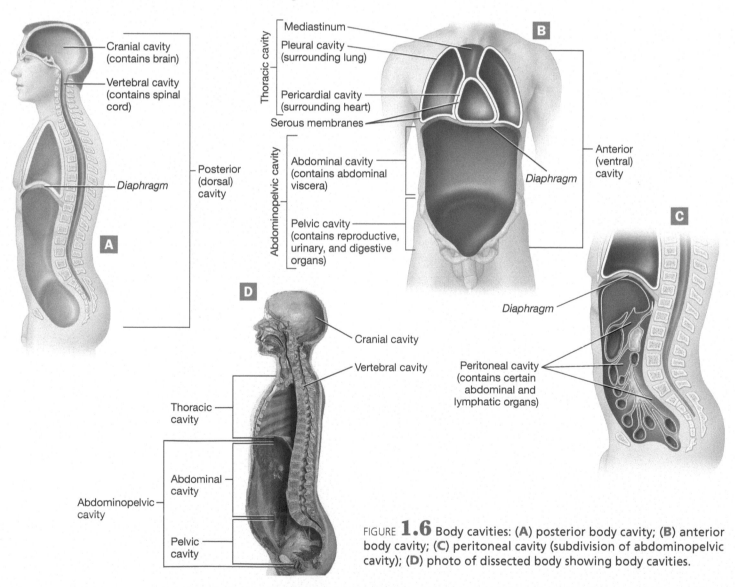

FIGURE **1.6** Body cavities: (**A**) posterior body cavity; (**B**) anterior body cavity; (**C**) peritoneal cavity (subdivision of abdominopelvic cavity); (**D**) photo of dissected body showing body cavities.

2. **Anterior (ventral) body cavity.** The **anterior body cavity** is largely on the anterior side of the body (Fig. 1.6B). It has two main divisions: the *thoracic* and *abdominopelvic cavities*, each of which is divided into smaller subcavities.

   a. **Thoracic cavity.** The **thoracic cavity** (thoh-RAEH-sik) is located superior to a muscle called the diaphragm and encompasses the area encased by the ribs. We find the following smaller cavities within the thoracic cavity:

      (1) **Pleural cavities.** Each **pleural cavity** (PLOO-ruhl) surrounds one of the lungs. The pleural cavities are located between two layers of a serous membrane called the **pleural membrane**. The **parietal pleurae** are attached to the body wall and the surface of the diaphragm, and the **visceral pleurae** are attached to the surface of the lungs. Between the two layers we find a thin layer of serous fluid.

      (2) **Mediastinum.** The cavity between the pleural cavities, called the **mediastinum** (mee-dee-uh-STY-num), contains the great vessels, the esophagus, the trachea and bronchi, and other structures. It houses another cavity formed by a serous membrane called the **pericardial membrane** (pehr-ee-KAR-dee-uhl). The **parietal pericardium** is attached to surrounding structures, and the inner **visceral pericardium** is attached to the heart muscle. Between the two layers we find the **pericardial cavity**, which is filled with a thin layer of serous fluid.

   b. **Abdominopelvic cavity.** The **abdominopelvic cavity** (ab-dom-ih-noh-PEL-vik) is located inferior to the diaphragm and extends into the bony pelvis. There are three subcavities within the abdominopelvic cavity:

      (1) **Abdominal cavity.** The area superior to the bony pelvis, called the **abdominal cavity**, houses many organs including the liver, gallbladder, small intestine, stomach, pancreas, kidneys, adrenal glands, spleen, and much of the colon (large intestine).

      (2) **Pelvic cavity.** The cavity housed within the bony pelvis, the **pelvic cavity**, contains certain sex organs, the urinary bladder, the rectum, and part of the colon.

      (3) **Peritoneal cavity.** The third subcavity of the abdominopelvic cavity, shown in Figure 1.6C, is formed by a serous membrane called the **peritoneal membrane** (pehr-ih-toh-NEE-uhl). The outer **parietal peritoneum** is attached to the body wall and surrounding structures, and the inner **visceral peritoneum** is attached to the surface of many of the abdominal and pelvic organs. Between these two layers of peritoneal membrane we find the **peritoneal cavity**, which is filled with serous fluid. Organs that are within the peritoneal cavity are **intraperitoneal** (in-trah-pehr-ih-toh-NEE-uhl), and include the liver, most of the small intestine, much of the colon, and part of the pancreas. Those organs that are posterior to the peritoneal cavity are said to be **retroperitoneal** (reh-troh-pehr-ih-toh-NEE-uhl), and include the kidneys, adrenal glands, the sex organs, the urinary bladder, part of the colon, and part of the pancreas.

We often divide the abdominopelvic cavity into four surface quadrants: the right upper, right lower, left upper, and left lower quadrants, as shown in Figure 1.7. Take note in the figure of which organs are located in each quadrant, as you will be identifying them in an upcoming procedure. Be aware, though, that certain structures are deep to the labeled organs and, so, aren't visible in the figure. These deeper organs are listed to the side of each quadrant.

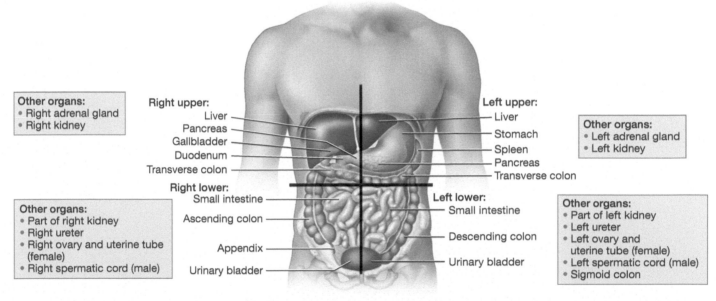

FIGURE **1.7** Quadrants of the abdominopelvic cavity.

Procedure 1 Body Cavities

In this procedure, you will use a human torso model or a preserved small mammal, such as a cat, fetal pig, or rat, to examine the body cavities. If you are using a preserved small mammal, you may use the following procedure to open the body cavities. *Note that you will not open the animal fully in this procedure to preserve structures for future dissections.*

1 Place the animal in the dissecting tray with its ventral (abdominal) side facing you. Use a scalpel to make shallow incisions, following the guide in Figure 1.8.

2 Using a blunt dissection probe, peel back the skin of the abdomen carefully to expose the abdominopelvic cavity. Examine its contents.

3 Gently peel back the skin of the chest and neck incisions. Use scissors or a scalpel to carefully cut through the sternum to expose the thoracic cavity. Examine its contents.

When you have opened the anterior body cavity of your preserved small mammal or human torso model, locate and identify each cavity, and do the following:

4 Identify the organs of the thoracic, abdominal, and pelvic cavities, and list these organs in Table 1.2. See Figures 1.9 (p. 16) and 1.10 (p. 17) for reference.

5 Mark each quadrant of the abdominopelvic cavity with a pin or marking tape (if working with a torso model). Note which organs are located in each quadrant.

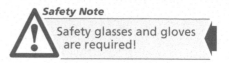

Safety Note

Safety glasses and gloves are required!

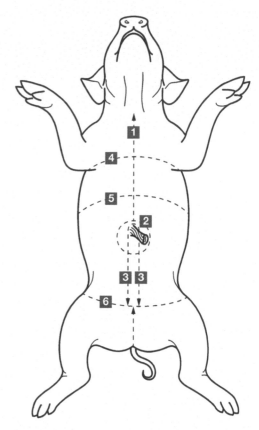

FIGURE **1.8** Incisions for a fetal pig dissection.

TABLE **1.2** Divisions of the Anterior Body Cavity and Quadrants of the Abdominopelvic Cavity

Cavity	Organ(s)
1. Thoracic cavity	
a. Pleural cavities	
b. Mediastinum	
(1) Pericardial cavity	
2. Abdominopelvic cavity	
a. Subdivisions	
(1) Abdominal cavity	
(2) Pelvic cavity	
(3) Peritoneal cavity	
b. Regions	
(1) Right upper quadrant	
(2) Right lower quadrant	
(3) Left upper quadrant	
(4) Left lower quadrant	

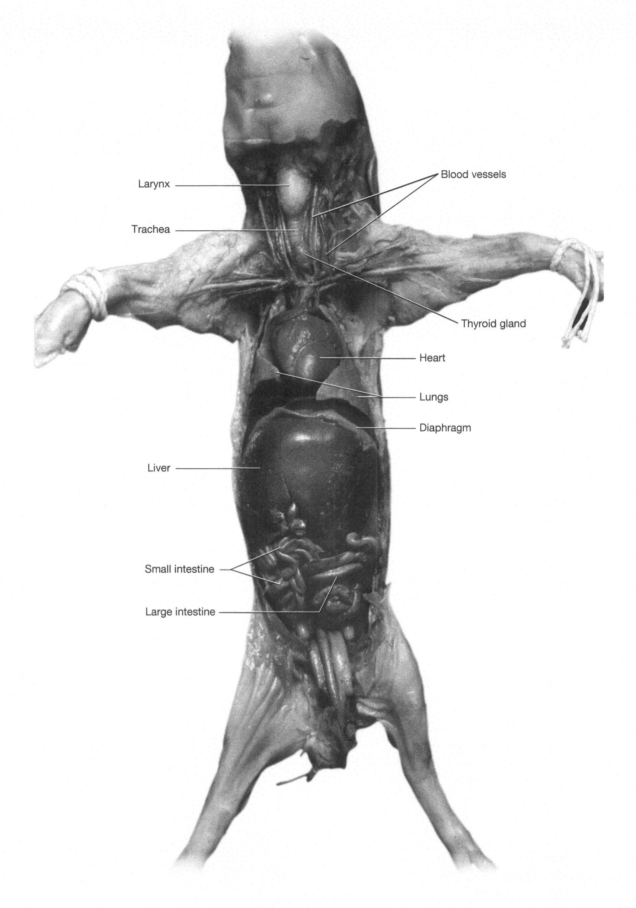

Larynx

Trachea

Blood vessels

Thyroid gland

Heart

Lungs

Diaphragm

Liver

Small intestine

Large intestine

FIGURE **1.9** Ventral view of the fetal pig.

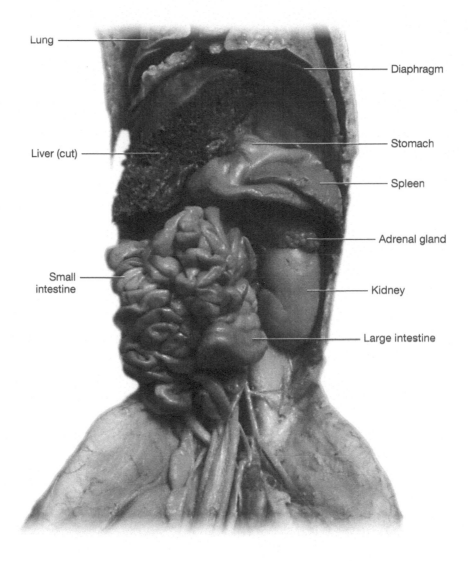

Lung

Diaphragm

Liver (cut)

Stomach

Spleen

Adrenal gland

Small
intestine

Kidney

Large intestine

FIGURE **1.10** Abdominopelvic cavity of the fetal pig.

Procedure **2** Serous Membranes

Part A

Serous membranes are best examined on a preserved specimen such as a fetal pig, because their structure is difficult to appreciate on a model. As you dissect the preserved mammal, look for the serous membranes listed in Table 1.3. Take care not to tear the fragile membranes, which consist of just a few layers of cells. As you identify each membrane, name the structure to which the membrane is attached (the lungs, heart, abdominal wall, etc.).

TABLE **1.3** Serous Membranes

Membrane	Cavity	Structure
Parietal pleura		
Visceral pleura		
Parietal pericardium		
Visceral pericardium		
Parietal peritoneum		
Visceral peritoneum		

Part B

Draw in the body cavities on your laminated outlines of the human body using water-soluble markers or on Figure 1.11. Label each body cavity, as well as the serous membranes surrounding the cavity, where applicable.

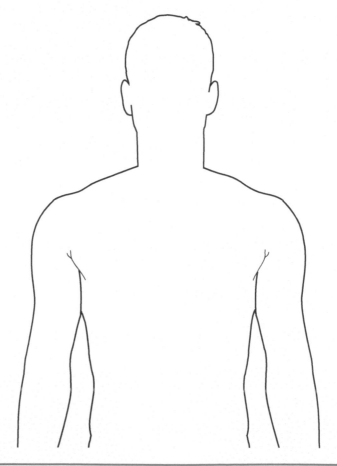

FIGURE **1.11** Anterior view of the human torso.

Procedure 3 Applications of Terms, Cavities, and Membranes

Remember this from the unit opener?

The bullet entered the right posterior scapular region, 3 centimeters lateral to the vertebral region, 4 centimeters inferior to the cervical region, and penetrated deep to the muscle and bone, but superficial to the parietal pleura.

Now it's your turn to combine the different anatomical terms you have learned in order to describe a "bullet wound." Assume that you are acting as coroner, and you have a victim with three gunshot wounds. In this scenario, your victim will be your fetal pig or a human torso model that your instructor has "shot." For each "bullet wound," describe the location of the wound using at least three directional terms and as many regional terms as possible, following this example. As coroner, remember to keep your patient in anatomical position and that you need to be as specific as possible.

Shot 1	
Shot 2	
Shot 3	

Exercise 1-3

Planes of Section

MATERIALS
☐ Modeling clay
☐ Knife or scalpel

Often in anatomy and physiology, we need to obtain different views of the internal structure of an organ or a body cavity. These views are obtained by making an **anatomical section** along a specific plane. The commonly used planes of section, shown in Figures 1.12 and 1.13, are as follows:

1. **Sagittal plane.** A section along the **sagittal plane** (SAJ-ih-tuhl) divides the specimen into right and left parts. The sagittal section has two variations:

   a. **Midsagittal sections** divide the specimen into equal right and left halves.

   b. **Parasagittal sections** divide the specimen into unequal right and left parts.

2. **Frontal plane.** The **frontal plane**, also known as the **coronal plane** (koh-ROH-nuhl), divides the specimen into an anterior (front) part and a posterior (back) part.

3. **Transverse plane.** The **transverse plane**, also known as a **cross section** or the **horizontal plane**, divides the specimen into a superior (or proximal) part and an inferior (or distal) part.

Note that although there is only a single midsagittal plane, there are a near infinite number of possible parasagittal, frontal, and transverse planes. In your study of anatomy and physiology, you will see many different examples of these planes of section.

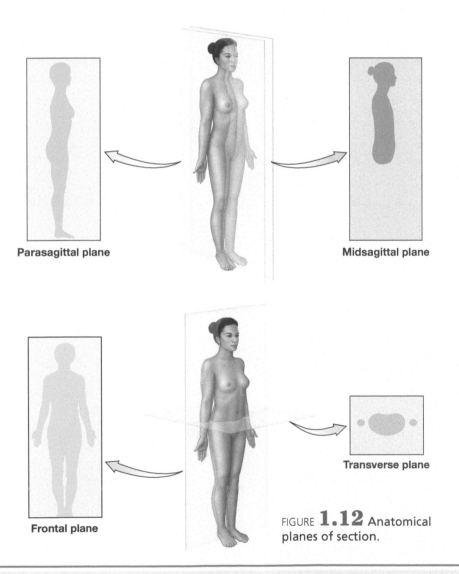

Parasagittal plane

Midsagittal plane

Transverse plane

Frontal plane

FIGURE **1.12** Anatomical planes of section.

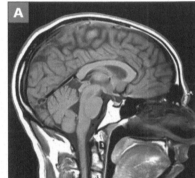

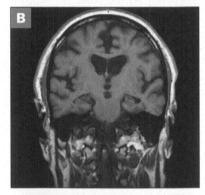

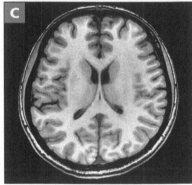

FIGURE **1.13** Planes of anatomical section visible on computed tomography scans of the brain: (**A**) midsagittal section; (**B**) frontal section; (**C**) transverse section.

Procedure 1 Sectioning Along Anatomical Planes

Mold a ball of modeling clay into the shape of a head and draw eyes on the head to denote anterior and posterior sides. Then, use a scalpel to cut the ball in each of the numbered anatomical planes. When you have finished cutting the planes, sketch a model "head" in the space provided, and draw each of the anatomical planes. Use Figures 1.12 and 1.13 for reference.

1. Sagittal
   a. Midsagittal
   b. Parasagittal
2. Frontal
3. Transverse

Exercise 1-4

Organs and Organ Systems

MATERIALS

❑ Human torso models

❑ Fetal pigs (or other preserved small mammals)

❑ Dissection kits/dissection trays

The organs that you have identified in this lab are organized into functional groups called **organ systems**. The human body has 11 organ systems, each with specific organs and functions (Fig. 1.14). In this exercise, you will examine the organ systems and identify their major organs.

Procedure 1 Identifying Organs

Identify the following organs on your preserved mammal specimen or human torso models. Check off each organ as you identify it, and record the organ system to which it belongs in Table 1.4, using Figure 1.14 for reference. Remember that some organs may function in more than one system.

❑ Adrenal glands   ❑ Large intestine   ❑ Spinal cord

❑ Blood vessels   ❑ Larynx   ❑ Spleen

❑ Bones   ❑ Liver   ❑ Stomach

❑ Brain   ❑ Lungs   ❑ Testes (male) or ovaries (female)

❑ Esophagus   ❑ Lymph nodes   ❑ Thymus

❑ Gallbladder   ❑ Pancreas   ❑ Thyroid gland

❑ Heart   ❑ Skeletal muscles   ❑ Trachea

❑ Joints   ❑ Skin   ❑ Urinary bladder

❑ Kidneys   ❑ Small intestine   ❑ Uterus

TABLE **1.4** Organs and Organ Systems

Organ System	Major Organ(s)
Integumentary system	
Skeletal system	
Muscular system	
Nervous system	
Endocrine system	
Cardiovascular system	
Lymphatic system	
Respiratory system	
Digestive system	
Urinary system	
Reproductive systems	

Procedure **2** Researching Organ Systems

The body's organ systems are illustrated in Figure 1.14. Fill in the blanks next to each organ system to identify the major organs shown. Then, use your textbook or other resources to research the principal functions of each system.

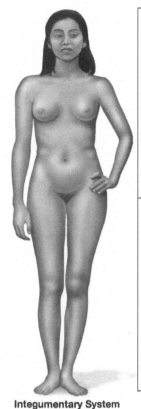

Main Organs:

Main Functions:

Integumentary System

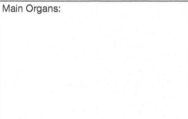
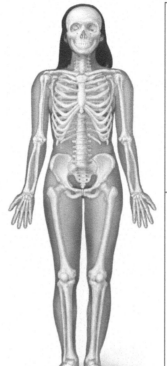

Main Organs:

Main Functions:

Skeletal System

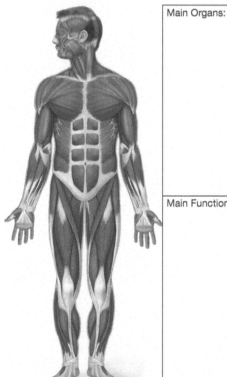

Main Organs:

Main Functions:

Muscular System

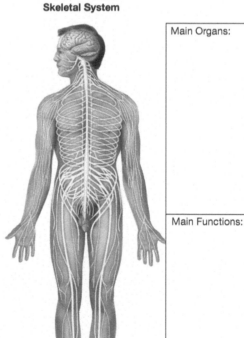

Main Organs:

Main Functions:

Nervous System

FIGURE **1.14** Organ systems of the body *(continues)*

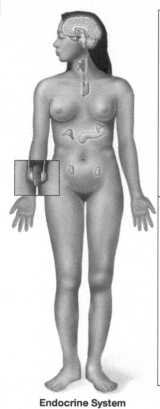

Main Organs:

Main Functions:

Endocrine System

Main Organs:

Main Functions:

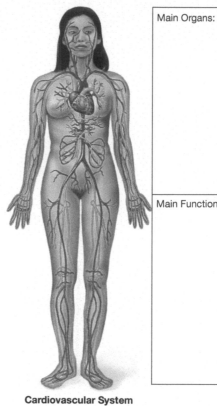

Cardiovascular System

Main Organs:

Main Functions:

Lymphatic System

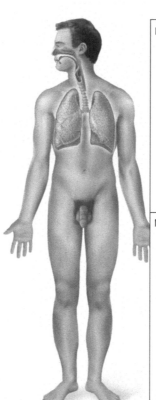

Main Organs:

Main Functions:

Respiratory System

FIGURE **1.14** *(cont.)* Organ systems of the body *(continues)*

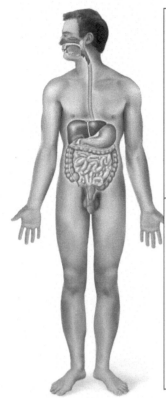

Main Organs:

Main Functions:

Digestive System

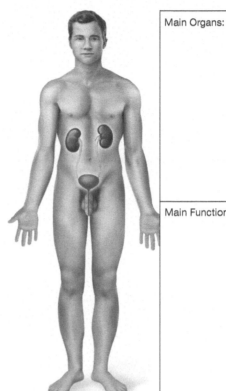

Main Organs:

Main Functions:

Urinary System

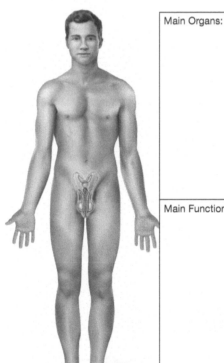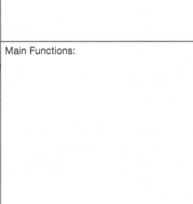

Main Organs:

Main Functions:

Male Reproductive System

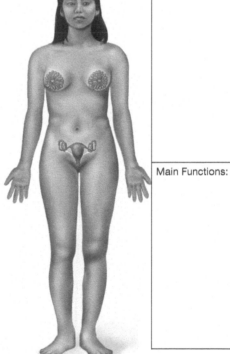

Main Organs:

Main Functions:

Female Reproductive System

FIGURE **1.14** *(cont.)* Organ systems of the body.

1 Which of the following best describes anatomical position?
a. Body facing forward, toes pointing forward, palms facing backward.
b. Body, toes, and palms facing backward.
c. Body facing forward, arms at the sides, palms facing forward.
d. Body facing backward and palms facing outward.

2 *Matching:* Match the directional term with its correct definition.

_____ Distal   A. Away from the surface/toward the body's interior

_____ Lateral   B. Toward the back of the body

_____ Anterior   C. Closer to the point of origin (e.g., of a limb)

_____ Proximal   D. Away from the body's midline

_____ Inferior   E. Toward the head

_____ Deep   F. Farther from the point of origin (e.g., of a limb)

_____ Superficial   G. Toward the body's midline

_____ Posterior   H. Away from the head/toward the tail

_____ Medial   I. Toward the front of the body

_____ Superior   J. Toward the surface/skin

3 Which of the following is a correct use of a directional term?
a. The ankle is inferior to the knee.
b. The sternum is distal to the abdomen.
c. The bone is superficial to the muscle.
d. The ear is lateral to the mouth.

4 *Fill in the blanks:* A serous membrane secretes _____, which

_____ the organs in certain _____ body cavities.

5 Label the following anatomical regions on **Figure 1.15**.

- ❏ Arm
- ❏ Axillary region
- ❏ Brachial region
- ❏ Carpal region

- ❏ Cervical region
- ❏ Femoral region
- ❏ Forearm
- ❏ Leg

- ❏ Otic region
- ❏ Tarsal region
- ❏ Thigh
- ❏ Vertebral region

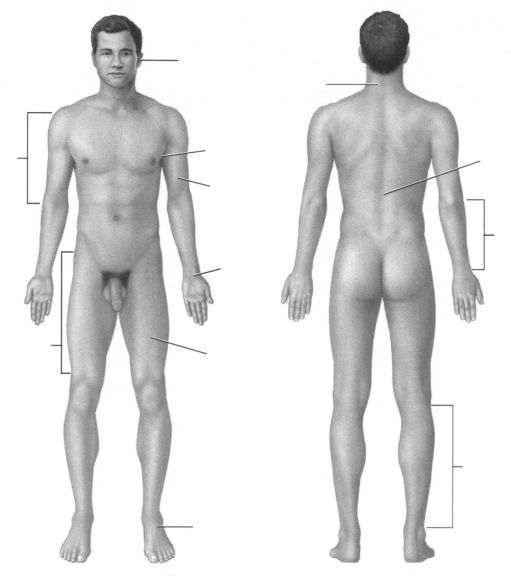

FIGURE **1.15** Anterior and posterior views of the body.

6 Anatomical position and specific directional and regional terms are used in anatomy and physiology to
a. standardize units of measure.
b. provide a standard that facilitates communication and decreases the chances for errors.
c. provide a standard used to develop drug delivery systems.
d. make students' lives difficult.

UNIT 1 QUIZ
(continued)

Name _____

Section _____ Date _____

7 Label the following body cavities on **Figure 1.16**, and indicate with an asterisk (*) which cavities are surrounded by serous membranes.

❑ Abdominopelvic cavity

❑ Cranial cavity

❑ Mediastinum

❑ Pericardial cavity

❑ Peritoneal cavity

❑ Pleural cavity

❑ Vertebral cavity

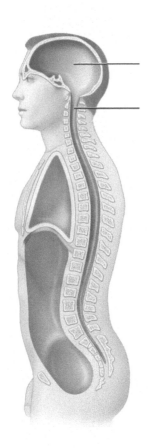

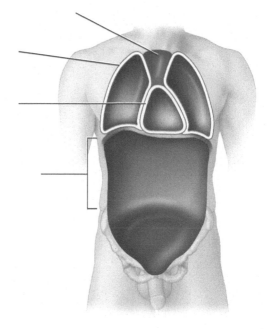

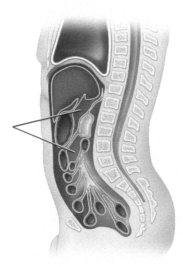

FIGURE **1.16** Body cavities.

8 Define the following planes of section:

a Midsagittal plane _____

b Parasagittal plane _____

c Frontal plane _____

d Transverse plane _____

9 The following organs belong to the _____ system: spleen, lymph nodes, and thymus.

    a. integumentary

    b. reproductive

    c. lymphatic

    d. digestive

10 **Figure 1.17** is not in anatomical position. List all the deviations from anatomical position.

_____

_____

_____

_____

FIGURE **1.17**
Figure not in
anatomical position.

11 Locate the following wounds, and mark them on the body illustrated in **Figure 1.18**.

a The wound is located in the right superior, posterior acromial region. It is 8 centimeters lateral to the vertebral region and 3 centimeters superior to the scapular region.

b The wound is located on the left anterior, medial femoral region, 10 centimeters distal to the inguinal region.

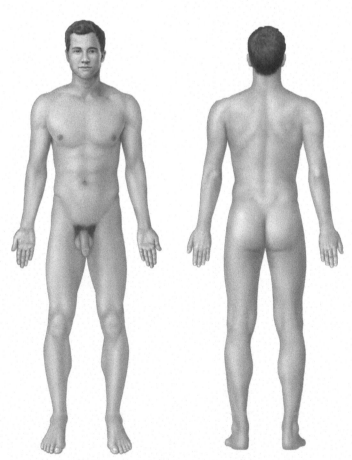

FIGURE **1.18** Anterior and posterior views of the body.

UNIT **1** QUIZ
(continued)

12 You are reading a surgeon's operative report. During the course of the surgery, she made several incisions. Your job is to read her operative report and determine where the incisions were made. Draw and label the incisions on **Figure 1.19**.

    a The first incision was made in the right anterior sternal region, 3 centimeters inferior to the cervical region. The cut extended vertically in an inferior direction, ending 2 centimeters superior to the umbilical region.

    b The second incision began in the left anterior, lateral inguinal region and extended horizontally in a medial direction for 4 centimeters to the pubic region. At the pubic region, the cut turned and extended vertically in a distal direction for 4 centimeters and ended in the proximal femoral region.

    c The third incision was made in the left posterior, lateral femoral region, 2 centimeters proximal to the popliteal region. The cut extended in a distal direction to 2 centimeters proximal to the tarsal region.

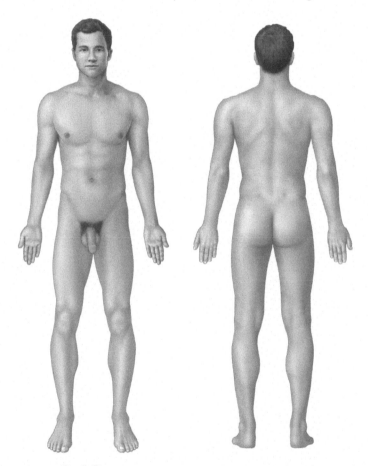

FIGURE **1.19** Anterior and posterior views of the body.

The Chemical Level of Organization

When you have completed this unit, you should be able to:

1 Demonstrate the proper interpretation of pH paper.

2 Describe and apply the pH scale.

3 Describe the purpose and effects of a buffer.

4 Explain the purpose of an enzyme.

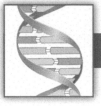

Name _____ Section _____ Date _____

PRE-LAB EXERCISES

Complete the following exercises prior to coming to lab, using your lab manual and textbook for reference.

Pre-Lab Exercise 2-1

✎ Key Terms

You should be familiar with the following terms before coming to lab.

Term	Definition
pH	
Acid	
Base	
Buffer	
Chemical reaction	
Enzyme	

Pre-Lab Exercise 2-2

The pH Scale and –logarithms

The **pH scale** is a numerical scale from 0 to 14 that represents the hydrogen ion concentration of a solution. Solutions with a pH of 0 have the highest hydrogen ion concentration, and solutions with a pH of 14 have the lowest hydrogen ion concentration. This is because the pH isn't the actual hydrogen ion concentration; instead, it is a negative logarithm of a solution's hydrogen ion concentration. This actually isn't as difficult as it sounds. Grab your calculator and enter the following:

–log(0.01)   =   _____   –log(0.0001)   =   _____

–log(0.001)   =   _____   –log(0.00001)   =   _____

Notice the general trend here: *As the number gets smaller, its negative log gets bigger.*

Here's a quick example using hydrogen ions: Solution X has a hydrogen ion concentration of 0.05 M. Solution Y has a hydrogen ion concentration of 0.0002 M.

❚ Which solution has more hydrogen ions? (*Hint:* It's just the solution with the bigger number.) _____

❚ What is the –log of solution X? _____   ❚ What is the –log of solution Y? _____

As you can see, the solution with the *higher* hydrogen ion concentration has the *lower* –log. This means that:

❚ Solution X is the more _____ (acidic/basic) solution and has a _____ (lower/higher) pH.

❚ Solution Y is the more _____ (acidic/basic) solution and has a _____ (lower/higher) pH.

EXERCISES

Molecular model of lumazine synthase, an enzyme involved in riboflavin (vitamin B₂) biosynthesis.

Including a chemistry unit with an anatomy and physiology course may seem odd, but consider for a moment the simplest level of organization—the chemical level. Our cells, tissues, and organs, as well as our extracellular environments, are all composed of chemicals that undergo countless chemical reactions every second. So, to be able to understand our anatomy and physiology, we first must understand the most basic structures in our bodies—chemicals. The following exercises introduce you to the world of chemistry with procedures pertaining to the pH scale, buffers, and enzymes.

Safety Note

⚠ Many of the chemicals you will use in this lab are poisons and can burn your skin, eyes, and mucous membranes. Take care in handling all chemicals. Do not handle any chemicals without wearing safety glasses, gloves, and a lab coat.

Exercise 2-1

pH, Acids, and Bases

MATERIALS

- ❏ 3 glass test tubes
- ❏ Test tube rack
- ❏ Graduated cylinder
- ❏ Dropping pipette
- ❏ pH paper
- ❏ Samples of various acids and bases
- ❏ 0.1 M hydrochloric acid (HCl)
- ❏ Samples of Tums, Rolaids, and Alka-Seltzer antacid tablets
- ❏ Glass stirring rods
- ❏ Sharpie

The **pH** represents the concentration of hydrogen ions present in a solution. As the hydrogen ion concentration increases, the solution becomes more **acidic**. As the hydrogen ion concentration decreases, the solution becomes more **alkaline**, or **basic**. Notice that the **pH scale**, shown in Figure 2.1, ranges from 0 (the most acidic) to 14 (the most basic). As you learned in Pre-Lab Exercise 2-2, the pH is actually a negative logarithm, so the lower the pH, the higher the hydrogen ion concentration. A pH of 7 is considered **neutral**, which is neither acidic nor basic, because the number of hydrogen ions equals the number of base ions in the solution.

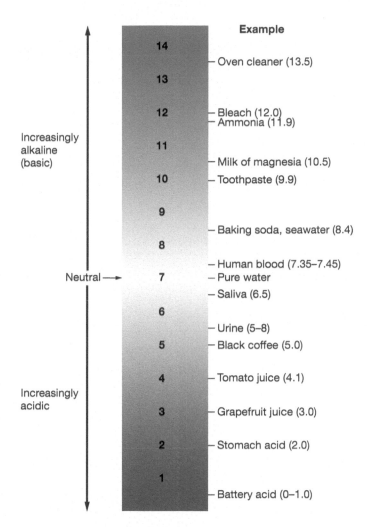

FIGURE 2.1 pH scale and examples of solutions with different pH values.

Acids and bases are some of the most important chemicals in the human body: They are found in the stomach, in the blood and other extracellular body fluids, in the cytosol, and in the urine, and are released by cells of the immune system. The normal pH for the blood is about 7.35 to 7.45. The body regulates this pH tightly because swings of even 0.2 points in either direction can cause serious disruptions to homeostasis.

Procedure 1 Reading the pH

A simple way to measure pH is to use pH paper (Fig. 2.2). To test the pH with pH paper, place one or two drops of the sample solution on the paper with a dropping pipette, and compare the color change with the colors on the side of the pH paper container. The pH is read as the number that corresponds to the color the paper turned.

Safety Note

⚠ Safety glasses and gloves are required!

1 Obtain two samples each of known acids and bases, and record their molecular formulae in Table 2.1 (the molecular formula should be on the side of each bottle).

2 Measure the pH of each acid and base using pH paper and a dropping pipette, and record their pH values in Table 2.1.

3 Obtain two randomly selected unknown samples, and measure their pH using pH paper and a dropping pipette.

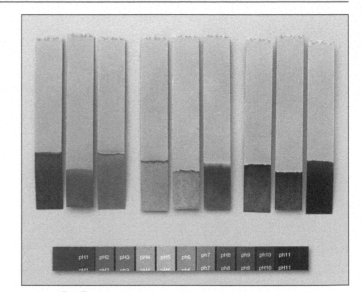

FIGURE 2.2 pH paper and indicator strip.

4 Record the pH values in Table 2.1, and determine whether each substance is an acid, a base, or is neutral.

TABLE 2.1 Samples of Acids and Bases

Samples	Molecular Formula	pH
Acid #1		
Acid #2		
Base #1		
Base #2		
Unknown Samples	**pH**	**Acid, Base, or Neutral?**
Unknown #1		
Unknown #2		

Procedure 2 pH Applications

Now let's apply the pH scale to physiological systems. The stomach contains concentrated hydrochloric acid (HCl), and the pH of the stomach contents ranges between 1 and 3. **Antacids** are medications that raise the pH of the stomach contents to treat a variety of conditions, including *gastroesophageal reflux*, commonly known as *heartburn*. In this activity, you will compare the effectiveness of three widely available antacids in neutralizing concentrated hydrochloric acid.

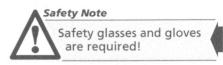

Safety Note

Safety glasses and gloves are required!

1 Obtain three glass test tubes, and label them 1, 2, and 3 with a Sharpie.

2 Obtain a bottle of 0.1 M hydrochloric acid (HCl), and record its pH in Table 2.2.

3 In each tube, place 2 mL of the 0.1 M HCl.

4 Add about one-fourth of one crushed Tums tablet to Tube 1. Stir the contents with a glass stirring rod.

5 Add about one-fourth of one crushed Rolaids tablet to Tube 2. Stir the contents with a glass stirring rod.

6 Add about one-fourth of one crushed Alka-Seltzer tablet to Tube 3. Stir the contents with a glass stirring rod.

7 Allow the tubes to sit undisturbed for 3 minutes.

8 Measure the pH of the contents of each tube, and record the values in Table 2.2.

9 What effect did the antacids have on the acid? _____

TABLE **2.2** Results of Antacid Test

Substance Tested	Molecular Formula (or Active Ingredient)	pH of Solution
Acid alone		
Tums + acid		
Rolaids + acid		
Alka-Seltzer + acid		

10 Based upon your observations, which antacid is the most effective at neutralizing the acid?

_____

_____

Exercise 2-2

Buffers

MATERIALS

- ❑ Well plates
- ❑ Stirring rods
- ❑ Dropping pipettes
- ❑ 0.1 M hydrochloric acid (HCl)
- ❑ Buffered solution
- ❑ Distilled water
- ❑ 0.1 M sodium hydroxide (NaOH)
- ❑ pH paper
- ❑ Sharpie

A **buffer** is a chemical that resists large or dramatic changes in pH; a solution to which a buffer has been added is called a **buffered solution.** When acid is added to a buffered solution, the buffer binds the added hydrogen ions and removes them from the solution. Similarly, when a base is added to a buffered solution, the buffer releases hydrogen ions into the solution. Both effects minimize pH changes that otherwise would occur in the solution if the buffer were not present.

It's important to note that a buffer does not necessarily make the pH of a solution *neutral*. If the starting pH of a buffered solution is 2 (very acidic), it will remain around 2 when acid or base is added. Similarly, if a buffered solution's starting pH is 11 (very basic), it will remain around 11 when you add acid or base. The buffer simply resists changes in pH—it does not neutralize the solution.

Procedure 1 Testing Buffered Solutions

In this experiment, you will examine the effects of adding an acid or a base to buffered solutions and nonbuffered solutions. You will use distilled water as your nonbuffered solution, and your instructor will choose an appropriate buffered solution for you to use.

1 Obtain a well plate, and number four wells 1, 2, 3, and 4 with a Sharpie.

2 Fill Wells 1 and 2 about half-full of distilled water. Measure the pH of the distilled water, and record that value in Table 2.3.

3 Fill Wells 3 and 4 about half-full of buffered solution. Measure the pH of the buffered solution, and record that value in Table 2.3.

4 Add two drops of 0.1 M HCl to Wells 1 and 3. Stir the solutions with a stirring rod (or a toothpick), and measure the pH of each well. Record the pH in Table 2.3.

5 Add two drops of 0.1 M NaOH (a base) to Wells 2 and 4. Stir the solutions, and measure the pH of each well. Record the pH in Table 2.3.

6 Interpret your results. What effect did the buffer have on the pH changes you saw?

Safety Note

Safety glasses and gloves are required!

TABLE 2.3 Buffered and Nonbuffered Solutions

Well	Initial Solution Contents	pH	Solution After Adding Acid or Base Contents	pH
1	Water		Water + 0.1 M HCl	
2	Water		Water + 0.1 M NaOH	
3	Buffered solution		Buffered solution + 0.1 M HCl	
4	Buffered solution		Buffered solution + 0.1 M NaOH	

Exercise 2-3

Enzymes and Chemical Reactions

MATERIALS

- ❑ 6 glass test tubes
- ❑ Test tube rack
- ❑ Lipase
- ❑ Boiled lipase
- ❑ Bile salts
- ❑ Vegetable oil
- ❑ 0.1 M sodium hydroxide (NaOH)
- ❑ Ice-water bath
- ❑ Warm-water bath (set to 37°C)
- ❑ Phenol red
- ❑ Sharpie

Often, when atoms, molecules, or compounds interact, chemical bonds are formed, broken, or rearranged, or electrons are transferred between reactants. These interactions are called **chemical reactions**. Most chemical reactions can proceed spontaneously, but they often take an extremely long time.

One factor that can alter the rate at which a reaction takes place is *temperature*. Generally, when the temperature increases (up to a point), molecules move faster, and they collide and react at a faster rate. The opposite happens when temperature decreases.

Another factor that can affect reaction rate is the addition of a substance called a **catalyst** to a reaction. Catalysts increase the rate of a reaction, but are not consumed in the reaction and may be reused after it has completed.

In the body, biological catalysts called **enzymes**, nearly all of which are **proteins**, speed up essentially all of our chemical reactions. They work by reducing the amount of energy required for a reaction to proceed (called the *activation energy*). In order to do this, an enzyme must bind specifically to its unique set of reactants. This means that if the enzyme is damaged and loses its shape, then it will not be able to function. Enzymes can be damaged by the same processes that can damage all proteins, including extreme heat and extreme pH swings. An enzyme that has lost its shape as a result of such damage is said to be **denatured** and is not able to catalyze reactions.

Procedure 1 Testing Enzymatic Activity

In the following procedure, you will be comparing the ability of two solutions—lipase and boiled lipase—to digest vegetable oil at three different temperatures. **Lipase** (LY-payz) is an enzyme found in the human digestive tract that catalyzes the digestion of fats called **triglycerides** (try-GLISS-er-aye′dz). Triglycerides consist of three long hydrocarbon chains called *fatty acids* bound to a sugar alcohol called *glycerol* (Fig. 2.3). Lipase catalyzes the reactions that break a triglyceride into two free fatty acids and a *monoglyceride*.

You will check for the presence of digestion using an indicator called **phenol red**. Phenol red appears pink at an alkaline (basic) pH, changes to an orange-red color at a neutral pH, and changes to a yellow color when the pH becomes acidic. If triglycerides have been digested, fatty acids will be released that will decrease the pH of the contents of your tube and turn them orange-red or yellow. You may interpret your results in the following way (Fig. 2.4):

- ❚ Yellow color = the pH is acidic, and significant triglyceride digestion occurred.

- ❚ Red-orange color = the pH is neutral, and some triglyceride digestion occurred.

- ❚ Pink color = the pH is basic, and no (or limited) triglyceride digestion occurred.

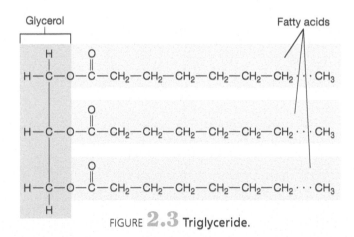

FIGURE 2.3 Triglyceride.

FIGURE 2.4 Possible results from the lipid digestion experiment.

Note that you also must add another component called **bile** to your mixture for digestion to occur. Bile is not an enzyme itself, but it does increase the ability of lipase to catalyze triglyceride digestion.

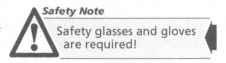

2

1 Obtain six glass test tubes, and label them 1 through 6 with a Sharpie.

2 Add 3 mL of vegetable oil to each test tube.

3 Add 8 to 10 drops of the pH indicator phenol red to each test tube. The oil now should appear pink. If it does not, add drops of 0.1 M NaOH (a base) to each tube until the indicator turns pink.

4 Add the following ingredients to each test tube:

- Tubes 1, 3, and 5: 1 mL lipase, 1 mL bile.
- Tubes 2, 4, and 6: 1 mL boiled lipase, 1 mL bile.

5 Place Tubes 1 and 2 in an ice-water bath, and leave them to incubate for 30 minutes.

6 Leave Tubes 3 and 4 in your test tube rack to incubate at room temperature for 30 minutes.

7 Place Tubes 5 and 6 in a warm-water bath set to 37°C, and leave them to incubate for 30 minutes.

8 After 30 minutes have passed, remove the tubes from the ice-water and warm-water baths, and place them in your test tube rack.

9 Record the color and interpretations of your results for each tube in Table 2.4.

10 Answer the following questions about your results:

a What effect does temperature have on enzyme activity?

_____

_____

b Were your results with the boiled lipase different from the results with nonboiled lipase? Why?

_____

_____

_____

TABLE **2.4** Enzymatic Activity of Lipase

Tube Number	Color	pH (Acidic, Neutral, or Alkaline)	Amount of Digestion that Occurred
1			
2			
3			
4			
5			
6			

Name _____

Section _____ Date _____

1 *Fill in the blanks:* An acidic solution has a pH value of _____,

a neutral solution has a pH value of _____, and a basic (alkaline)

solution has a pH value of _____.

2 The pH is a measure of a solution's
   a. hydrogen concentration.
   b. oxygen concentration.
   c. cation concentration.
   d. hydrogen ion concentration.

3 What is the pH detected by the pH paper in **Figure 2.5**?

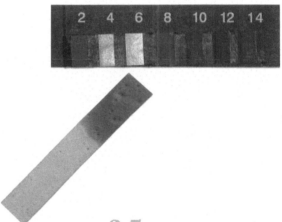

FIGURE **2.5** pH paper and indicator strip.

4 What does a buffer do when hydrogen ions are added to a solution?
   a. The buffer keeps the pH of a solution neutral.
   b. The buffer releases base ions.
   c. The buffer releases additional hydrogen ions.
   d. The buffer binds the hydrogen ions to remove them from the solution.

5 A chemical reaction takes place when
   a. chemical bonds are formed.
   b. chemical bonds are broken.
   c. chemical bonds are rearranged.
   d. electrons are transferred between reactants.
   e. All of the above.

6 *Fill in the blanks:* Increasing the temperature generally _____ the rate of reactions, and decreasing the temperature generally _____ the rate of reactions.

7 How does an enzyme affect the rate of a chemical reaction, and how does it accomplish this?

8 An enzyme that has lost its shape
   a. is denatured and no longer functional.
   b. is denatured but still functional.
   c. is not different from a regular enzyme because shape is not important.
   d. can be restored by extreme temperatures or pH swings.

9 Your lab partner argues with you that if you add base to a solution, the hydrogen ion concentration should decrease and, therefore, the pH should decrease. Is he correct? What do you tell him?

10 The same lab partner tells you that if you add a buffer to a solution, it will make the solution neutral. Is he correct? What do you tell him?

11 Ms. Young presents to the emergency department in respiratory distress. The respiratory therapist measures the pH of her blood and determines that it is 7.15. Is the pH of her blood more acidic or more basic than normal? Has the number of hydrogen ions in her blood increased or decreased? Explain.

Introduction to the Cell and the Microscope

When you have completed this unit, you should be able to:

1 Identify the major parts of the microscope and demonstrate its proper use.

2 Define the magnification of high, medium, and low power and depth of focus.

3 Identify parts of the cell and organelles.

4 Describe the process of diffusion.

5 Describe the effects of hypotonic, isotonic, and hypertonic environments on cells.

6 Identify the stages of the cell cycle and mitosis.

PRE-LAB EXERCISES

Complete the following exercises prior to coming to lab, using your lab manual and textbook for reference.

3

Pre-Lab Exercise 3-1

✎ Key Terms

You should be familiar with the following terms before coming to lab.

Term	Definition
Cell Structures and Organelles	
Plasma membrane	
Cytoplasm	
Nucleus	
Mitochondrion	
Ribosome	
Peroxisome	
Smooth endoplasmic reticulum (SER)	
Rough endoplasmic reticulum (RER)	
Golgi apparatus	
Lysosome	
Centrosome	

Cilia _____

Flagella _____

Membrane Transport

Diffusion _____

Osmosis _____

Tonicity _____

Cell Cycle and Mitosis

Cell cycle _____

Interphase _____

Mitosis _____

The Parts of the Cell

Color the parts of the cell in Figure 3.1, and label them with the following terms. Use Exercise 3-2 (p. 54) in this unit and your text for reference.

❑ Plasma membrane

❑ Cytoplasm
  ☐ Ribosome
  ☐ Mitochondrion
  ☐ Rough endoplasmic reticulum (RER)
  ☐ Smooth endoplasmic reticulum (SER)
  ☐ Golgi apparatus
  ☐ Lysosome
  ☐ Centrioles

❑ Nucleus
  ☐ Nuclear envelope
  ☐ Nuclear pores
  ☐ Chromatin
  ☐ Nucleolus

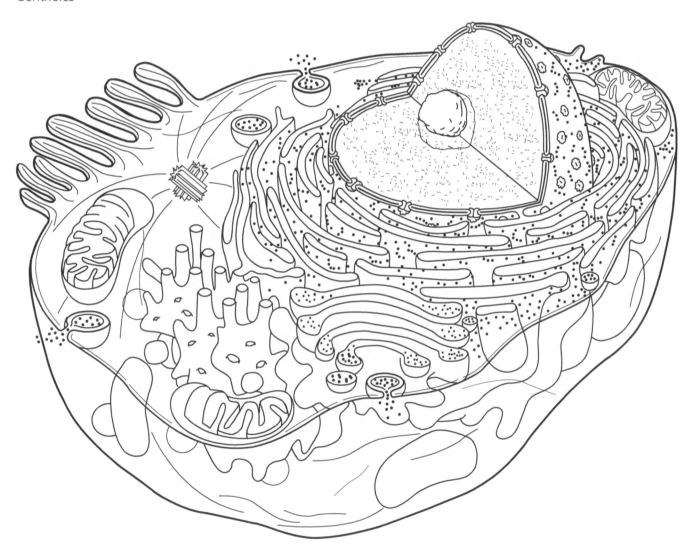

FIGURE **3.1** Generalized cell.

Blood sample being examined by a light microscope.

A basic principle we will revisit repeatedly in our study of anatomy and physiology is:

Form follows function.

This principle may be stated in a variety of ways and is alternatively called the "principle of complementarity of structure and function." Essentially, this means the structure's anatomy (the *form*) is always suited for its physiology (the *function*). This is obvious at the organ level: Imagine if the heart were solid rather than composed of hollow chambers, or if the femur were pencil-thin rather than the thickest bone in the body. These organs wouldn't be able to carry out their functions of pumping blood and supporting the weight of the body very well, would they? But this principle is applicable even at the chemical and cellular level, which we will see in this unit.

We begin with an examination of the instrument we use to look at the cell: the microscope. We then turn our attention to the cell, its structures, and two critical processes that occur at the cellular level: *diffusion* and *osmosis*. In the final exercise, we explore the *cell cycle* and the process of cell division.

Note that this is the first lab in which you will use a *Model Inventory*—something you will use throughout this lab manual. In this inventory, you will list the anatomical models or diagrams you use in lab (if the model is not named, make up a descriptive name for it), and record in a table the structures you are able to locate on each model. This is particularly helpful for study purposes because it allows you to return to the proper models to locate specific structures.

Exercise 3-1

Introduction to the Microscope

MATERIALS

❏ Light microscopes with three objective lenses
❏ Introductory slides (letter "e" and three colored threads)
❏ Fetal pigs (or other preserved small mammals)
❏ Sterile cotton swabs
❏ Methylene blue dye
❏ Glass slide (blank)
❏ Distilled water
❏ Coverslip
❏ Colored pencils

Working with the microscope and slides seems to be one of the least favorite tasks of anatomy and physiology students. But with a bit of help, a fair amount of patience, and a lot of practice, the use of microscopes becomes progressively easier with each unit.

The microscopes that you will use in this lab are called **light microscopes** (Fig. 3.2). This type of microscope shines light through the specimens to illuminate them, and the light is refracted through objective lenses to magnify the image. Light microscopes have the following components:

■ **Ocular lens.** The **ocular lens** (AWK-yoo-lur) is the lens through which you look to examine the slide. It generally has a 10× power of magnification, meaning it magnifies the image you are examining 10 times. The microscope may have one ocular lens (a **monocular** microscope) or two ocular lenses (a **binocular** microscope). Many ocular lenses have pointers that can be moved by rotating the black **eyepiece.** The area of the slide visible when you look into the ocular is known as the **field of view.**

■ **Objective lenses.** The **objective lenses** are lenses with various powers of magnification. Most microscopes have low- (4×), medium- (10×), and high-power (40×) objective lenses. Note that sometimes the 4× objective is referred to as the *scan* objective. On such microscopes, the 10× objective is actually called the low-power objective. The objective lenses are attached to the **nosepiece,** which allows the user to switch between objectives. Certain microscopes have a higher power objective (100×), called the **oil-immersion lens,** which requires that a drop of oil be placed between the slide and objective lens. Without this drop of oil, you will not be able to focus the image.

■ **Stage.** The **stage** is the surface on which the slide sits. It typically has stage clips to hold the slide in place. The stages of many microscopes are moveable thanks to the **mechanical stage adjustment knob.** On such microscopes, you can move the slide by turning this knob. Other microscopes have stages that are not adjustable, and these require you to move the slide manually.

■ **Arm.** The **arm** supports the body of the microscope and typically houses the adjustment knobs.

- **Coarse adjustment knob.** The large knob on the side of the arm is the **coarse adjustment knob.** Turning it moves the stage up and down to change the distance of the stage from the objective lenses. This allows gross focusing of the image.

- **Fine adjustment knob.** The smaller knob underneath the coarse adjustment knob is the **fine adjustment knob.** Turning it allows fine-tuning of the image's focus. Note that occasionally the fine adjustment knob is situated on top of the coarse adjustment knob.

- **Lamp.** The **lamp,** also called the *illuminator,* provides the light source. It rests on the base of the microscope. Light from the lamp is focused on the specimen by the **condenser,** a lens that sits under the stage.

- **Iris diaphragm.** On the underside of the stage is an adjustable wheel called the **iris diaphragm** that controls the amount of light allowed to pass through the slide. It is important to check the iris diaphragm when you are focusing on an image. Too much light will wash the image out, and too little light will make it harder to see details. You may also be able to adjust the amount of light coming into the image with a **light adjustment dial** (or knob), which is found on the arm or base of the microscope. This dial allows you to control the brightness of the lamp.

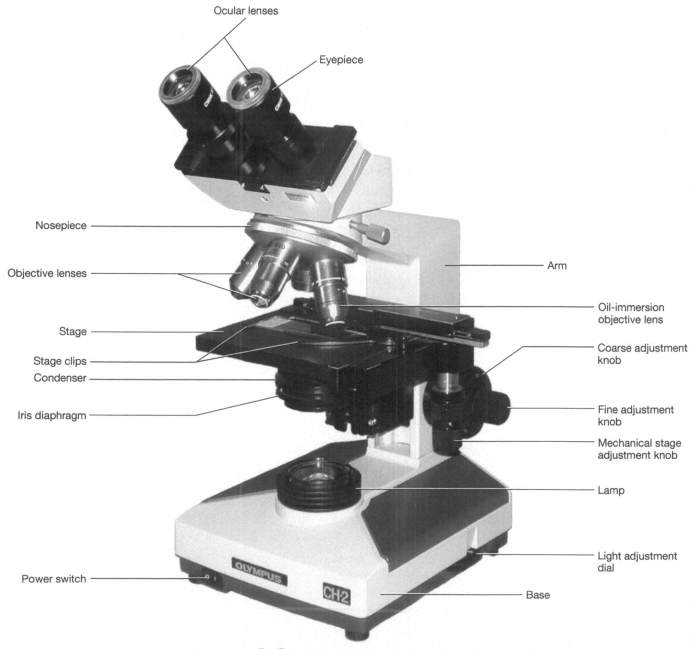

FIGURE **3.2** Compound light microscope.

As I'm certain your professor will tell you, microscopes are expensive! Care must be taken to ensure that the microscopes stay in good working condition. Taking proper care of microscopes makes the histology sections of your labs run more smoothly, and also ensures that you stay on your lab instructor's good side. Keeping that in mind, following are some general guidelines for handling the microscopes:

▮ Use two hands to support the microscope when carrying it—one hand to hold the arm and the other hand to support the base.

▮ Gather up the cord so it does not dangle off the lab table after you have plugged in the microscope. This will help to prevent people from tripping over loose cords.

▮ Clean the lenses with lens paper only, as paper towels and cloth will scratch its surface.

▮ Make sure the *lowest* power objective is in place before you begin. This will prevent you from cracking the slide and, possibly, the lens.

▮ Get the image in focus on low power, then switch to higher power and adjust with the **fine** adjustment knob. Be careful not to use the coarse adjustment knob with the high-power objective in place, because you could break the slide and damage the lens.

▮ When you are finished with the microscope, turn the nosepiece to the lowest power objective, and remove the slide. Be sure to clean off the objective if you used oil, because oil left on the objective tends to harden. Turn off the power to the microscope, and unplug it. This will decrease the chances of a blown bulb or fuse next time the microscope is used. Before putting the microscope away, wrap the electrical cord around the base or secure with a Velcro strap, and cover it with a dust cover.

If you follow these general guidelines, you can rest assured that the microscopes (and your grade) will not suffer any harm.

Procedure 1 Magnification

Light is refracted through two lenses to obtain magnification—the ocular lens and the objective lens.

Magnification of the ocular lens is usually 10×. This means that if you were to view a slide with only the ocular lens, the image would be magnified 10 times. Magnification of the objective lenses varies, but typically is 4× for low power, 10× for medium power, and 40× for high power. (To verify that this is the case for your microscope, look at the side of the objective lens, which usually is labeled with its magnification.) Remember that oil immersion provides even greater magnification at 100×. When calculating the total magnification, you must multiply the magnification of the ocular lens (10) by the power of the objective lens.

Fill in Table 3.1 to determine total magnification at each of the different objective lenses on your microscope. Remember that the magnification of the objective lens is usually printed on the side of the lens itself.

TABLE **3.1** Total Magnification at Each Power

Magnification of Ocular Lens	Power	Magnification of Objective Lens	Total Magnification
	Low		
	Medium		
	High		
	Oil immersion		

Procedure 2 Focusing the Microscope

Now that we know how to handle the microscope properly, let's practice using it.

1 Obtain a slide of the letter "e."

2 Examine the letter "e" slide macroscopically (with the naked eye) before placing it on the stage. How is the "e" oriented

on the slide? Is it right side up, upside down, backward, etc.? _____

3 Switch the nosepiece to the low-power objective, place the slide on the stage, and secure it with the stage clips. Move the slide using the stage adjustment knob until the "e" is over the condenser.

4 Use the coarse adjustment knob to bring the slide into focus slowly. Once it is grossly in focus, use the fine adjustment knob to sharpen the focus. Note that you might need to adjust the iris diaphragm to allow more light to pass through the specimen, as the newsprint is relatively thick. How is the "e" oriented in the field of view? Is it different from the

way it was when you examined it in item 2? _____

5 Move the nosepiece to medium power. You should only have to adjust the focus with the fine adjustment knob; no adjustment of the coarse focus should be necessary.

6 Once you have examined the slide on medium power, move the nosepiece to high power. Again, focus only with the fine adjustment knob.

Wasn't that easy?

Procedure 3 Depth of Field

At times you will look at a slide and see something brown or black and kind of neat-looking with interesting swirls and specks. What is this fascinating discovery you've made? It's dirt on top of the slide. This confusion happens because students tend to focus the objective on the first thing they can make out, which usually is the top of the coverslip on the slide, and this has a tendency to be dirty.

These "dirt discoveries" can be avoided by appreciating what is known as **depth of field**. Also called the *depth of focus*, the depth of field is the thickness of a specimen that is in sharp focus. Thicker specimens will require you to focus up and down in order to see all levels of the specimen. This takes practice and skill, so let's get some practice doing this. Here we will view a slide that has three differently colored threads stacked on top of one another.

1 Obtain a slide with three colored threads. The threads on the slide are stacked on top of one another, and you will examine each thread individually to practice focusing on all levels of a specimen.

2 Examine the slide macroscopically prior to putting it on the stage.

3 Switch the nosepiece to the low-power objective, place the slide on the stage, and secure it with the stage clips. Move the slide using the stage adjustment knob until the threads are in your field of view.

4 Use the coarse adjustment knob to get the slide into focus on low power.

5 Switch to medium power, and use the fine adjustment knob to sharpen the focus. How many threads are in focus?

6 Move the objective up and down slowly with the coarse adjustment knob, focusing on each individual thread. Figure out which color thread is on the bottom, in the middle, and on the top, and write the color order here:

Bottom _____

Middle _____

Top _____

How to Approach Microscopy

Okay, so you can focus on newsprint and threads, but what about cellular structures and tissue sections? Well, those certainly are more difficult, but if you keep the following hints in mind, the task becomes much simpler:

- ℹ **Always start on low power.** You are supposed to start on low power, anyway, to avoid damaging the objective lenses. Sometimes, though, students forget, and jump straight to medium or high power. This risks damaging the lenses and also makes it harder on you. Bear in mind that most slides will have more than one histological or cellular structure on each slide. Starting on low power allows you to scroll through a large area of the slide, and then focus in on the desired part of the section.

- ℹ **Make sure the specimen is over the condenser.** This tip may sound simplistic, but I can't count the number of times that my students have declared in frustration that the slide is blank because they either forgot to move the stage around or the specimen is small and they couldn't find it. The easy solution to this problem is to examine the slide with the naked eye first, and then place it on the stage so that the specimen is right over the condenser (the light). This way, you are guaranteed to see something when you look through the ocular.

- ℹ **Beware of too much light.** It is easy to wash out the specimen with too much light. If you are having difficulty making out details, first adjust the focus with the fine adjustment knob. If this doesn't help, use the iris diaphragm to reduce the amount of light illuminating the specimen. This will increase the contrast and allow you to observe more details. It also helps to reduce headaches and eyestrain.

- ℹ **Keep both eyes open.** It is tempting to close one eye when looking through a monocular microscope. Admittedly, keeping both eyes open isn't easy at first, but it helps to reduce eyestrain and headaches.

- ℹ **Compare your specimen to the photos in your lab manual.** Although the slides you are examining will not necessarily be identical to the photos in this book, they should be similar in appearance. Generally speaking, if you are looking at something that is vastly different from what is in this book, you probably should move the slide around a bit or change to a different power objective to find the correct tissue or cell type on the slide. Other good sources for micrographs include atlases, your textbook, and images on the Internet.

- ℹ **Remember that the slides aren't perfect.** Not all slides will clearly demonstrate what you need to see. Some aren't stained adequately or properly. Some are sectioned at a funny angle. Some don't contain all of the tissue or cell types you need to see. What should you do about this? See the next hint for the answer.

- ℹ **Look at more than one slide of each specimen.** This will help you in the face of subpar slides and also will assist you overall in gaining a better understanding of the specimens you are examining.

- ℹ **Draw what you see.** Although you may tend to resist drawing, even the most basic picture is helpful for two reasons. First, it allows you to engage more parts of your brain in the learning process. The more areas of your brain engaged, the better are the chances you will retain the information. Also, drawing is helpful in that you actually have to look at the specimen long enough to draw it.

- ℹ **Have patience!** It really does get easier. Don't get frustrated, and don't give up. By the end of the semester, you may come to appreciate the microscope and the fascinating world it reveals.

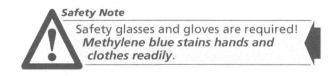

Procedure 4 Preparing a Cell Smear

Now let's do some actual laboratory microscopy. In this procedure, you will prepare what is known as a cell smear (Fig. 3.3). To prepare a cell smear, obtain a sample of cells with a cotton applicator swab and "smear" these cells on a blank slide. However, cells and cell structures aren't visible unless they are first stained, so the next step in preparing a smear is to apply a stain to the sample. In our procedure, we will be using a cell sample obtained from the mouth and applying the stain methylene blue.

Safety Note

⚠ Safety glasses and gloves are required! *Methylene blue stains hands and clothes readily*.

1 Obtain a blank slide and a coverslip.

2 Clean the slide with lens paper.

3 Swab the inside of your cheek with a sterile cotton swab. (As an alternative, swab the inside cheek of a fetal pig or other preserved small mammal.) Do not get large chunks of tissue on the swab, or individual cells will not be visible.

4 Wipe the swab with the cheek cells on the blank slide. If you are using your own cheek cells, dispose of the swab in a biohazard bag.

5 Place one drop of methylene blue dye onto the slide. Wait 1 minute.

6 Rinse the dye off the slide with distilled water and pat dry. The blue dye should be barely visible on the slide. If you see large areas of blue, rinse the slide again. If the blue areas remain, obtain a new sample of cheek cells and repeat the procedure.

7 Place a coverslip over the stained area, and place the slide on the stage of a microscope. Focus the image grossly on low power, then switch to the high-power objective lens to find individual cells (use oil immersion if available).

8 Identify and draw some of the cells from your smear in the space provided with colored pencils.

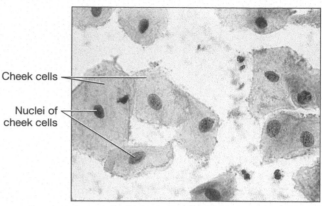

Cheek cells

Nuclei of cheek cells

FIGURE **3.3** Cell smear.

Exercise 3-2

Organelles and Cell Structures

MATERIALS

❏ Cell models and diagrams

❏ Light microscope

❏ Cell slides (with red blood cells, skeletal muscle cells, and sperm cells)

❏ Colored pencils

Most cells in the body are composed of three basic parts—the plasma membrane, the cytoplasm, and the nucleus.

1. **Plasma membrane.** The **plasma membrane** is the outer boundary of the cell (Fig. 3.4). It is a dynamic, fluid structure that acts as a selectively permeable barrier, meaning that it only allows certain solutes to pass into or out of the cell. The main component of the plasma membrane is a **phospholipid bilayer** (FAHS-foh-lip-id), where the two rows of phospholipids form a "sandwich": They line up so that their nonpolar fatty acid tails face one another, and their polar phosphate heads face the water-containing fluids inside and outside of the cell. There are multiple components mixed in with the phospholipids, including proteins, cholesterol, and carbohydrates. In parts of the body where rapid absorption is necessary, the plasma membrane is folded into projections called **microvilli** (my-kroh-VIL-aye), which increase its surface area.

2. **Cytoplasm.** The **cytoplasm** (SY-toh-plaz-m) is the material inside the cell. It consists of three parts: cytosol, the cytoskeleton, and organelles (ohr-gan-ELLZ; Fig. 3.5).

   a. **Cytosol** (SY-toh-sahl) is the fluid portion of the cytoplasm; it contains water, solutes, RNA, enzymes, and other proteins.

   b. The **cytoskeleton** is a collection of protein filaments including actin filaments, intermediate filaments, and microtubules. **Actin filaments** are small protein strands, many of which are located along the plasma membrane and in the core of microvilli, that help maintain the shape of the cell and function in cell movement. The larger **intermediate filaments** are ropelike structures that help maintain the shape of the organelles and the nucleus and give the cell mechanical strength. **Microtubules**, the largest filaments, are hollow tubes that maintain the shape of the cell, hold organelles in place, move substances within the cell, and function in cell division. In addition, microtubules form the core of motile extensions from the cell called cilia and flagella. **Cilia** (SIL-ee-uh) are small, hairlike extensions that beat rhythmically together to propel substances past the cell. **Flagella** (flah-JEL-uh) are single extensions that propel the cell itself (sperm cells are the only flagellated cells in the human body).

   c. **Organelles** are specialized cellular compartments that carry out a variety of functions. The organelles we cover in this unit include the following:

   - **Ribosomes.** The small, granular **ribosomes** (RY-boh-zohmz) are composed of two subunits. Some ribosomes float freely in the cytosol, whereas others are bound to the membrane of another organelle or the nucleus. They are one of the few organelles that are not enclosed by a membrane. Ribosomes are the sites of protein synthesis in the cell.

   - **Peroxisomes.** The small, vesicular organelles known as **peroxisomes** (per-AWKS-ih-zohmz) are small, round membrane-enclosed sacs. They metabolize fatty acids, synthesize certain phospholipids, and contain enzymes that catalyze reactions to detoxify chemicals produced by cellular reactions.

   - **Mitochondria.** The bean-shaped **mitochondria** (my-toh-KAHN-dree-ah) produce the bulk of the cell's ATP (energy). Notice in Figure 3.6 that they are surrounded by a double plasma membrane that encloses a central space called the **matrix**. Within the matrix, we find numerous enzymes, circular mitochondrial DNA, and ribosomes.

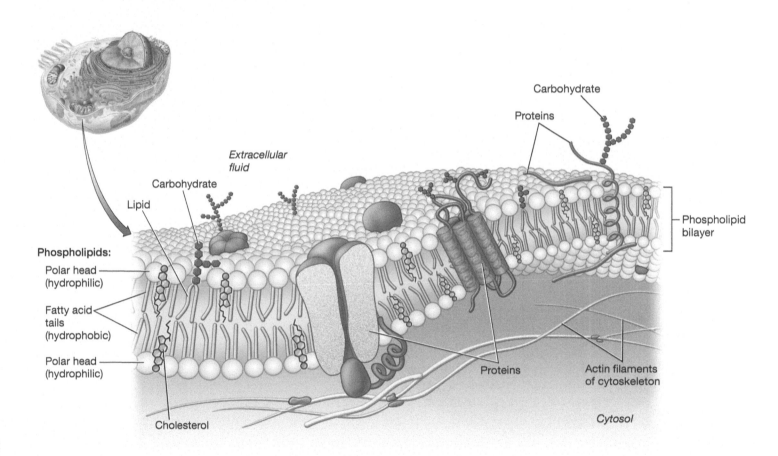

FIGURE **3.4** Plasma membrane.

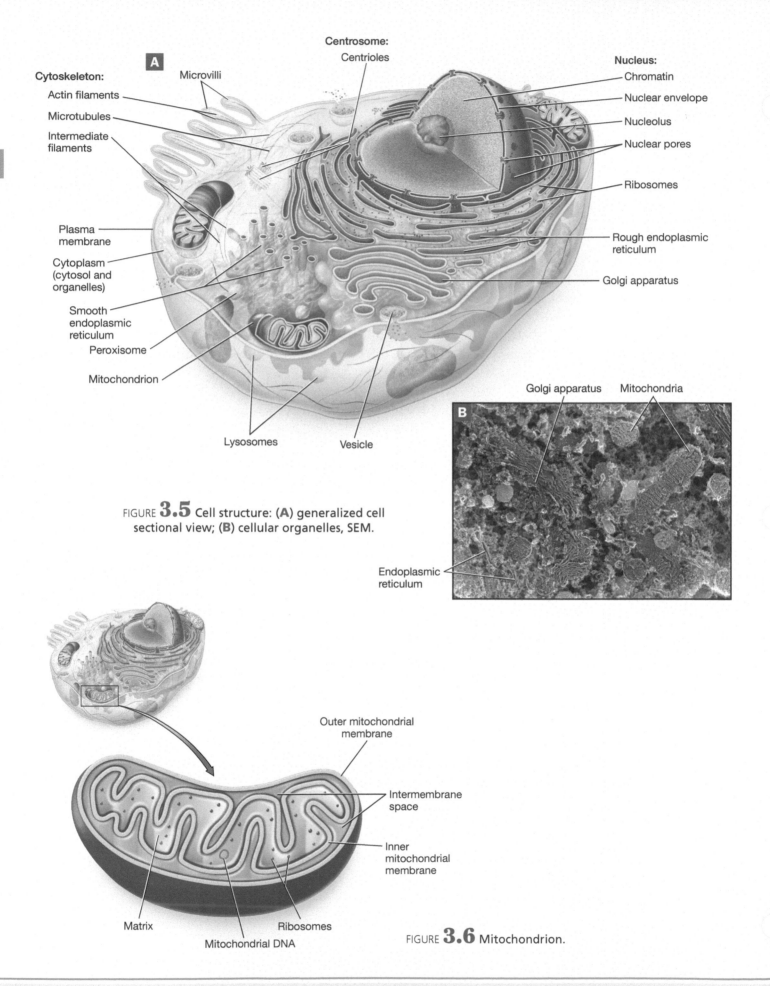

Cytoskeleton:
Actin filaments
Microtubules
Intermediate filaments

A Microvilli

Centrosome:
Centrioles

Nucleus:
Chromatin
Nuclear envelope
Nucleolus
Nuclear pores

Ribosomes

Plasma membrane

Cytoplasm (cytosol and organelles)

Smooth endoplasmic reticulum

Peroxisome

Mitochondrion

Rough endoplasmic reticulum

Golgi apparatus

Lysosomes

Vesicle

FIGURE **3.5** Cell structure: (**A**) generalized cell sectional view; (**B**) cellular organelles, SEM.

B

Golgi apparatus   Mitochondria

Endoplasmic reticulum

Outer mitochondrial membrane

Intermembrane space

Inner mitochondrial membrane

Matrix

Ribosomes

Mitochondrial DNA

FIGURE **3.6** Mitochondrion.

- **Endoplasmic reticulum.** The series of membrane-enclosed sacs known as the **endoplasmic reticulum** (en-doh-PLAZ-mik reh-TIK-yoo-lum) may be of two types: **rough endoplasmic reticulum,** or **RER,** which has ribosomes on its surface, and **smooth endoplasmic reticulum,** or **SER,** which lacks ribosomes. The RER modifies proteins that the ribosomes have made. The SER has multiple functions, including lipid synthesis and detoxification reactions.
- **Golgi apparatus.** The **Golgi apparatus** (GOHL-jee) is a stack of flattened sacs near the RER. Its membrane-enclosed sacs receive products in vesicles from the RER and other places in the cell. Enzymes within the sacs catalyze reactions that process, modify, and sort these products, most of which then exit the Golgi apparatus via new vesicles.
- **Lysosomes.** Lysosomes (LY-soh-zohmz) are membrane-enclosed vesicular organelles filled with digestive enzymes. These enzymes catalyze many types of reactions, including those that break down substances brought into the cell, components within the cell such as old and worn-out organelles, and even the cell itself.
- **Centrioles.** Centrioles (SEN-tree-ohlz) are paired organelles composed primarily of microtubules. They are located in the central area of the cell, called the **centrosome** (SEN-troh-sohm), and appear to be microtubule-organizing centers that are important in facilitating the assembly and disassembly of microtubules.

3. **Nucleus.** The third component in nearly all cells is a specialized structure called the **nucleus** (NOO-klee-us). The nucleus is the cell's biosynthetic center, which directs the synthesis of nearly all the body's proteins, as well as certain nucleic acids. The nucleus is surrounded by a double membrane called the **nuclear envelope,** which contains holes called **nuclear pores.** Within the nucleus we find **chromatin** (KROH-mah-tin), a ball-like mass of tightly coiled DNA and proteins; RNA; and a dark-staining region called the **nucleolus** (noo-klee-OH-lus). The nucleolus contains a type of RNA called **ribosomal RNA** and is the "birthplace" of ribosomes.

Note that the cell shown in Figure 3.5 is a **generalized cell** that contains each organelle in average numbers. Most cells in the body don't look like this and instead are specialized so their structure follows their functions. For example, the cells of the liver contain a large amount of smooth endoplasmic reticulum, and immune cells called phagocytes house many lysosomes.

Procedure **1** Model Inventory for the Cell

Identify the following structures of the cell on models and diagrams, using this unit and your textbook for reference. As you examine the anatomical models and diagrams, record the name of the model and the structures that you were able to identify on the model inventory in Table 3.2.

1. Plasma membrane
   a. Phospholipid bilayer
   b. Microvilli

2. Cytoskeleton
   a. Actin filaments
   b. Intermediate filaments
   c. Microtubules
   d. Cilia
   e. Flagella

3. Cytoplasmic organelles
   a. Ribosomes
   b. Peroxisomes
   c. Mitochondria
   d. Rough endoplasmic reticulum
   e. Smooth endoplasmic reticulum
   f. Golgi apparatus
   g. Lysosomes
   h. Centrioles

4. Nucleus
   a. Nuclear envelope
   b. Nuclear pores
   c. Chromatin
   d. Nucleolus

TABLE **3.2** Cellular Structures Model Inventory

Model/Diagram	Structures Identified

Procedure **2** Time to Draw

In the space below, draw, color, and label one of the cell models that you examined. In addition, write the function of each organelle that you label.

Procedure **3** Examining Cellular Diversity with Microscopy

The structures of different cell types can vary drastically. Cells differ not only in size and shape but also in the types and prevalence of organelles in the cell. In this activity, you will examine prepared microscope slides of red blood cells, sperm cells, and skeletal muscle cells. Use the techniques you learned in Exercise 3-1: Begin your observation on low power, and advance to high power for each slide. Note that sperm cells can be difficult to find, so an oil-immersion lens is helpful for finding the tiny cells. Draw, color, and label the cellular structures and organelles you see on each slide. You may wish to look at Figures 4.9 (p. 87) for red blood cells, 4.10A (p. 93) for skeletal muscle cells, and 16.14 (p. 446) for sperm cells.

Figures 4.9 (p. 87) ... 4.10A (p. 93) ... 16.14 (p. 446)

HINTS & TIPS

How to Draw Useful Micrograph Diagrams

You don't need to be a great artist to draw useful micrograph diagrams. Following are some tips for producing effective drawings that can help you learn the material and study for practical exams.

ℹ️ First look at a diagram of the cell or tissue in question in this book or your textbook to get oriented and get an idea of what you should look for in the field of view.

ℹ️ Make note of the image's magnification. As always, start on low power, and then advance to higher-powered objectives as needed. You should aim for a magnification about the same as the magnification in the diagram.

ℹ️ Reproduce as closely as possible what you see in the field of view using a pencil. This doesn't have to be a work of art, but you should take care to draw the shape(s) of the cells, the way the cells are organized, and the components of the extracellular matrix. You may draw in the circles in this manual, or on a piece of white paper for larger diagrams.

ℹ️ Add color to your drawing with colored pencils, using the slide as a guide. Color is a critical component, particularly for study purposes. Some slides will have uniquely colored stains or staining patterns, and studying the colored images can prove quite helpful. In addition, the slides your lab uses might have different stains than the images in your book. In these cases, it is useful to have a drawing with those specific stain colors.

ℹ️ The final step is to label your drawing. Use your lab manual, textbook, and other resources as a guide to ensure your labels are accurate.

1 Red blood cells

2 Sperm cells

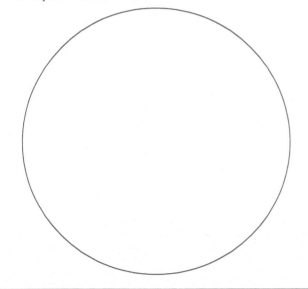

3 Skeletal muscle cells

Exercise 3-3

Diffusion

MATERIALS

- ❑ 2 100 mL beakers
- ❑ Food coloring
- ❑ Hot water
- ❑ Ice water
- ❑ Ruler
- ❑ Sharpie

Diffusion is defined as the movement of solute molecules from a high concentration to a low concentration until a state of equilibrium is reached (Fig. 3.7). Diffusion is a **passive process**—one that requires no net input of energy by a cell—because the energy for diffusion comes from a **concentration gradient**. A concentration gradient is defined as a situation in which two connected areas have different concentrations of a solute. **Figure** 3.7 illustrates a concentration gradient in the panels "Time 1" and "Time 2."

The rate at which diffusion takes place depends upon several factors, including the steepness of the concentration gradient, the temperature, and the size of the particles. Generally, smaller particle size, a steeper concentration gradient, and/or a higher temperature will increase the rate of diffusion. Diffusion will continue until the concentration of the substance is distributed evenly, a condition known as **equilibrium**. At this point, net diffusion ceases, although molecules continue moving. In **Figure** 3.7, equilibrium is illustrated in "Time 3."

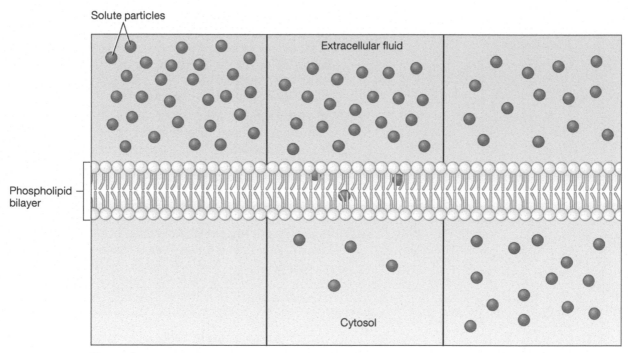

Solute particles

Extracellular fluid

Phospholipid bilayer

Cytosol

Time 1: Steep concentration gradient

Time 2: Diffusion begins

Time 3: Equilibrium

FIGURE **3.7** Diffusion across a plasma membrane.

Procedure 1 Measuring Rates of Diffusion

Diffusion is a process that occurs all around us and, therefore, is easy to witness in action (Fig. 3.8). In this experiment, you will place food coloring in water to examine the effects of temperature on the rate of diffusion. The setup here is very simple, and requires only food coloring and water at two different temperatures.

1 Obtain two 100 mL glass beakers. Label one beaker "cold" and the other "hot" using a Sharpie. Fill the "cold" beaker with ice water (taking care not to get ice in the beaker, though), and fill the "hot" beaker with water that has been heated.

2 Add two drops of food coloring to each beaker.

3 Observe the beakers from the side, and use a ruler to measure the distance that the dye spreads each minute for 5 minutes. Record your results in Table 3.3.

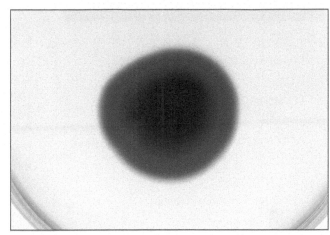

FIGURE **3.8** Example of diffusion as seen with potassium permanganate dye in agar. The dye diffuses outward over time from a high concentration to a low concentration.

TABLE **3.3** Diffusion Results for Food Coloring in Water

Time	Distance of Dye Diffusion: Cold Water	Distance of Dye Diffusion: Hot Water
1 minute		
2 minutes		
3 minutes		
4 minutes		
5 minutes		

4 Interpret your results. What effect does temperature have on the rate of diffusion?

_____

_____

_____

Exercise 3-4

Osmosis and Tonicity

MATERIALS

- ❏ Animal or human blood cells
- ❏ 3 glass slides (blank)
- ❏ Coverslips
- ❏ Dropper
- ❏ Wooden applicator stick
- ❏ 5% dextrose in water solution
- ❏ Deionized water
- ❏ 25% NaCl in water solution
- ❏ Lens paper/paper towel
- ❏ Light microscope with 40× objective
- ❏ Sharpie

For alternate procedure:

- ❏ Pieces of potato cut into equal sizes
- ❏ 3 400 mL beakers
- ❏ Calipers
- ❏ Wax marker

Whereas diffusion refers to the movement of solute, the passive process called **osmosis** (oz-MOH-sis) refers to the movement of *solvent*. Specifically, osmosis is the movement of solvent (usually water in biological systems) from a solution with a lower solute concentration to a solution with a higher solute concentration (Fig. 3.9). A key feature of osmosis is that it only occurs through a *selectively permeable membrane*; in this case, the membrane allows water to cross it but blocks the movement of certain solute particles. For this reason, it is the solvent that crosses the membrane and moves to the more concentrated solution. Notice in the figure that this results in a change in volume of the container—the more dilute solution on the left loses water and so has a lower volume when osmosis has completed. The more concentrated solution on the right gains water and so has a higher volume at the end of the process.

We can compare the solute concentration of two solutions—for example, the cytosol and the extracellular fluid (ECF)—with the concept of **tonicity**. A simple way to imagine the concept of tonicity is as a comparison of the ability of two solutions to cause osmosis. The ECF surrounding a cell can have three variations in tonicity:

- **Isotonic.** An **isotonic** ECF has the same ability to cause osmosis as the cytosol. As neither the ECF nor the cytosol has a stronger "pull" on water, there is no net movement of water into or out of a cell in an isotonic solution (Fig. 3.10A).

- **Hypotonic.** A **hypotonic** ECF has a lower ability to cause osmosis than the cytosol (*hypo* = less). This means that the cytosol has a greater ability to cause osmosis, and so will pull water into the cell. This may cause the cell to swell, which is shown in Figure 3.10B. If a cell is placed in an extremely hypotonic environment, it may swell so much that it ruptures.

- **Hypertonic.** A **hypertonic** ECF has a greater ability to cause osmosis than the cytosol. This causes the ECF to pull water molecules out of the cytosol by osmosis. The cell may shrivel or crenate as it loses water to the ECF, which you can see in Fig. 3.10C.

The following experiment will allow you to watch osmosis in action by placing cells—in this case, red blood cells or *erythrocytes*—into solutions of different tonicity. If your lab doesn't have access to blood products, skip to the alternate procedure.

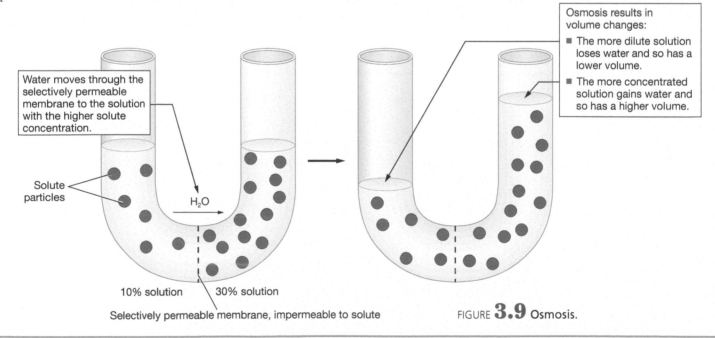

Water moves through the selectively permeable membrane to the solution with the higher solute concentration.

Solute particles

H₂O

10% solution   30% solution

Selectively permeable membrane, impermeable to solute

Osmosis results in volume changes:
- The more dilute solution loses water and so has a lower volume.
- The more concentrated solution gains water and so has a higher volume.

FIGURE **3.9** Osmosis.

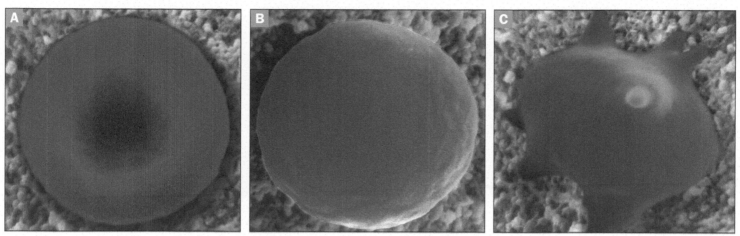

FIGURE **3.10** Effects of solutions of different tonicities on cells:
(**A**) isotonic solution; (**B**) hypotonic solution; (**C**) hypertonic solution.

Procedure **1** Watching Osmosis in Action, Blood Version

1 Obtain three blank slides, and number them 1, 2, and 3 with a Sharpie.

2 Place one small drop of animal blood on each slide with a dropper. Take care to keep your droplet fairly small, otherwise you won't be able to see individual cells.

3 Gently spread the droplet around the center of the slide with a wooden applicator stick, and place a coverslip on it.

4 On Slide 1, place a drop of 5% dextrose solution on one side of the coverslip. On the other side of the coverslip, hold a piece of lens paper or a paper towel. The lens paper will draw the fluid under the coverslip.

5 Observe the cells under the microscope on high power.

6 Repeat steps 3 through 5 with the 25% NaCl solution on Slide 2 and distilled water on Slide 3.

7 Draw and describe what you see with each slide. Then, move on to Procedure 3 to interpret your results.

Slide 1

Description:

Slide 2

Description:

_____

_____

_____

Slide 3

Description:

_____

_____

_____

Procedure **2** Alternate: Watching Osmosis in Action, Potato Version

1 Obtain three pieces of potato of equal size. Mark them as 1, 2, and 3 with a Sharpie.

2 Measure the length and width of each piece of potato with calipers. Record these data in Table 3.4.

3 Obtain three 400 mL beakers and label them as 1, 2, and 3 with a wax marker.

4 Fill Beaker 1 about half-full with distilled water; fill Beaker 2 about half-full with 5% dextrose; and fill Beaker 3 about half-full with 25% NaCl.

5 Place each piece of potato in its corresponding numbered beaker, and leave it in the solution undisturbed for one hour.

6 After an hour has passed, remove the potato pieces and measure their length and width with calipers. Record these data in Table 3.4. After you have recorded your data, move on to Procedure 3.

TABLE **3.4** Results for Potato Osmosis Experiment

Sample	Sample Contents	Length Before	Width Before	Length After	Width After
1	Distilled water				
2	5% dextrose				
3	25% NaCl				

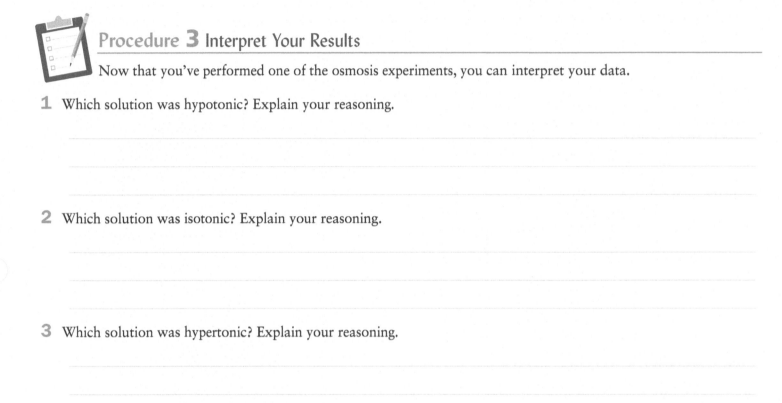

Procedure 3 Interpret Your Results

Now that you've performed one of the osmosis experiments, you can interpret your data.

1 Which solution was hypotonic? Explain your reasoning.

2 Which solution was isotonic? Explain your reasoning.

3 Which solution was hypertonic? Explain your reasoning.

Exercise 3-5

Mitosis and the Cell Cycle

MATERIALS
- ☐ Cell models or diagrams
- ☐ Mitosis models
- ☐ Mitosis slides
- ☐ Light microscope
- ☐ Colored pencils

Many cells go through a continual cycle of growth and replication called the **cell cycle**. The cell cycle consists of four phases (Fig. 3.11):

1. G_1 is the initial growth phase during which the cell grows, develops, and carries out activities that are specific to that cell type. Some cells are *amitotic*, meaning that they never divide, and remain in G_1 indefinitely. These cells are said to be in phase G_0 (pronounced "G not") of the cell cycle.

2. **S phase** is the period during which the cell's DNA is replicated. In this case, the "S" stands for "synthesis" because another copy of the DNA is synthesized.

3. G_2 is the second growth phase during which the cell makes its final preparations for division.

4. **M phase** is the phase during which the cell, sometimes called the *mother cell*, undergoes the process of **mitosis** (my-TOH-sis) and divides its replicated DNA among two identical **daughter cells**. Also occurring during M phase is the process of **cytokinesis** (sy-toh-kin-EE-sis), during which the mother cell's cytoplasm is distributed between the two daughter cells. Each daughter cell has the same exact genetic and structural characteristics as the original mother cell.

The portions of the cycle from G_1 to G_2, when the cell is not dividing, are collectively called **interphase**, which is shown in Figure 3.12A. In G_1, the cell has only a single copy of its DNA. As the cell enters S phase, DNA replication occurs. Human cells have 23 pairs of **homologous chromosomes** (KROH-moh-sohmz): one set from the mother, and one set from the father. After the DNA is replicated, each homologous chromosome exists in a set of "identical twins" called **sister chromatids**. Note that in Figure 3.12A we only show two pairs of homologous chromosomes for simplicity. Note also that the DNA through all of interphase is still in the form of chromatin and you can't see individual chromosomes or sister chromatids—we have shown them here in the illustration for teaching purposes only.

When S phase has completed, the cell enters G_2, and when G_2 is finished, mitosis begins. Mitosis proceeds in the four general stages shown in Figure 3.12B.

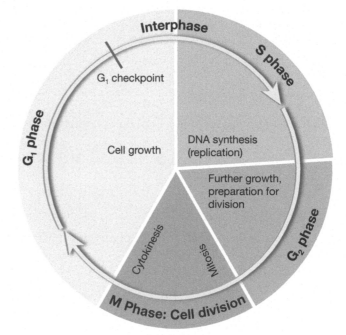

FIGURE 3.11 Cell cycle.

1. **Prophase.** During **prophase**, the nuclear membrane starts to degenerate, and the chromatin condenses into individual **chromosomes**. Also during this stage, we see a structure called the **mitotic spindle** organizing around the centrioles. By the end of late prophase, the centrioles have migrated toward the opposite poles of the cell. Microtubules called spindle fibers emanate from each side of the mitotic spindle and attach to a structure known as the **centromere** (SIN-troh-meer) that joins the two sister chromatids—one spindle fiber attaches to each side of the centromere.

2. **Metaphase.** In **metaphase**, we see the sister chromatids, attached to the spindle fibers, line up along the equator of the cell.

3. **Anaphase.** During **anaphase**, we see the spindle fibers shorten, which pulls the centromeres apart, causing the sister chromatids to migrate toward the opposite poles of the cell. In addition, cytokinesis begins, and the cell elongates.

4. **Telophase.** In the final phase of mitosis, **telophase** (TEL-oh-phayz), a divot called a **cleavage furrow** forms between the two cells (Fig 3.13). As the cleavage furrow progressively narrows, the cell is pinched into two identical daughter cells and cytokinesis is completed. In addition, during this stage, the nuclear membranes begin to reassemble, the mitotic spindle becomes less visible, and the DNA returns to its chromatin form.

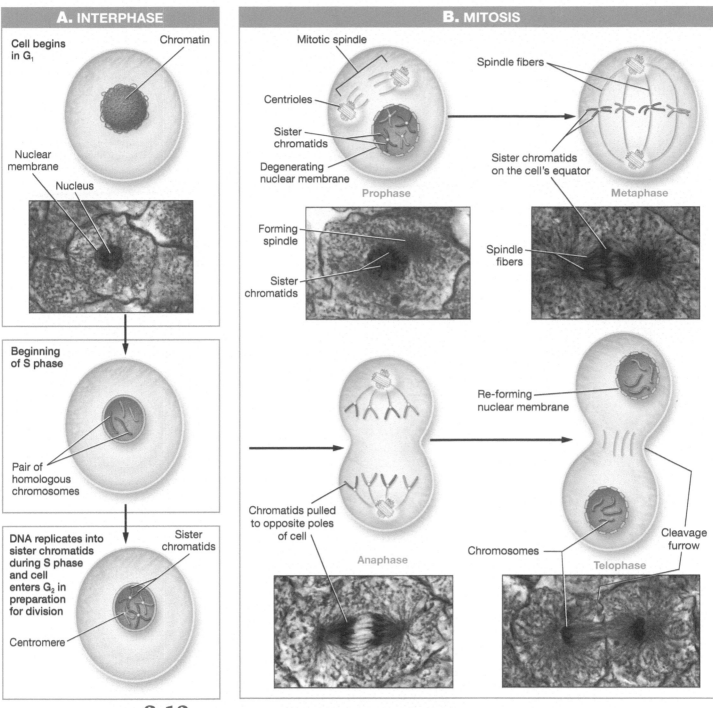

A. INTERPHASE

Cell begins in G₁

Chromatin

Nuclear membrane

Nucleus

Beginning of S phase

Pair of homologous chromosomes

DNA replicates into sister chromatids during S phase and cell enters G₂ in preparation for division

Sister chromatids

Centromere

B. MITOSIS

Mitotic spindle

Centrioles

Sister chromatids

Degenerating nuclear membrane

Prophase

Spindle fibers

Sister chromatids on the cell's equator

Metaphase

Forming spindle

Sister chromatids

Spindle fibers

Chromatids pulled to opposite poles of cell

Anaphase

Re-forming nuclear membrane

Chromosomes

Cleavage furrow

Telophase

FIGURE **3.12** Cell during the stages of the cell cycle: (**A**) interphase; (**B**) stages of mitosis.

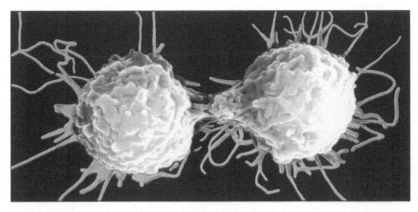

FIGURE **3.13** White blood cell in telophase of mitosis.

Procedure 1 Model Mitosis

Arrange models of the cell cycle and mitosis in the proper order.

1. Interphase

2. Mitosis
   a. Prophase
   b. Metaphase
   c. Anaphase
   d. Telophase

Procedure 2 Microscopy of the Cell Cycle

Examine the five phases of the cell cycle on prepared whitefish mitosis slides using the highest-power objective. Note that every stage of the cell cycle may not be visible on one single slide, so you may have to use more than one slide. Note also that most of the cells you see will be in interphase.

Draw what the cell looks like during each phase of the cell cycle, and then describe what you see in each phase in Table 3.5. Use Figure 3.12 for reference.

1 Interphase

2 Mitosis

   a Prophase

   b Metaphase

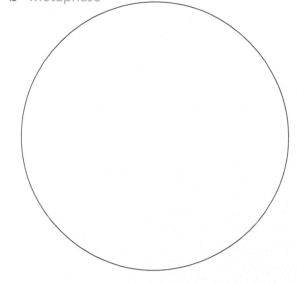

   c Anaphase

d Telophase

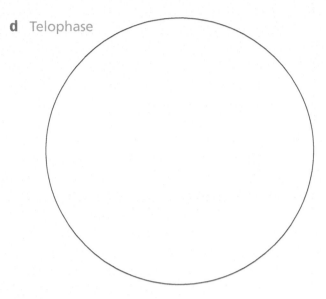

TABLE **3.5** Stages of the Cell Cycle

Stage of the Cell Cycle	Events Taking Place in the Cell	Cell Appearance
Interphase		
Mitosis		
Prophase		
Metaphase		
Anaphase		
Telophase		

UNIT 3

QUIZ

1 The lens that provides 100× magnification is the
   a. low-power lens.
   b. medium-power lens.
   c. high-power lens.
   d. oil-immersion lens.

2 Which knob should you use to get the image in focus when using the higher-power objective lenses?
   a. The coarse adjustment knob.
   b. The fine adjustment knob.
   c. The iris diaphragm.
   d. The light adjustment dial.

3 If you are having trouble making out details on your slide, what can you do?
   a. Adjust the focus with the fine adjustment knob.
   b. Reduce the amount of light with the iris diaphragm.
   c. Both a and b.
   d. Neither a nor b.

4 Label the following parts of the cell on **Figure 3.14**.
   ❏ Golgi apparatus
   ❏ Mitochondrion
   ❏ Nuclear envelope
   ❏ Nucleolus
   ❏ Plasma membrane
   ❏ Rough endoplasmic reticulum
   ❏ Smooth endoplasmic reticulum

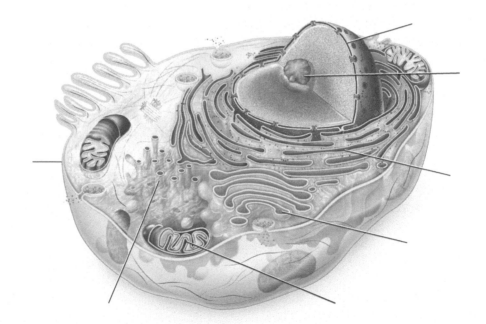

FIGURE **3.14** Generalized cell.

5 Which of the following is not a basic component of most cells?

a. Microvilli.
b. Plasma membrane.
c. Nucleus.
d. Cytoplasm.

6 *Matching:* Match the following organelles and cell structures with the correct definitions.

_____ Plasma membrane   A. Biosynthetic center of the cell; houses the cell's DNA

_____ Smooth ER   B. Produce(s) the bulk of the cell's ATP

_____ Mitochondria   C. Contain(s) digestive enzymes

_____ Ribosomes   D. Stack of flattened sacs that modify and sort proteins

_____ Rough ER   E. Barrier around the cell; composed of a phospholipid bilayer

_____ Nucleus   F. Membrane-enclosed sacs with ribosomes on the surface

_____ Lysosome   G. Membrane-enclosed sacs that detoxify substances and synthesize lipids

_____ Golgi apparatus   H. Granular organelles that are the sites of protein synthesis

7 *Fill in the blanks:* Diffusion is a _____ process where a solute moves from a

_____ concentration to a _____ concentration.

8 Which of the following factors influence the rate at which diffusion takes place?

a. Size of the particles.
b. Temperature.
c. Steepness of the concentration gradient.
d. All of the above.

9 *Fill in the blanks:* Osmosis is the movement of _____ from a solution

with a _____ solute concentration to a solution with a

_____ solute concentration.

10 How do isotonic, hypertonic, and hypotonic solutions differ?

_____

_____

_____

11 Label the stages of mitosis and the cell cycle on **Figure 3.15**.

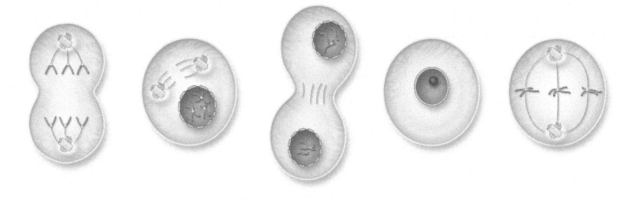

FIGURE **3.15** Cells in different stages of the cell cycle.

12 The mitotic spindle fibers are composed of
   a. microtubules.
   b. microvilli.
   c. intermediate filaments.
   d. actin filaments.

13 Autolysis is a process in which damaged or dying cells digest themselves via reactions catalyzed by their own digestive enzymes due to rupture of a specific organelle. Which organelle do you think is involved in autolysis? Explain.

14 Isotonic saline and 5% dextrose in water are solutions considered isotonic to human blood. What effect on red blood cells would you expect if a patient were given these fluids intravenously? A solution of 10% dextrose in water is hypertonic to human blood. What would happen if you were to infuse your patient with this solution?

15 Many anticancer drugs inhibit the formation of the mitotic spindle. What impact will this have on cell division? Why?

_____

_____

_____

_____

3

Histology: The Tissue Level of Organization

When you have completed this unit, you should be able to:

1 Relate tissue structure to tissue function, and describe how organs are formed from two or more tissue types.

2 Identify epithelial tissues by number of layers, cell shape, and specializations.

3 Identify and describe connective tissues.

4 Identify and describe muscle and nervous tissues.

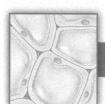

PRE-LAB EXERCISE

Complete the following exercise prior to coming to lab, using your lab manual and textbook for reference.

Pre-Lab Exercise **4-1**

4

✎ Key Terms

You should be familiar with the following terms before coming to lab.

Term	Definition

Epithelial Tissue

Simple epithelial tissue _____

Stratified epithelial tissue _____

Squamous cell _____

Cuboidal cell _____

Columnar cell _____

Connective Tissue

Loose connective tissue _____

Dense connective tissue _____

Cartilage _____

Bone _____

Blood _____

Muscle Tissue

Striated _____

Skeletal muscle tissue _____

Cardiac muscle tissue _____

Smooth muscle tissue _____

Nervous Tissue

Neuron _____

Neuroglial cell _____

SEM of a fibroblast and collagen fibers.

EXERCISES

The histology labs can be some of the more intimidating and frustrating labs for beginning anatomy and physiology students. The subjects are somewhat abstract and unfamiliar and require use of a complicated tool—the microscope. The best way to approach this subject is to be systematic, and to let your lab manual walk you through it step-by-step. If you get confused, don't despair. With the help of this book, your lab instructor, and a little patience, you can do it! Before you begin, you may wish to review the "Hints and Tips" boxes on pages 51 and 57.

The exercises in this unit introduce you to the four basic types of tissue: **epithelial tissue** (ep-ih-THEE-lee-uhl), **connective tissue, muscle tissue,** and **nervous tissue** (Fig. 4.1). All four of these tissue types have two main components: cells, which are specialized for each tissue type, and the **extracellular matrix (ECM)**, which is the material around the tissue's cells that is largely produced by the cells themselves. As you can see in Figure 4.1, all tissues have ECM, although its prominence, composition, and location vary in the different tissue types.

ECM consists of two components: ground substance and protein fibers (Fig. 4.2). **Ground substance** is a gelatinous material that contains water, ions, nutrients, and large polysaccharides, as well as *proteoglycans*, which are polysaccharides bound to a protein core. Notice in Figure 4.2 that thousands of proteoglycans come together to form huge structures called *proteoglycan aggregates* that resemble bottle brushes. The polysaccharides, proteoglycans, and proteoglycan aggregates trap water in the ECM and make it firmer, which helps a tissue resist compression.

Protein fibers are found within ground substance. The three types of protein fiber include:

1. **Collagen fibers.** Cells make many different types of **collagen fibers** (known as type I, type II, type III, etc.), which

Epithelial Tissue

Epithelial tissue lines all external and internal body surfaces.

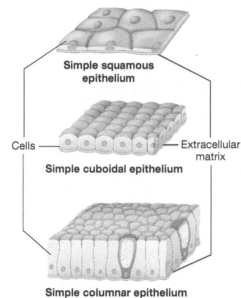

Simple squamous epithelium

Cells ———— Extracellular matrix

Simple cuboidal epithelium

Simple columnar epithelium

Connective Tissue

Connective tissue has binding, supporting, and transport functions.

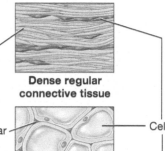
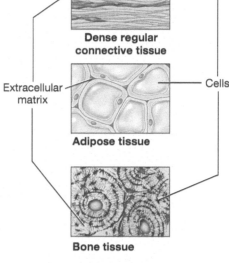

Dense regular connective tissue

Extracellular matrix ———— Cells

Adipose tissue

Bone tissue

Muscle Tissue

Muscle tissue is adapted to contract.

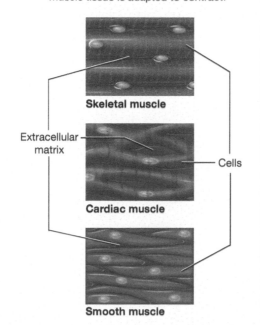

Skeletal muscle

Extracellular matrix

———— Cells

Cardiac muscle

Smooth muscle

Nervous Tissue

Nervous tissue receives stimuli and transmits signals from one part of the body to another.

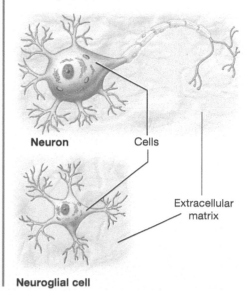

Neuron   Cells

Extracellular matrix

Neuroglial cell

FIGURE **4.1** The four different types of tissue.

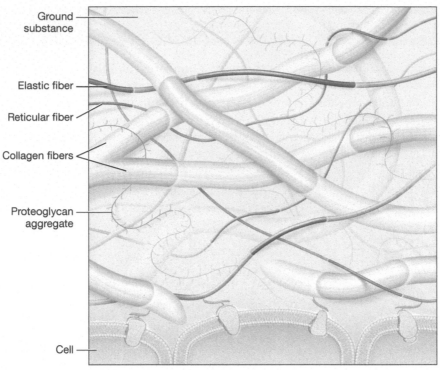

Ground substance

Elastic fiber

Reticular fiber

Collagen fibers

Proteoglycan aggregate

Cell

FIGURE **4.2** Composition of the extracellular matrix.

are composed of multiple entwining strands of the thick protein **collagen**. Imagine collagen fibers as the steel cables of a tissue—just as steel cables give a structure the ability to withstand tension, collagen fibers give a tissue tensile strength (i.e., the ability to resist stretching forces).

2. **Elastic fibers.** Cells may produce another type of fiber called **elastic fibers**, which are composed of the protein **elastin**. These protein strands perform the opposite function of collagen fibers—whereas collagen fibers resist stretch, elastic fibers allow a tissue to be stretched, which is a property known as *distensibility*. Elastic fibers also give a tissue the property of *elasticity*, meaning that the tissue returns to its original shape and size when the stretching force is removed.

3. **Reticular fibers.** The final fiber type made by cells is thin **reticular fibers**. These fibers were long thought to be a unique fiber type, but are now understood to be a special type of collagen fiber composed of collagen proteins. The thin structure of reticular fibers allows them to interweave to form "nets" that support blood vessels, nerves, and other structures. Indeed, the word root *reticul-* means "netlike."

Let's now begin our exploration of this fascinating level of organization, starting with epithelial tissues.

Exercise 4-1

Epithelial Tissue

MATERIALS
- ❑ Epithelial tissue slides
- ❑ Light microscope
- ❑ Colored pencils

Epithelial tissues are our covering and lining tissues. They are found covering body surfaces, lining body passageways, lining body cavities, and forming glands. Epithelia contain mostly cells, called **epithelial cells**, with little visible ECM. In fact, the majority of their ECM is located underneath the cells in a thin layer called the **basal lamina** (BAY-zul LAM-in-uh). The basal lamina performs an important function: It adheres to another layer of ECM produced by the connective tissues deep to the epithelium (known as the *lamina reticularis*), which effectively "glues" the epithelium in place. Together, the basal lamina and lamina reticularis are called the **basement membrane**.

Epithelial tissues are all **avascular**, which means that they have no blood vessels to supply them directly. As a result, epithelial cells must obtain oxygen and nutrients that have diffused up from blood vessels supplying deeper tissues. For this reason, epithelial tissues can be only a certain number of cell layers in thickness. If they are too thick, oxygen and nutrients will not reach the more superficial cells, and these cells will die.

The different types of epithelia are classified according to the number of layers of cells they have: **Simple epithelia** have only one layer of cells, and **stratified epithelia** have two or more cell layers. Epithelia are also classified by their predominant cell shape: **squamous** (SKWAY-muss) **cells** are flat, **cuboidal cells** are about as tall are they are wide, and **columnar cells** are taller than they are wide. These two criteria give us the following classes of epithelia (see Fig. 4.3):

1. **Simple epithelia:**

   a. **Simple squamous epithelium. Simple squamous epithelium,** shown in Figure 4.3A, consists of a single layer of flat cells with a centrally located, flattened nucleus. We often find simple squamous epithelium in places where substances have to cross the epithelium quickly, such as the air sacs of the lungs.

   b. **Simple cuboidal epithelium.** Note in Figure 4.3B that the cells of **simple cuboidal epithelium** are short and have a spherical, central nucleus. Simple cuboidal epithelium is found lining glands such as the thyroid gland, certain respiratory passages, and in the kidneys.

   c. **Simple columnar epithelium.** The cells of **simple columnar epithelium,** shown in Figure 4.3C, have spherical nuclei generally located near the base of the cell. These cells line certain respiratory passages and much of the digestive tract. The plasma membranes of simple columnar epithelial cells often contain cilia or are folded into microvilli.

d. **Pseudostratified ciliated columnar epithelium.** **Pseudostratified** (SOO-doh-strat-ih-fy'd) **epithelium**, seen in Figure 4.3D, looks as if it has many cell layers but actually has only one layer of cells (*pseudo* = false). Notice in the figure that the nuclei of the cells are at different heights, which gives the epithelium the appearance of having many layers. However, the cells are all attached to the same basal lamina, and so exist in a single layer. This type of epithelium usually has cilia, and the cell shape is always columnar. It is found lining the nasal cavity and much of the respiratory tract, where it is often called *respiratory epithelium*.

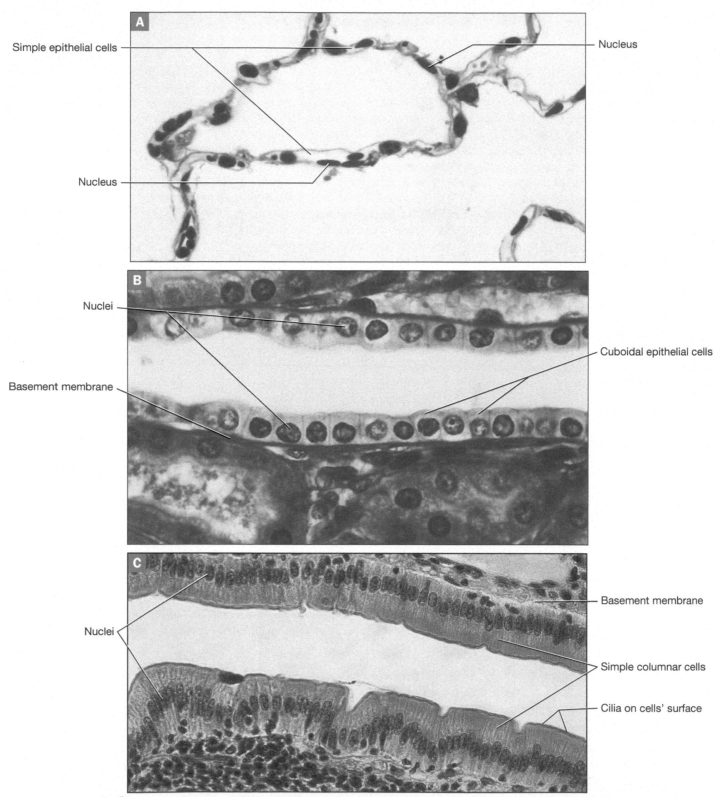

FIGURE **4.3** Simple epithelial tissues: (**A**) simple squamous epithelium from the lungs; (**B**) simple cuboidal epithelium from the kidney; (**C**) simple ciliated columnar epithelium from the uterine tube; *(continues)*

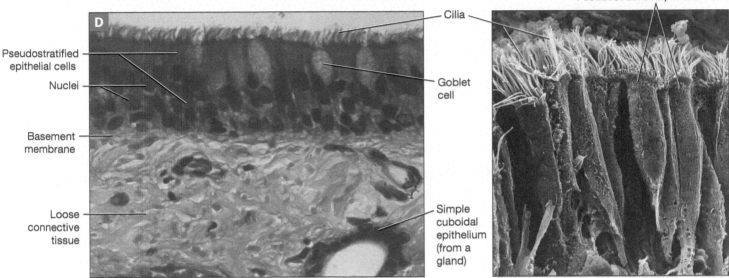

FIGURE **4.3** Simple epithelial tissues *(cont.)*: **(D)** pseudostratified ciliated columnar epithelium from the trachea, light micrograph and scanning electron micrograph.

2. **Stratified epithelia:**

   a. **Stratified squamous epithelium.** This type of epithelium consists of many layers of flattened cells and has two variants. The first, shown in Figure 4.4A, is **stratified squamous keratinized epithelium,** which consists of epithelial cells called **keratinocytes** (kehr-ah-TIN-oh-syt'z) that produce the protein **keratin** (KEHR-ah-tin). The more superficial flaky-looking cells are dead because they are too far away from the blood supply in the deeper tissues and, as a result, they harden and die. Stratified squamous keratinized epithelial cells resist mechanical stresses and so are found in the superficial layer of the skin where the body is subject to multiple environmental stresses. **Stratified squamous nonkeratinized epithelium** (Fig. 4.4B) contains no keratin and is located in places subject to lesser degrees of mechanical stress, such as the oral cavity, the pharynx (throat), the anus, and the vagina. This type of epithelium is not as thick and so the superficial cells are alive. As you can see, this gives them a much different appearance than the superficial cells of keratinized epithelium.

   b. **Stratified cuboidal epithelium** and **stratified columnar epithelium.** Both of these types of epithelium are rare in the human body (they are not included in Figure 4.4) and are found lining the ducts of certain glands.

   c. **Transitional epithelium. Transitional epithelium,** shown in Figure 4.4C, is stratified but is not classified by its shape because its cells can change shape. Typically, the apical or surface cells are dome-shaped, but when the tissue is stretched, they flatten and are squamous in appearance. Transitional epithelium is found lining the urinary bladder and ureters, and, for this reason, is also sometimes known as *urinary epithelium.*

HINTS & TIPS

Identifying Epithelial Tissue

The first step in locating and identifying a tissue is to figure out if you are looking at epithelial, connective, muscle, or nervous tissue. Fortunately, it is relatively easy to determine whether you are seeing epithelium. A few unique features can help you distinguish epithelial tissues from other tissues:

- ℹ Epithelial tissues are all avascular, so you won't see any blood vessels in epithelium.

- ℹ Epithelial tissues are often on the outer edge of the slide or, at least, on the outer edge of the tissue. Keep in mind that most slides have several tissues in each section. To find the epithelial tissue, often you need to scroll to one end of the slide or the other.

- ℹ Epithelial tissues consist mostly of cells. As we have established, everything in a tissue is either cells or ECM. In the case of epithelium, the vast majority of the ECM is restricted to a thin layer deep to the cells. So, you will be looking for cells—structures with dark purple nuclei—that are tightly packed together.

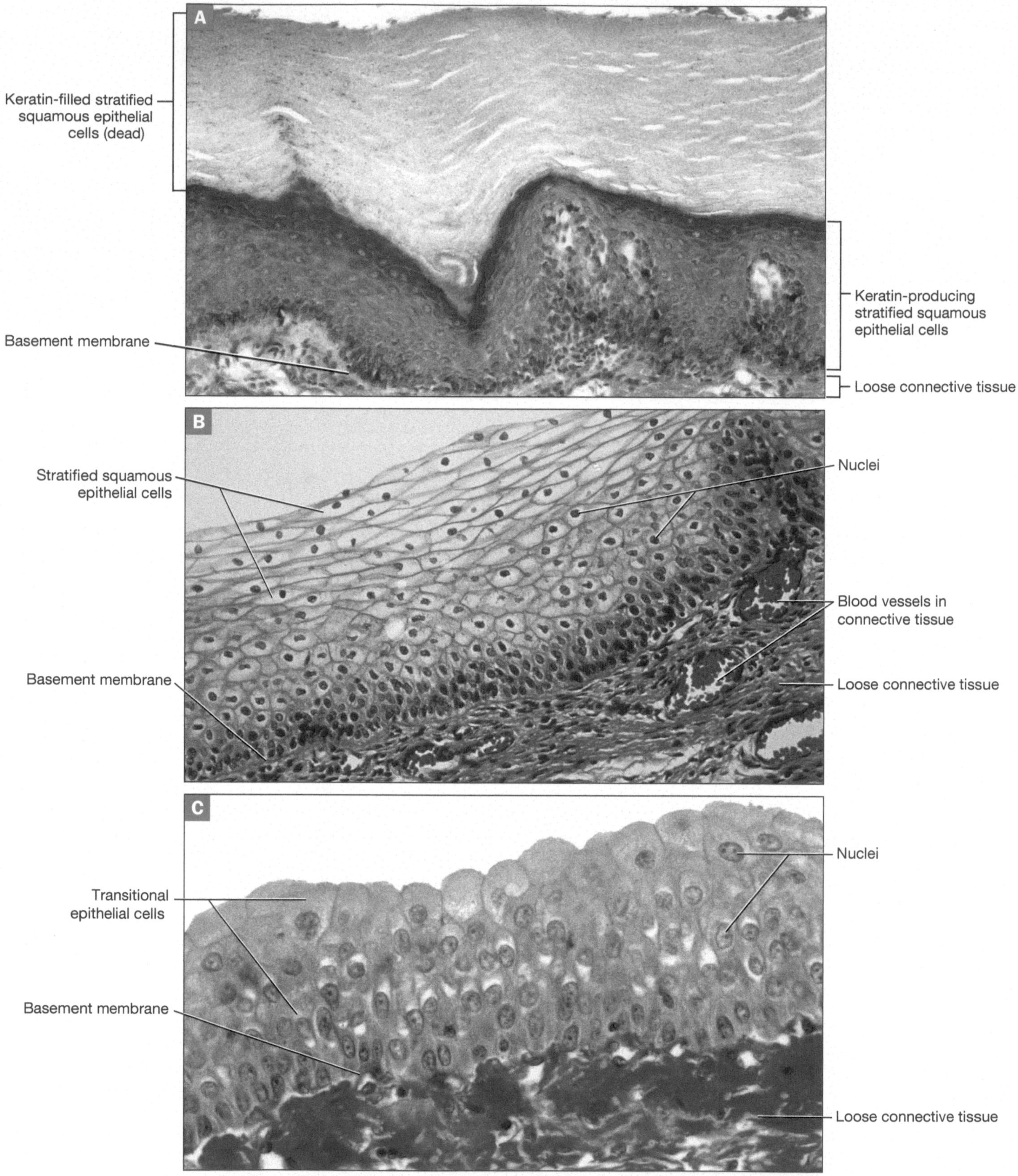

Keratin-filled stratified squamous epithelial cells (dead)

Basement membrane

Keratin-producing stratified squamous epithelial cells

Loose connective tissue

Stratified squamous epithelial cells

Nuclei

Basement membrane

Blood vessels in connective tissue

Loose connective tissue

Transitional epithelial cells

Nuclei

Basement membrane

Loose connective tissue

FIGURE 4.4 Stratified epithelial tissues: (A) stratified squamous keratinized epithelium from the skin; (B) stratified squamous nonkeratinized epithelium from the vagina; (C) transitional epithelium from the urinary bladder.

Procedure 1 Microscopy of Epithelial Tissue

Examine prepared slides of the following epithelial tissues. Use colored pencils to draw what you see under the microscope, and label your drawings with the terms from Figures 4.3 and 4.4. Then (a) describe what you see, and (b) give examples of locations in the body where this tissue is found.

1 Simple squamous epithelium

a _____

_____

b _____

_____

2 Simple cuboidal epithelium

a _____

_____

b _____

_____

3 Simple columnar epithelium

a _____

_____

b _____

_____

4 **Stratified squamous nonkeratinized epithelium**

a _____

_____

b _____

_____

5 **Transitional epithelium**

a _____

_____

b _____

_____

6 **Pseudostratified ciliated columnar epithelium**

a _____

_____

b _____

_____

Exercise 4-2

Connective Tissue

MATERIALS
❑ Connective tissue slides
❑ Light microscope
❑ Colored pencils

Connective tissues are found throughout the body. They have a variety of functions, most of which serve to *connect*, as their name implies. All connective tissues stem from a common embryonic tissue called **mesenchyme** (MEZ-en-ky′m). Connective tissues are distinguished easily from epithelial tissues by the prominence of their extracellular matrices. Typically, connective tissues contain few cells and have an extensive ECM.

The four general types of connective tissue (CT) are as follows:

1. **Connective tissue proper. Connective tissue proper,** the most widely distributed class of connective tissue in the body, consists of scattered cells called **fibroblasts** (FY-broh-blastz) that secrete an extensive ECM filled with many types of protein fibers. This tissue is highly vascular with an extensive blood supply. The subclasses of CT proper include the following:

❚ **Loose (areolar) CT.** You can see in Figure 4.5A that the primary element in **loose CT** is ground substance, which gives it a "loose" appearance on a slide. All three types of protein fibers are scattered in loose CT ground substance, although the slide is stained for only collagen and elastic fibers. Loose CT is found as part of the basement membrane and in the walls of hollow organs, where it helps to "glue" together the tissue layers of these organs.

❚ **Reticular CT.** As you can guess by the name, **reticular CT** consists of many reticular fibers produced by cells called **reticular cells** (Fig. 4.5B). It is located in the spleen and lymph nodes, where the fine reticular fibers interweave to form "nets" that trap pathogens and foreign cells. Reticular CT also is located around blood vessels and nerves, where it forms supportive networks.

❚ **Adipose tissue.** Notice in Figure 4.5C that **adipose tissue** (fat tissue) has a much different appearance than the other types of CT proper. It consists mostly of huge cells called **adipocytes** (AD-ih-poh-syt′z) with collagen fibers in the ECM. Each adipocyte contains a large lipid droplet that occupies most of its cytoplasm. The nucleus and other organelles are barely visible, because they are pushed to the periphery of the cell against the plasma membrane. Adipose tissue is distributed widely throughout the body under the skin and around organs.

❚ **Dense regular collagenous CT.** The difference between loose and dense CT is obvious in Figure 4.6A. **Dense regular collagenous CT** consists primarily of collagen fibers arranged in parallel bundles with little ground substance and few cells. It is exceptionally strong and makes up structures that require tensile strength in a single plane, such as tendons and ligaments.

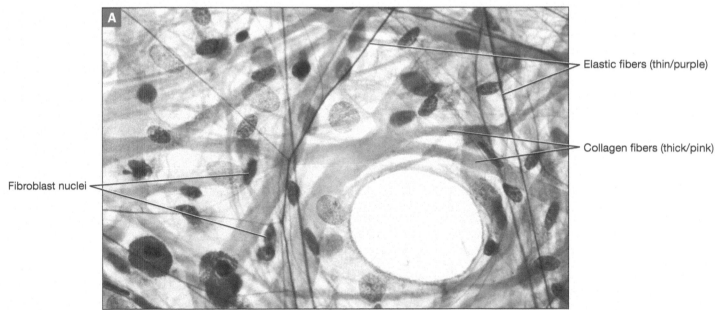

Elastic fibers (thin/purple)

Collagen fibers (thick/pink)

Fibroblast nuclei

FIGURE **4.5** Loose connective tissues: **(A)** loose (areolar) CT; *(continues)*

- **Dense irregular collagenous CT.** Like dense regular collagenous CT, **dense irregular collagenous CT** consists of bundles of collagen fibers. Notice in Figure 4.6B, however, that these collagen bundles are arranged in an irregular, haphazard fashion without a consistent pattern. You can see the difference clearly in the scanning electron micrographs of the two tissues. Like dense regular collagenous CT, dense irregular collagenous CT also is quite strong. The arrangement of its collagen bundles makes it well-suited to support structures that require tensile strength in multiple planes, such as the dermis of the skin, joint capsules, and organ capsules.

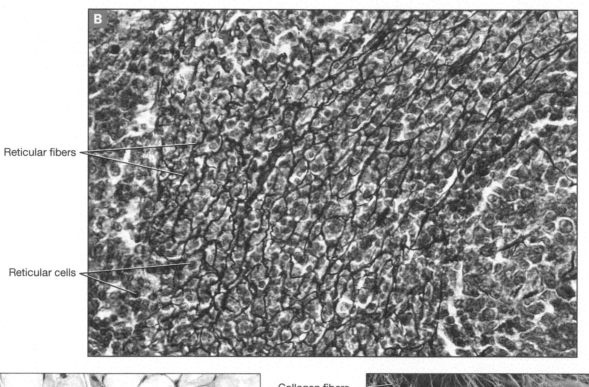

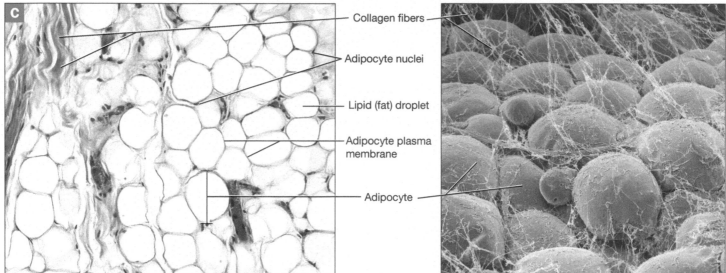

FIGURE **4.5** *(cont.)* Loose connective tissues: (**B**) reticular CT from the spleen; (**C**) adipose tissue, light micrograph and SEM.

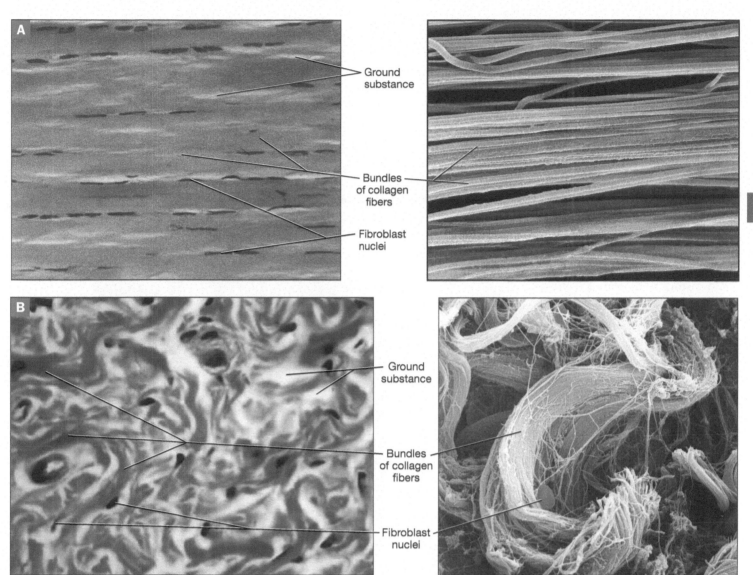

Ground substance

Bundles of collagen fibers

Fibroblast nuclei

Ground substance

Bundles of collagen fibers

Fibroblast nuclei

FIGURE **4.6** Dense connective tissues: (**A**) dense regular collagenous CT from a tendon, light micrograph and SEM; (**B**) dense irregular collagenous CT from the dermis, light micrograph and SEM.

2. **Cartilage. Cartilage** is a tough but flexible tissue that is resistant to tension, twisting, and compressive forces. It consists of cells called **chondrocytes** (KAHN-droh-syt′z), which are embedded in the ECM in cavities called **lacunae** (lah-KOO-nee). Cartilage is notable among the connective tissues for being avascular. Each of the three types of cartilage has a different ECM composition.

- **Hyaline cartilage.** Notice in Figure 4.7A that **hyaline cartilage** (HY-ah-lin) contains mostly chondrocytes scattered in ground substance with few visible protein fibers. This gives hyaline cartilage a smooth, glassy appearance and makes it an ideal tissue to cover the ends of bones where they form joints with another bone. The smooth texture of hyaline cartilage provides a nearly frictionless surface on which bones can articulate. Hyaline cartilage also is found connecting the ribs to the sternum, lining certain respiratory passages, and forming the framework for the nose.

- **Fibrocartilage.** As you can see in Figure 4.7B, **fibrocartilage** is named appropriately because its ECM is full of protein fibers (mostly collagen). This makes fibrocartilage tough and extremely strong but not very smooth (think of the surface of fibrocartilage like a flannel sheet, with the cotton fibers representing the protein fibers). For this reason, fibrocartilage does not form articular cartilage, but it does reinforce ligaments and form *articular discs*, tough structures that improve the fit of two bones. In addition, fibrocartilage is found in intervertebral discs, structures between two vertebrae that help to support the weight of the vertebral column and absorb shock.

- **Elastic cartilage.** The final type of cartilage, **elastic cartilage,** is shown in Figure 4.7C. Its ECM is filled with elastic fibers that allow it to stretch and recoil. Elastic cartilage is found in the ear and in the epiglottis.

3. **Bone.** Bone tissue, also called **osseous tissue** (AHS-see-us), consists of bone cells called **osteocytes** (AHS-tee-oh-syt′z) encased in an ECM that contains collagen fibers and calcium hydroxyapatite crystals. Note in Figure 4.8 that the ECM is arranged in concentric layers called **lamellae** (lah-MELL-ee), with the osteocytes sandwiched between them. This structure makes bone one of the hardest tissues in the body and very resistant to mechanical stresses.

4. **Blood.** Blood (Fig. 4.9) consists of a liquid ECM called **plasma,** within which we find cells called **erythrocytes** (eh-RITH-roh-syt′z; red blood cells) and **leukocytes** (LOO-koh-syt′z; white blood cells), and cellular fragments called **platelets.** Its main role is to transport oxygen, nutrients, electrolytes, wastes, and many other substances through the body. Note that blood features the only cell type in the body that lacks a nucleus: the mature erythrocyte.

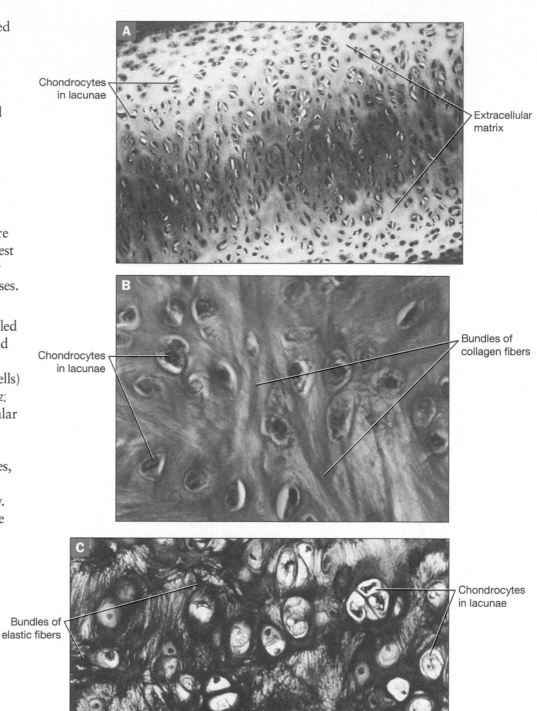

FIGURE **4.7** Cartilaginous connective tissues: (**A**) hyaline cartilage from a joint; (**B**) fibrocartilage from an articular disc; (**C**) elastic cartilage from the ear.

Osteocytes
in lacunae

Lamellae containing
collagen fibers and
calcium salts in the
extracellular matrix

FIGURE **4.8** Bone tissue.

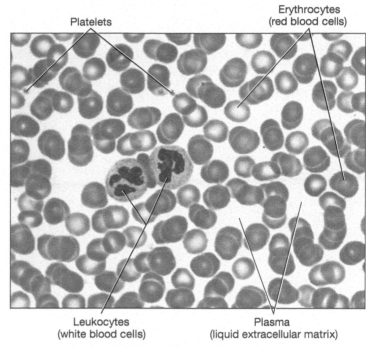

Platelets

Erythrocytes
(red blood cells)

Leukocytes
(white blood cells)

Plasma
(liquid extracellular matrix)

FIGURE **4.9** Blood.

HINTS & TIPS

Identifying Connective Tissue

Connective tissue really isn't as confusing as it seems. If you approach these slides systematically, use the figures in this manual as a guide, and, of course, don't panic, you will find they are surprisingly simple. Remember, everything in the field of view is only one of two things: cells or ECM.

The following points will help you to identify the various connective tissues and differentiate them from other tissue types:

ⓘ Usually, in connective tissue, you will see a large amount of space between the cells because they aren't packed tightly together (adipose tissue is an exception). Remember to look for the nucleus to find the cells.

ⓘ You will often see a lot of straight or wavy lines running through a connective tissue section. These are just protein fibers. Many types of CT can be distinguished by the types of fibers they contain:

• Reticular fibers are the thinnest fibers, and they require a special stain to be visible. As you saw in Figure 4.5B (p. 84), this stain makes reticular fibers brownish-black. In general, only reticular tissue is prepared with this stain, so if you see brownish-black reticular fibers, you know you are looking at reticular tissue.

• Collagen fibers are thick fibers that usually stain pink or light purple. Look for collagen fibers in fibrocartilage, dense regular collagenous CT, dense irregular collagenous CT, and loose CT. The arrangement of the collagen fibers can tell you even more about the tissue. Are the fibers arranged haphazardly (basically, does your slide look like an undefined mess)? Then it's loose CT. Are they arranged in straight rows? Then it's dense regular collagenous CT. Are there small, wavy bundles of collagen? Then it's dense irregular collagenous CT.

• Elastic fibers are thinner than collagen fibers and may have a wavy appearance. Their color ranges from purple-black to blue, depending on the stain used. Look for them in tissues such as elastic cartilage and loose CT.

ⓘ Cartilage is easy to differentiate from CT proper by looking at the shape of the cells. Fibroblasts are generally small and flat, and often appear as just purple dots. Chondrocytes are also fairly small, but they sit in lacunae, which creates a clear space around the cells. This makes the cells much easier to see and makes them appear to be bigger.

ⓘ Blood and bone are perhaps the two easiest tissues you will examine in this lab. They should look much like they do in Figures 4.8 and 4.9, and no other tissues resemble them.

Procedure 1 Microscopy of Connective Tissue Proper

View prepared slides of each type of connective tissue proper. Use colored pencils to draw pictures of what you see under the microscope, and label your drawings with the terms from Figures 4.5 and 4.6. Then (a) describe what you see, and (b) give examples of locations in the body where this tissue is found.

4

1 Loose (areolar) CT

a _____

_____

b _____

_____

2 Reticular CT

a _____

_____

b _____

_____

3 Adipose tissue

a _____

_____

b _____

_____

4 Dense regular collagenous CT

a _____

_____

b _____

_____

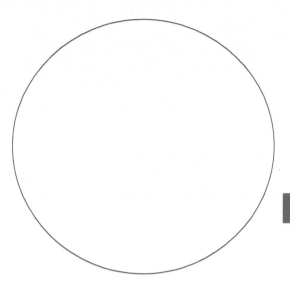

5 Dense irregular collagenous CT

a _____

_____

b _____

_____

Procedure 2 Microscopy of Cartilage

View prepared slides of the three types of cartilage. Use colored pencils to draw pictures of what you see under the microscope, and label your drawings with the terms from Figure 4.7. Then (a) describe what you see, and (b) give examples of locations in the body where this tissue is found.

1 Hyaline cartilage

a _____

_____

b _____

_____

2 Fibrocartilage

a _____

_____

b _____

_____

3 Elastic cartilage

a _____

_____

b _____

_____

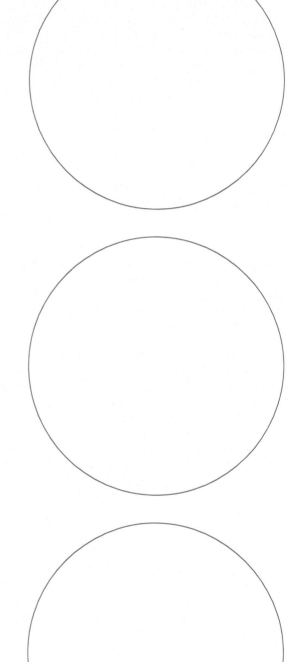

Procedure 3 Microscopy of Bone and Blood

View prepared slides of bone and blood. Use colored pencils to draw pictures of what you see under the microscope, and label your drawings with the terms in Figures 4.8 and 4.9. Then (a) describe what you see, and (b) give examples of locations in the body where this tissue is found.

1 Bone

a _____

_____

b _____

_____

2 Blood

a _____

_____

b _____

_____

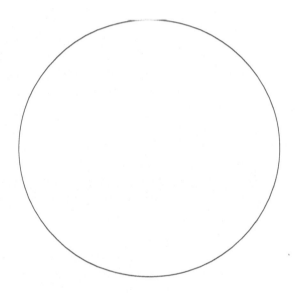

Exercise 4-3

Muscle Tissue

MATERIALS

❑ Muscle tissue slides

❑ Light microscope

❑ Colored pencils

4

Muscle tissue is located in skeletal muscles, the walls of hollow organs, the heart, and other locations such as the iris of the eye. It consists of muscle cells, sometimes called **myocytes** (MY-oh-syt′z) or **muscle fibers,** and a small amount of ECM called the **endomysium** (en-doh-MY-see-um). Note that the endomysium blends with connective tissue surrounding groups of muscle fibers and is often called connective tissue as a result. As you can see in Figures 4.10A–C, myocytes aren't shaped like the cells of epithelial or connective tissues. This makes muscle tissue easy to tell apart from most other tissue types.

There are three types of muscle tissue:

1. **Skeletal muscle tissue.** The muscle fibers of **skeletal muscle tissue** are long, tubular, and have multiple nuclei (Fig. 4.10A). They are also **striated** (STRY-ayt-ed; striped) in appearance, which results from the arrangement of proteins called **myofilaments** within the muscle fiber. This type of muscle tissue is *voluntary*, which means that its contraction is under conscious control. For this reason, it has a close relationship with the nervous system and nervous tissue, because skeletal muscle fibers must be stimulated by a nerve cell to contract.

2. **Cardiac muscle tissue.** The myocytes of **cardiac muscle tissue,** located in the heart, are short, wide, striated, and tend to be branching (Fig. 4.10B). Cardiac myocytes typically have only one nucleus, but some may have two or more. Adjacent cardiac myocytes are linked by specialized junctions called **intercalated discs** (in-TUR-kuh-layt-ed) that contain desmosomes and gap junctions. Intercalated discs link all cardiac myocytes physically and electrically so that the heart may contract as a unit. Cardiac muscle tissue is *involuntary* and *autorhythmic,* meaning that it requires no outside stimulus from the nervous system to contract. (Note, however, that the nervous system can influence the rate and force of cardiac muscle contraction.)

3. **Smooth muscle tissue.** The myocytes of **smooth muscle tissue** are flat with a single nucleus in the center of the cell (Fig. 4.10C). The arrangement of myofilaments within smooth muscle fibers differs from that of skeletal and cardiac muscle fibers, and, as a result, these cells lack noticeable striations (hence the name *smooth* muscle). Smooth muscle lines all hollow organs and is found in the skin, the eyes, and surrounding many glands. Like cardiac muscle tissue, smooth muscle tissue is involuntary. However, although there are populations of smooth muscle cells that are autorhythmic, much of our smooth muscle requires stimulation from the nervous system to contract.

HINTS & TIPS

Identifying Muscle Tissue

The three types of muscle tissue aren't that difficult to tell apart from one another. The only two that might give you troubles are skeletal and cardiac muscle tissues, but if you remember that cardiac muscle tissue has intercalated discs, you shouldn't have any problems. However, sometimes students mistake skeletal or smooth muscle tissue for an entirely different tissue type: dense regular collagenous CT.

Take a look back to **Figure 4.6A** (p. 85). Notice how the collagen bundles are arranged in a way that almost looks like the tubular skeletal muscle fibers? And notice also how the nuclei of the fibroblasts have the same shape and arrangement as those of smooth muscle cells? Both of these situations can cause students to sometimes mix up these tissue types, so careful examination is needed to prevent these mistakes:

ⓘ Look closely at the "tubes" to see if they have striations running through them. If there are striations, as in **Figure 4.10A,** the section is skeletal muscle tissue.

ⓘ If there are no striations, look more closely at the nuclei.

   ● Advance to a higher-power objective lens and look for plasma membranes or boundaries around each nucleus, which you can see in **Figure 4.10D.** If you can find plasma membranes around the nuclei, you are looking at smooth muscle tissue.

   ● If you can't see any boundaries around the nuclei, and the space between the nuclei looks slightly fuzzy rather than well-defined, you're probably looking at dense regular collagenous CT.

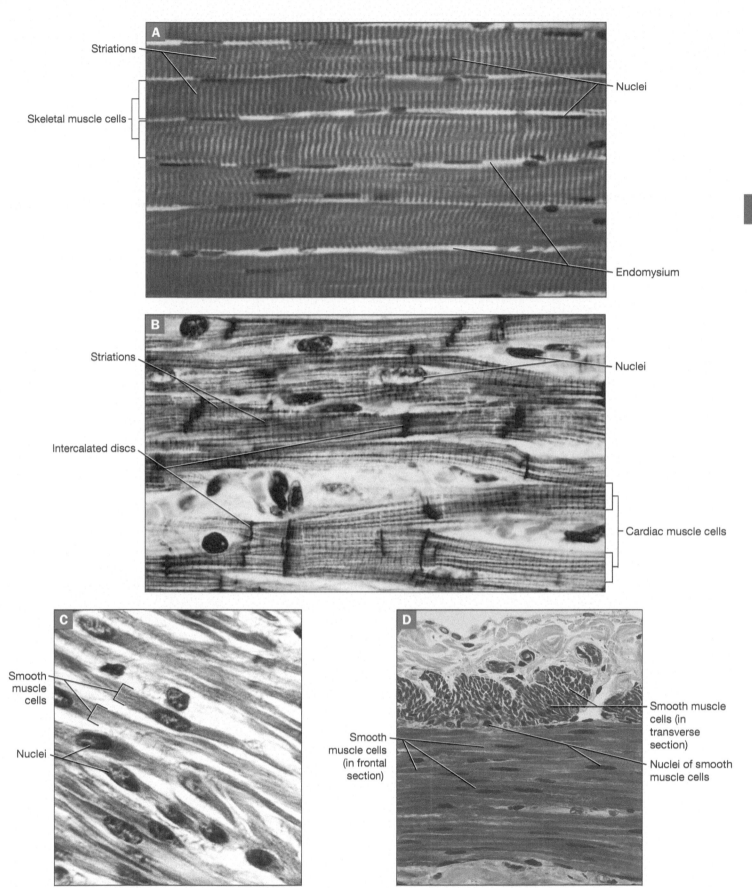

FIGURE **4.10** Muscle tissue: (**A**) skeletal muscle tissue; (**B**) cardiac muscle tissue; (**C**) teased smooth muscle cells; (**D**) smooth muscle cells in an organ.

Procedure 1 Microscopy of Muscle Tissue

View prepared slides of skeletal, smooth, and cardiac muscle tissue. Use colored pencils to draw what you see under the microscope, and label your drawing with the terms from Figure 4.10. Record your observations of each slide in Table 4.1.

4

1 Skeletal muscle

2 Cardiac muscle

3 Smooth muscle

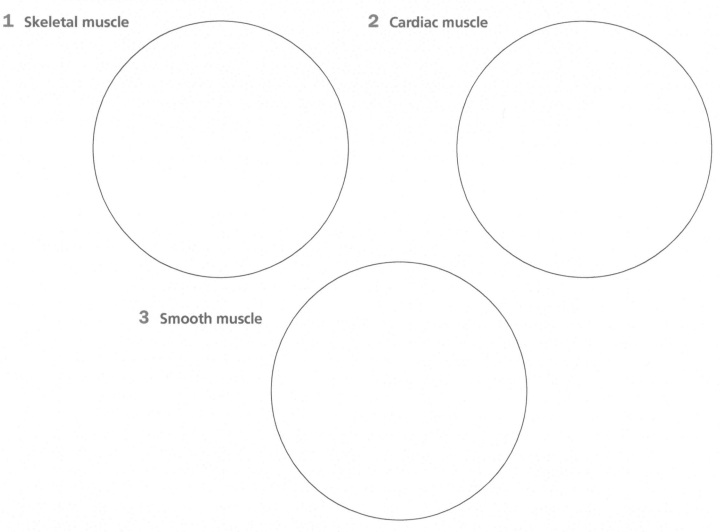

TABLE **4.1** Characteristics of Muscle Tissues

Muscle Tissue Type	Striated or Nonstriated	One or Multiple Nuclei	Size, Shape, and Special Features of Cells	Location in the Body
Skeletal muscle				
Cardiac muscle				
Smooth muscle				

Exercise 4-4

Nervous Tissue

MATERIALS
❑ Nervous tissue slides
❑ Light microscope
❑ Colored pencils

Nervous tissue (Fig. 4.11) is the primary component of the brain, the spinal cord, and the peripheral nerves. It consists of a unique ECM and two main cell types:

1. **Neurons.** The **neurons** (NOOR-ahnz) are responsible for sending and receiving messages within the nervous system. On your slide, they are the larger of the two cell types. The large, central portion of the neuron is called the **cell body.** Within the cell body, we find the nucleus and many of the neuron's organelles, including clusters of rough endoplasmic reticulum called **Nissl bodies.** Most neurons contain two types of long armlike processes extending from the cell body: (1) the **dendrites** (DEN-dryt'z), which receive messages from other neurons, and (2) the **axon** (AX-ahn), which sends messages to other neurons, muscle cells, or gland cells.

2. **Neuroglial cells.** The smaller and more numerous cells around the neurons are the **neuroglial cells** (noor-oh-GLEE-uhl). The six different types of neuroglial cells vary significantly in shape and appearance. Neuroglial cells in general perform functions that support the neurons or the ECM in some way.

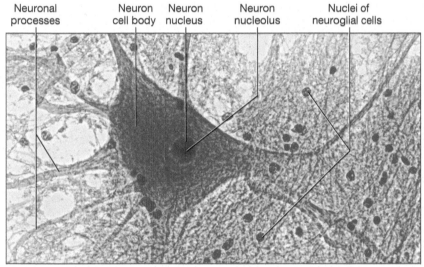

Neuronal processes · Neuron cell body · Neuron nucleus · Neuron nucleolus · Nuclei of neuroglial cells

FIGURE **4.11** Nervous tissue.

 Procedure 1 Microscopy of Nervous Tissue

View a prepared slide of nervous tissue (the slide might be called a "motor neuron smear"). Use colored pencils to draw a picture of what you see under the microscope, and label your drawing with the terms from Figure 4.11. Then (a) describe what you see, and (b) give examples of locations in the body where this tissue is found.

Nervous tissue

a _____

_____

b _____

_____

Exercise 4-5

Organology

All organs consist of two or more tissues that must work together to enable the organ to function properly. The study of the tissues that make up the body's organs is called **organology**. Most organs are made of layers of tissues stacked upon one another and "glued" together by proteins and other compounds in the ground substance. This exercise introduces you to organology, a topic we explore repeatedly in the remainder of this lab manual.

Procedure 1 Determine the Tissue Types of the Knee Joint

Following is an illustration of the knee joint (Fig. 4.12). Tissue sections from each part of the joint have been taken, and your task is to identify each tissue. Be specific about which type of muscle, epithelial tissue, and connective tissue you identify in the joint.

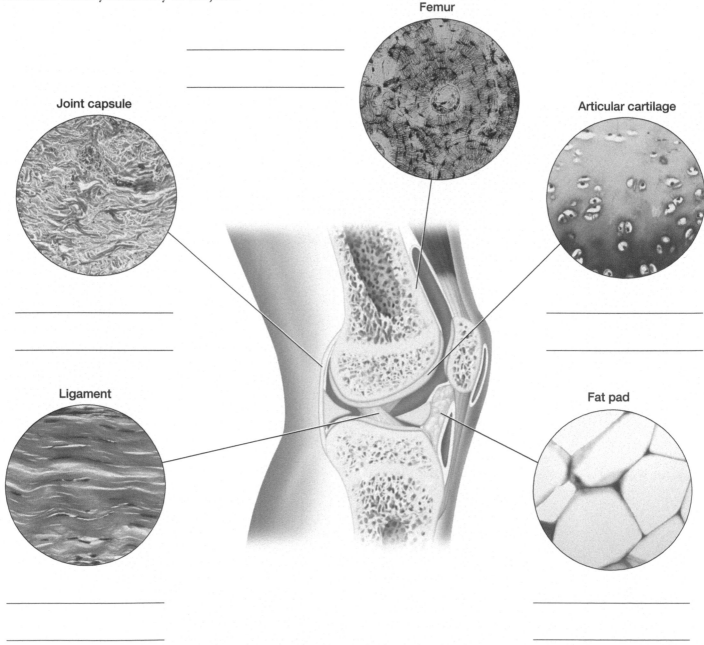

FIGURE **4.12** Tissues making up the knee joint.

Name _____

Section _____ Date _____

4

1 Identify each of the following tissues in Figure 4.13.

a _____   c _____

b _____   d _____

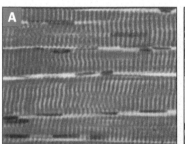
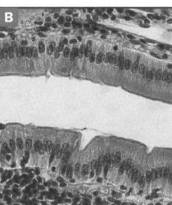
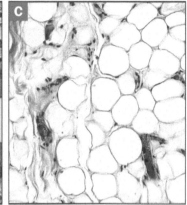
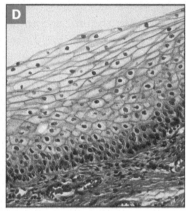

FIGURE **4.13** Unknown tissues for question 1.

2 What are the four basic tissue types?

_____   _____

_____   _____

3 *Fill in the blanks:* All tissues consist of two main components: _____

and _____ .

4 An epithelial tissue that has one layer of flat cells would be classified as
  a. simple keratinized epithelium.
  b. simple cuboidal epithelium.
  c. stratified columnar epithelium.
  d. simple squamous epithelium.

5 *True/False:* Mark the following statements concerning connective tissue proper and cartilage as true (T) or false (F). If the statement is false, correct it to make it a true statement.

_____ Adipocytes contain a large lipid droplet in their cytoplasm.

_____ The primary element in dense regular collagenous connective tissue is ground substance.

_____ Bundles of collagen fibers are arranged parallel to one another in dense irregular collagenous connective tissue.

_____ Loose connective tissue contains all three protein fiber types.

_____ Fibrocartilage has a smooth, glassy appearance due to a large amount of ground substance and few visible protein fibers.

6 Which of the following statements about muscle tissue is true?

   a. Skeletal muscle and cardiac muscle tissues have no striations.

   b. Smooth muscle tissue is found in the heart.

   c. The cells of skeletal muscle tissue are long, tubular, and multinucleated.

   d. Smooth muscle cells are joined by intercalated discs.

7 *Fill in the blanks:* Nervous tissue is composed of _____

and _____ .

4

8 The structure in the lungs known as the *respiratory membrane* is where gases are exchanged—oxygen leaves the lungs and enters the blood, and carbon dioxide leaves the blood and enters the lungs. The respiratory membrane is composed of two extremely thin layers of simple squamous epithelium. Explain how the structure of the respiratory membrane follows its function.

_____

_____

_____

_____

9 Predict what you think might happen if the respiratory membrane were instead composed of two layers of stratified squamous epithelium. Explain.

_____

_____

_____

_____

10 The formation of fibrocartilage is a common response to injury of hyaline cartilage. Do you think fibrocartilage would provide an articular surface (i.e., the cartilage in joints) as smooth as the original hyaline cartilage? Why or why not?

_____

_____

_____

11 When muscle tissue dies, it usually is replaced with dense irregular collagenous connective tissue. How do these tissues differ in structure? Will the muscle be able to function normally? Why or why not?

_____

_____

_____

_____

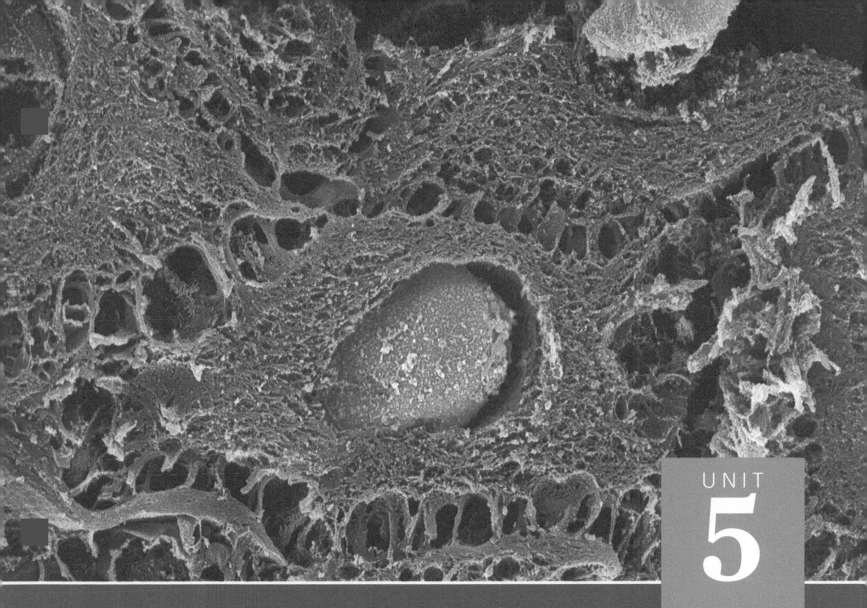

Integumentary System

When you have completed this unit, you should be able to:

1 Identify structures of the integumentary system.

2 Describe the gross and microscopic structure of thick and thin skin.

Name _____ Section _____ Date _____

PRE-LAB EXERCISES

Complete the following exercises prior to coming to lab, using your lab manual and textbook for reference.

Pre-Lab Exercise **5-1**

 Key Terms

You should be familiar with the following terms before coming to lab.

Term	Definition
Epidermal Structures	
Epidermis	
Keratinocyte	
Melanocyte	
Structures of the Dermis	
Dermis	
Dermal papillae	
Tactile corpuscle	
Lamellated corpuscle	
Accessory and Other Structures	
Sebaceous gland	
Sweat gland	
Hair follicle	

Arrector pili muscle

Nails

Hypodermis

Pre-Lab Exercise **5-2**

Skin Anatomy

Color the structures of the skin in Figure 5.1 and label them with the following terms from Exercise 5-1 (pp. 105–106). Use Exercise 5-1 in this unit and your text for reference.

❑ Epidermis
❑ Dermis
  ☐ Papillary layer
    ▪ Dermal papillae
  ☐ Reticular layer
  ☐ Blood vessels
❑ Hypodermis

Nerves

❑ Lamellated corpuscle
❑ Tactile corpuscles

Glands

❑ Sebaceous gland
❑ Sweat gland
  ☐ Sweat duct
  ☐ Sweat pore

Hair

❑ Hair shaft
❑ Hair root
❑ Hair follicle
❑ Arrector pili muscle

Nail

❑ Nail plate
❑ Nail fold (proximal and lateral)
❑ Nail matrix

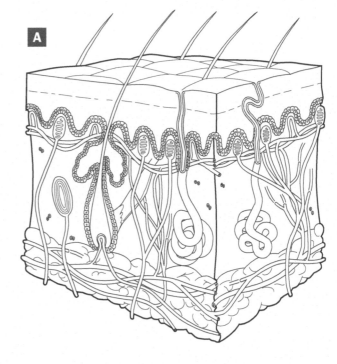

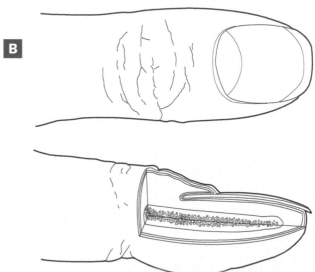

FIGURE **5.1** Structures of the integument: (**A**) skin section; (**B**) nail anatomy.

Colored SEM of a keratinocyte skin cell with a large central nucleus.

Although the skin is the largest organ in the body, most people don't realize that it is actually an organ. Like all organs, the skin or **integument** (in-TEG-yoo-ment) is composed of several tissue types, including epithelial tissue, connective tissue, muscle tissue, and nervous tissue. In the following exercises, you will examine the tissues of this organ, along with the other structures of the integumentary system.

Exercise 5-1

Skin Anatomy and Accessory Structures

MATERIALS
❏ Skin models and diagrams
❏ Colored pencils

The **integumentary system** (in-TEG-yoo-MEN-tuh-ree) is composed of the skin and its **accessory structures**: the *hair*, *glands*, and *nails*. The skin is composed of two general tissue layers: the *epidermis* and the *dermis*. The tissue beneath the dermis, called the **hypodermis** or the *subcutaneous tissue*, connects the skin to the underlying tissues and is not considered part of the integument. The hypodermis is richly supplied with blood vessels.

The **epidermis** contains layers (or *strata*) of stratified squamous keratinized epithelium. The predominant cell type found in the epidermis is the keratin-producing **keratinocyte** (kehr-ah-TIN-oh-sy't). Keratin is a hard protein that protects the skin from mechanical stresses. From superficial to deep, the layers of the epidermis are as follows (Fig. 5.2):

1. **Stratum corneum.** The superficial layer of cells, called the **stratum corneum** (STRAT-um KOHR-nee-um), is composed of dead keratinocytes. Under microscopic examination, the keratinocytes of the stratum corneum bear little resemblance to living cells and have a dry, flaky appearance. These cells typically appear dark purple or hot pink, which is due to the accumulation of keratin in their cytosol.

2. **Stratum lucidum.** The **stratum lucidum** (LOO-sid-um) is a single layer of translucent, dead keratinocytes found only in the skin on the palmar surfaces of the hands and the plantar surfaces of the feet.

3. **Stratum granulosum.** The third layer of keratinocytes is the **stratum granulosum** (gran-yoo-LOH-sum). Here the superficial keratinocytes are dead, but the deeper cells are alive. This layer is named for the cells' cytoplasmic granules, which contain keratin and a lipid-based substance. This lipid-based substance helps to keep the skin water-resistant.

4. **Stratum spinosum.** The first actively metabolizing cells are encountered in the fourth cell layer, the **stratum spinosum** (spin-OH-sum). In this layer, we find the pigment **melanin** (MEL-uh-nin), which protects keratinocytes from UV light and also decreases production of vitamin D (to prevent the body from overproducing it).

5. **Stratum basale.** The **stratum basale** (bay-SAY-lee) is the deepest layer and consists of a single row of actively dividing keratinocytes. Here we also find **melanocytes** (mel-AN-oh-syt'z), which produce melanin, and sensory receptor cells called **tactile discs** (also known as *Merkel cells*), which function in light touch and texture discrimination.

Why does the epidermis have so many dead cells? Recall that the epidermis is composed of epithelial tissue, and epithelial tissue is avascular (has no blood supply). All epithelial tissues require oxygen and nutrients to diffuse to them from the deeper tissues. In the case of the epidermis, this deeper tissue is the dermis. Only the cells of the deeper parts of the stratum granulosum, the stratum spinosum, and the stratum basale are close enough to the blood supply in the dermis to get adequate oxygen and nutrients for survival. For this reason, as the cells migrate farther away from the blood supply, they begin to die.

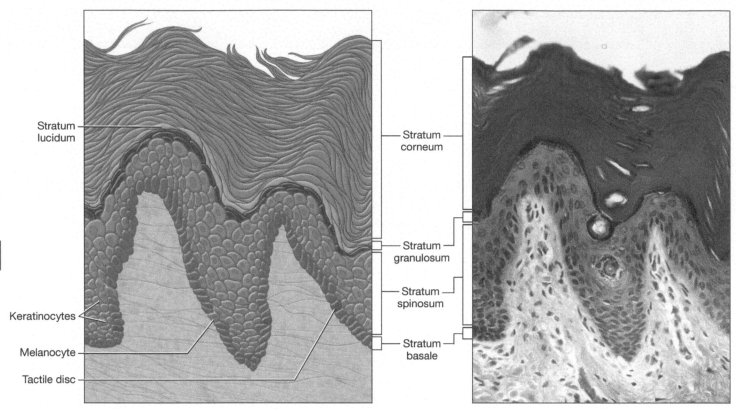

FIGURE **5.2** Layers of the epidermis.

Immediately deep to the stratum basale of the epidermis is the **basement membrane,** which holds the epidermis in place with molecular "glue" that prevents it from separating from the dermis. Deep to the basement membrane is the **dermis,** which is composed of highly vascular connective tissue. It has two layers (Fig. 5.3):

1. **Papillary layer.** The superficial **papillary layer** is composed of loose connective tissue. It contains fingerlike projections called the **dermal papillae** (pah-PILL-ee) that project into the epidermis, which you can see in **Figures 5.3A and B.** The dermal papillae contain touch receptors called **tactile corpuscles** that detect fine touch, and capillary loops that provide blood supply to the avascular epidermis.

2. **Reticular layer.** The thick **reticular layer** is composed of dense irregular collagenous connective tissue. It houses structures such as sweat glands, oil-producing *sebaceous glands* (seh-BAY-shuhs), blood vessels, and pressure receptors called **lamellated corpuscles.** Also, within the reticular layer, we find numerous *free nerve endings* that transmit pain sensations, which is why it hurts when you cut yourself down to the dermis.

Let's move on to accessory structures of the integumentary system. First are its glands. **Sebaceous glands** are exocrine glands that secrete **sebum** (SEE-bum; oil). Notice in Figure 5.3 that sebaceous glands are associated with hairs and, indeed, they release sebum directly into a hair follicle. **Sweat glands** are also exocrine glands; however, notice that they secrete their product, *sweat*, through a **sweat duct.** There are two types of sweat glands. The type shown in Figure 5.3, the *eccrine sweat gland*, is found all over the body. This type of gland releases sweat onto the surface of the skin through a small **sweat pore.** The other type, the *apocrine sweat gland*, is found in the axillae and genital region. This type releases sweat into a hair follicle like a sebaceous gland.

Both hair and nails are also accessory structures of the integumentary system (Fig. 5.4). A **hair** consists of two basic parts: (a) the long, slender **shaft** composed of dead keratinocytes that projects from the skin's surface, and (b) the **hair root** embedded in the dermis (Fig. 5.4A). The base of the root is indented by a projection from the dermis called the **hair papilla;** the papilla and root together are known as the **hair bulb.** The hair is embedded in a sheath known as the **hair follicle.** The inner lining of the hair follicle is epithelium that is continuous with the epidermis, and the outer lining is dermal connective tissue. A band of smooth muscle called an **arrector pili muscle** (ah-REK-tohr PIL-aye) attaches to this dermal sheath, and can pull the hair into an upright or erect position when it contracts.

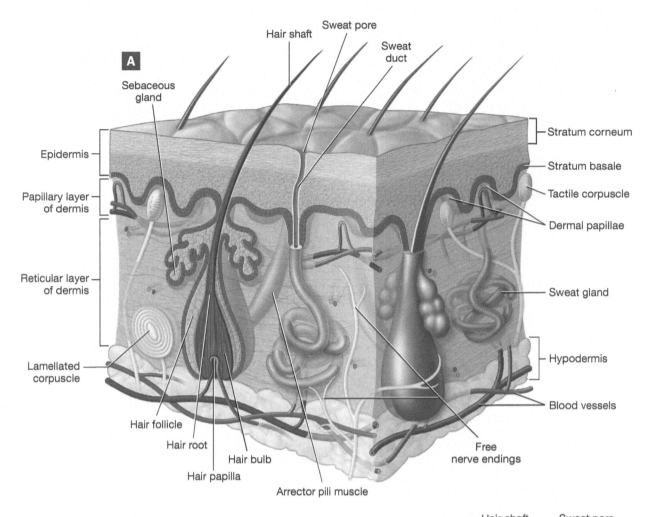

A

Hair shaft
Sweat pore
Sweat duct

Sebaceous gland

Epidermis

Papillary layer of dermis

Reticular layer of dermis

Lamellated corpuscle

Hair follicle

Hair root

Hair bulb

Hair papilla

Arrector pili muscle

Stratum corneum

Stratum basale

Tactile corpuscle

Dermal papillae

Sweat gland

Hypodermis

Blood vessels

Free nerve endings

Like hairs, **nails** are composed primarily of dead keratinocytes (Fig. 5.4B). A nail consists of a flat **nail plate** surrounded by folds of skin on all three sides, known as **nail folds**. Those on the sides of the nails are **lateral nail folds**, and the one at the base is the **proximal nail fold**. The stratum corneum of the proximal nail fold grows over the nail plate to form a thin structure called the **eponychium** (ep-oh-NIK-ee-um) or *cuticle*. Often, near the eponychium, we find a thickened area of the nail plate, the **lunula** (LOON-yoo-luh), which generally has a half-moon shape (note that many nails lack a lunula). Deep to the proximal nail fold is a group of dividing cells from which the nail plate grows. These cells are collectively known as the **nail matrix**.

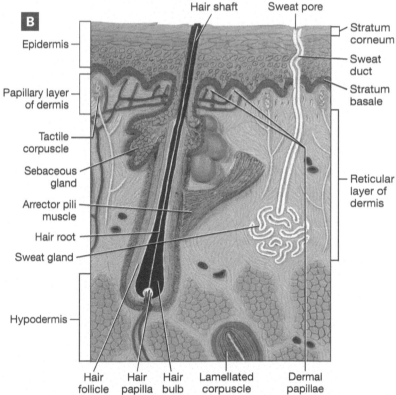

B

Hair shaft
Sweat pore

Epidermis

Papillary layer of dermis

Tactile corpuscle

Sebaceous gland

Arrector pili muscle

Hair root

Sweat gland

Hypodermis

Stratum corneum

Sweat duct

Stratum basale

Reticular layer of dermis

Hair follicle
Hair papilla
Hair bulb
Lamellated corpuscle
Dermal papillae

FIGURE **5.3** Skin section: (**A**) illustration; (**B**) anatomical model photo.

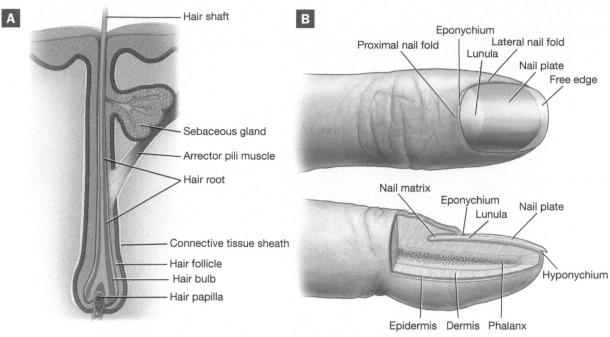

FIGURE **5.4** Accessory structures of the skin: (**A**) hair structure; (**B**) nail anatomy.

 Procedure 1 Model Inventory for the Integumentary System

Identify the following structures of the integumentary system on models and diagrams, using this unit and your textbook for reference. As you examine the anatomical models and diagrams, record the name of the model and the structures you were able to identify on the model inventory in Table 5.1.

Structures of the Skin

1. Epidermal layers
   a. Stratum corneum
   b. Stratum lucidum
   c. Stratum granulosum
   d. Stratum spinosum
   e. Stratum basale
2. Dermis
   a. Papillary layer
      (1) Dermal papillae
      (2) Tactile corpuscle
   b. Reticular layer
      (1) Lamellated corpuscle
3. Blood vessels

Accessory and Other Structures

1. Sebaceous gland
2. Sweat gland
   a. Sweat duct
   b. Sweat pore
3. Hair
   a. Hair shaft
   b. Hair root
   c. Hair follicle
   d. Arrector pili muscle
4. Nail
   a. Nail plate
   b. Nail fold (proximal and lateral)
   c. Nail matrix
5. Hypodermis

TABLE **5.1** Model Inventory for the Integumentary System

Model/Diagram	Structures Identified

Procedure **2** Time to Draw

In the space below, draw, color, and label one of the skin models that you examined. In addition, write down the main function of each structure that you label.

Exercise 5-2

Histology of Integument

MATERIALS
- ❏ Light microscope
- ❏ Slide of thick skin
- ❏ Slide of thin skin
- ❏ Colored pencils

In this exercise, we examine prepared slides of skin and hair. The skin sections are taken from different regions of the body so we can compare and contrast two types of skin: (a) **thick skin**, found on the palmar surfaces of the hands and the plantar surfaces of the feet, and (b) **thin skin**, found everywhere else (Fig. 5.5).

Before moving on, review the basics of microscopy from Unit 3. Remember to follow a step-by-step approach when examining the slides: Look at the slide with the naked eye first, then begin your examination on low power, and advance to higher power to see more details.

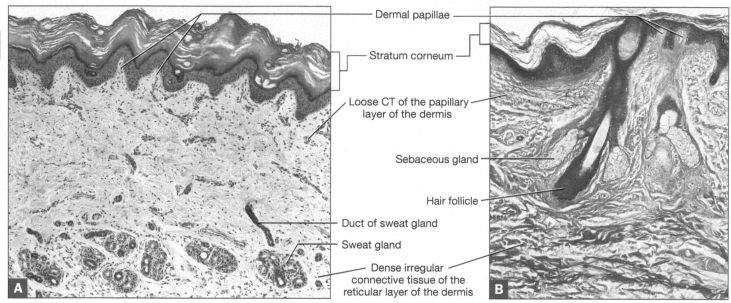

Dermal papillae

Stratum corneum

Loose CT of the papillary layer of the dermis

Sebaceous gland

Hair follicle

Duct of sweat gland

Sweat gland

Dense irregular connective tissue of the reticular layer of the dermis

FIGURE **5.5** Skin, photomicrographs: (**A**) thick skin; (**B**) thin skin.

Procedure **1** Microscopy of Thick Skin

Obtain a prepared slide of thick skin (which may be labeled "Palmar Skin"), and examine it with the naked eye to get oriented. After you are oriented, place the slide on the stage of the microscope, and scan it on low power. You should be able to see the epidermis with its superficial layers of dead cells and the dermis with its pink clusters of collagen bundles that make up the dense irregular collagenous connective tissue. Compare it to Figure 5.5A to make sure that you are looking at the right slide and using the right magnification. Advance to higher power to see the cells and associated structures in greater detail.

Use your colored pencils to draw what you see in the field of view (you will be able to see the most structures on low power). Label your drawing with the following terms, using Figure 5.5A as a reference. When you have completed your drawing, fill in Table 5.2.

1. Epidermis
   a. Stratum corneum
   b. Stratum lucidum
   c. Stratum granulosum
   d. Stratum spinosum
   e. Stratum basale

2. Dermis
   a. Dermal papillae
   b. Collagen bundles
   c. Sweat gland

Procedure 2 Microscopy of Thin Skin

Obtain a prepared slide of thin skin (which may be called "Scalp Skin"). As before, examine the slide with the naked eye, then scan the slide on low power, advancing to higher power as needed to see the structures more clearly. Compare what you see in your field of view to Figure 5.5B to ensure that you are looking at the correct region of the slide.

Use your colored pencils to draw what you see in the field of view (you will be able to see the most structures on low power). Label your drawing with the following terms, using Figure 5.5B as a reference. When you have completed your drawing, fill in the remainder of Table 5.2.

1. Epidermis
   a. Stratum corneum
   b. Stratum granulosum
   c. Stratum spinosum
   d. Stratum basale
2. Dermis
   a. Dermal papillae
   b. Collagen bundles
3. Hair follicle
4. Sebaceous gland
5. Sweat gland
6. Arrector pili muscle

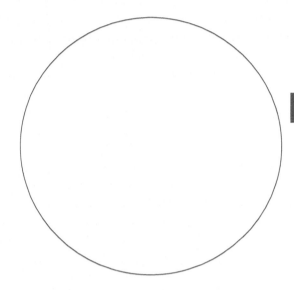

TABLE **5.2** Characteristics of Thick and Thin Skin

Characteristic	Thick Skin	Thin Skin
Thickness of stratum corneum		
Hair follicles present?		
Sebaceous glands present?		
Stratum lucidum present?		
Arrector pili muscles present?		

Name _____

Section _____ Date _____

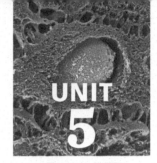

1 Label the following parts of the skin on **Figure 5.6**.
- ❏ Arrector pili muscle
- ❏ Dermal papillae
- ❏ Dermis
- ❏ Epidermis
- ❏ Hair follicle
- ❏ Hair shaft
- ❏ Hypodermis
- ❏ Sebaceous gland
- ❏ Sweat gland

5

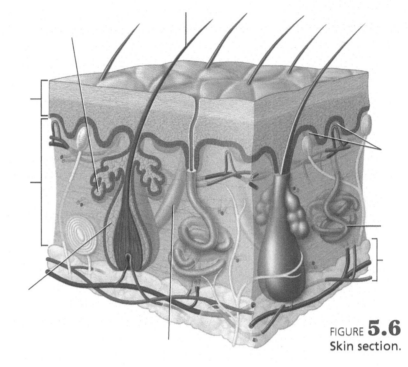

2 The primary protein produced by the main cell type in skin is
- a. actin.
- b. melanin.
- c. keratin.
- d. collagen.

FIGURE **5.6**
Skin section.

3 Number the layers of the epidermis, with 1 being the most superficial layer and 5 being the deepest layer.

_____ Stratum lucidum   _____ Stratum spinosum

_____ Stratum basale   _____ Stratum granulosum

_____ Stratum corneum

4 Which layers of the epidermis contain living cells?
- a. Stratum granulosum only.
- b. Stratum corneum, stratum granulosum, stratum lucidum.
- c. Stratum basale, stratum spinosum, stratum granulosum.
- d. All of the layers of the epidermis contain living cells.
- e. None of the layers of the epidermis contain living cells.

5 From where do the cells of the epidermis obtain oxygen and nutrients?
- a. From blood vessels in the epidermis.
- b. Diffusion from blood vessels in the dermis.
- c. Diffusion from the air.
- d. From blood vessels in other epithelial tissues.

6 Which of the following are characteristics of thick skin? *(Circle all that apply.)*
- a. Located over the palms and the soles of the feet.
- b. Contains hair and arrector pili muscles.
- c. Contains sweat glands.
- d. Very thick stratum corneum.
- e. Contains sebaceous glands.
- f. Contains a stratum lucidum.

7 *Matching:* Match the following terms with the correct description.

_____ Nail matrix

_____ Sebaceous gland

_____ Lamellated corpuscle

_____ Hair follicle

_____ Dermal papillae

_____ Hair shaft

_____ Eponychium

_____ Sweat gland

A. Secretes product through a pore

B. Pressure receptor in the dermis

C. Projections of the dermis that indent the epidermis

D. Location of the dividing cells of the nail

E. Sheath of epithelial and connective tissue around a hair

F. Thin growth of the proximal nail fold over the proximal nail plate

G. Secretes sebum (oil)

H. Portion of the hair that projects from the skin's surface

8 You develop a *callus*, which is a thickening of the epidermis, from a pair of ill-fitting shoes.

a To treat the callus, you decide to shave off some of the excess epidermis. While doing this, you notice that there is no pain or bleeding. Why wouldn't it cause pain to remove only the superficial part of the epidermis? Why wouldn't it bleed?

_____

_____

b Unfortunately, you go a bit overboard and feel a sudden pinch, after which you notice you are bleeding. What has happened? Are you still in the epidermis? Explain.

_____

_____

9 Shampoos and hair conditioners often claim to have nutrients and vitamins your hair must have to grow and be healthy.

a What is the composition of the hair shaft onto which shampoos and conditioners are applied?

_____

_____

b Taking into account the composition of the hair shafts, do you think these vitamins and nutrients will be beneficial? Why or why not?

_____

_____

10 The disease *bullous pemphigoid* results in the destruction of proteins within the basement membrane that hold the epidermis and dermis together. How would this likely affect the epidermis?

_____

_____

_____

11 *Carcinomas* are cancers of epithelial tissue, whereas *sarcomas* are cancers of connective tissue. Are epidermal cancers carcinomas or sarcomas? What about dermal cancers? Explain.

_____

_____

_____

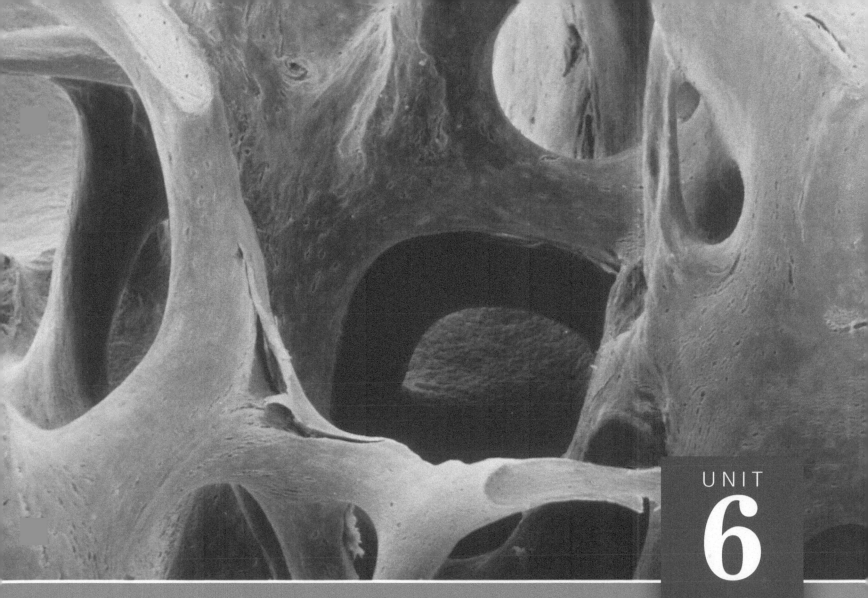

Skeletal System:
Bone Tissue, Bones, and Joints

When you have completed this unit, you should be able to:

1 Identify the structures and components of osseous tissue.

2 Classify bones according to their shape.

3 Identify the parts of a long bone.

4 Identify bones and markings of the skeleton.

5 Identify structures associated with synovial joints and knee joints.

6 Demonstrate and describe motions allowed at synovial joints.

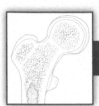

PRE-LAB EXERCISES

Complete the following exercises prior to coming to lab, using your lab manual and textbook for reference.

Pre-Lab Exercise **6-1**

 Key Terms

You should be familiar with the following terms before coming to lab.

Term	Definition
Bone Tissue	
Compact bone	
Spongy bone	
Osteon	
Lamellae	
Osteocytes	
Periosteum	
Bone Shapes	
Long bone	
Short bone	
Flat bone	
Irregular bone	
Structures of Long Bones	
Diaphysis	

Epiphysis

Epiphyseal plate/line

Axial Skeleton

Cranial bones

Suture

Facial bones

Hyoid bone

Vertebrae

Sternum

Ribs

Appendicular Skeleton—Upper Limb

Pectoral girdle

Humerus

Ulna

Radius

Carpals

Metacarpals

Phalanges

Appendicular Skeleton—Lower Limb

Pelvic girdle

Femur

Tibia

Fibula

Tarsals

Metatarsals

Classes of Joints

Fibrous

Cartilaginous

Synovial

Types of Movement in Synovial Joints

Flexion

Extension

Abduction

Adduction

Circumduction

Rotation

Pre-Lab Exercise **6-2**

Microscopic Anatomy of Compact Bone

Color the microscopic anatomy of compact bone tissue in **Figure 6.1**, and label it with the following terms from Exercise 6-1 (p. 123). Use Exercise 6-1 in this unit and your text for reference.

- ❏ Osteon
  - ◻ Central canal
  - ◻ Lamellae
  - ◻ Lacunae
    - ▪ Osteocytes
    - ▪ Canaliculi
  - ◻ Perforating canal
- ❏ Trabeculae

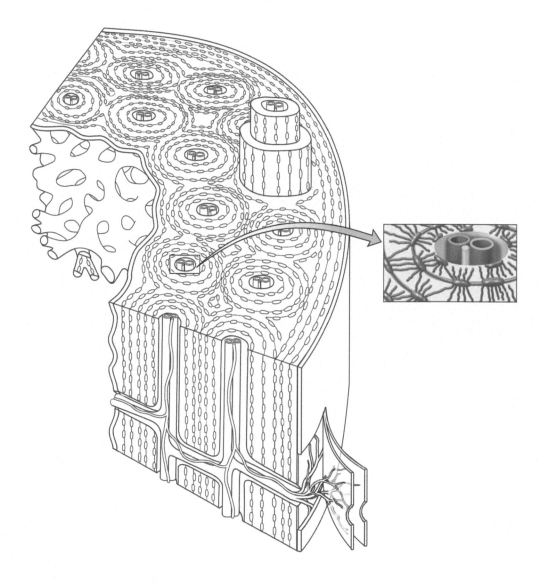

FIGURE **6.1** Microscopic anatomy of compact bone tissue.

Pre-Lab Exercise **6-3**

Bones of the Skull

Color the structures of the skull in Figure 6.2, and label them with the following terms from Exercise 6-3 (pp. 131, 132, and 135). Use Exercise 6-3 in this unit and your text for reference.

Cranial Bones

❑ Frontal bone
   ❑ Frontal sinuses
❑ Parietal bones
   ❑ Coronal suture
   ❑ Squamous suture
   ❑ Lambdoid suture

❑ Temporal bones
   ❑ Mastoid process
   ❑ External acoustic meatus
   ❑ Styloid process
❑ Occipital bone
   ❑ Foramen magnum

❑ Sphenoid bone
   ❑ Sphenoid sinus
   ❑ Sella turcica
❑ Ethmoid bone
   ❑ Perpendicular plate

Facial Bones and Other Structures

❑ Mandible
   ❑ Mandibular body
   ❑ Mandibular ramus
   ❑ Mandibular condyle

❑ Maxilla
❑ Lacrimal bones
❑ Nasal bones

❑ Vomer
❑ Palatine bones
❑ Zygomatic bones

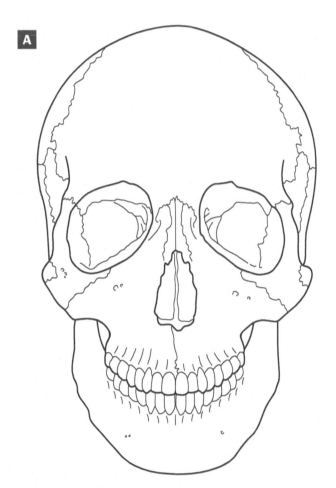

FIGURE **6.2** Skull: **(A)** anterior view; *(continues)*

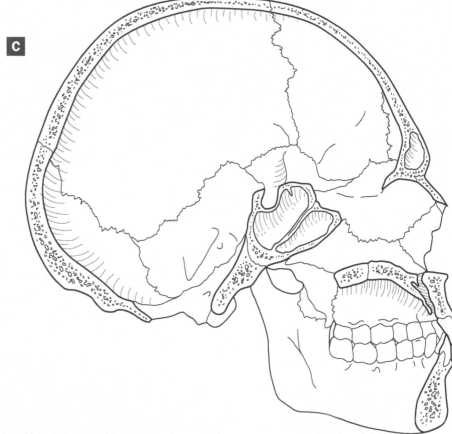

FIGURE **6.2** Skull *(cont.)*: **(B)** lateral view; **(C)** midsagittal section.

Pre-Lab Exercise **6-4**

Whole Skeleton

Color the structures of the skeleton in Figure 6.3, and label them with the following terms from Exercise 6-4 (p. 153). Use Exercise 6-4 in this unit and your text for reference.

Axial Skeleton

❏ Vertebrae
❏ Sacrum
❏ Coccyx
❏ Sternum
❏ Ribs

Pectoral Girdle and Upper Limb

❏ Scapula
❏ Clavicle
❏ Humerus
❏ Ulna
❏ Radius
❏ Carpals
❏ Metacarpals
❏ Phalanges

Pelvic Girdle and Lower Limb

❏ Ilium
❏ Ischium
❏ Pubis
❏ Femur
❏ Patella
❏ Tibia
❏ Fibula
❏ Tarsals
❏ Metatarsals
❏ Phalanges

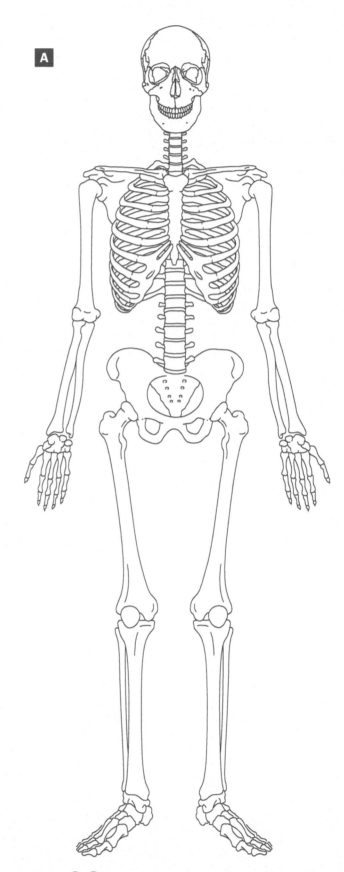

A

FIGURE **6.3** Skeleton: **(A)** anterior view; *(continues)*

6

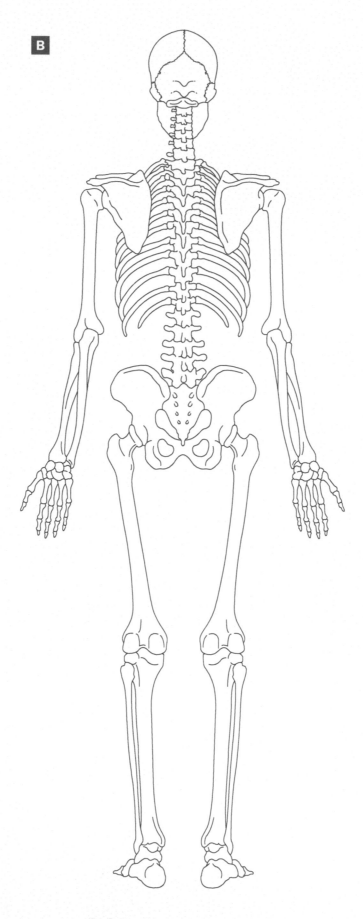

FIGURE **6.3** Skeleton *(cont.)*: **(B)** posterior view.

EXERCISES

The **skeletal system** consists of the bones, associated cartilages, and joints. At first, it might seem odd that a set of bones makes up an organ system, but remember that each bone is considered an organ. A bone consists of many tissue types, including osseous tissue, epithelial tissue, dense connective tissue, and adipose tissue. The following exercises will introduce you to these organs and the histology of osseous tissue.

Colored SEM of trabeculae in spongy bone tissue of the femur.

The two divisions of the skeletal system are the axial skeleton and the appendicular skeleton. The **axial skeleton** is composed of the bones of the head, neck, and trunk—specifically, the cranial bones, the facial bones, the vertebral column, the hyoid bone, the sternum, and the ribs. The **appendicular skeleton** consists of the bones of the upper limbs, the lower limbs, the pectoral girdle (the bones forming the shoulder joint), and the pelvic girdle (the bones forming the pelvis and the hip joint).

In this unit, we explore the anatomy of the bones and bone markings of the skeletal system, as well as the structure and function of joints. Note as you progress through the exercises that much of what you learn here will serve as a foundation for later units. For example, the radial and ulnar arteries parallel the radius and the ulna, and the frontal, parietal, temporal, and occipital lobes of the brain are named for the cranial bones under which they are located.

6

Exercise **6-1**

Histology of Osseous Tissue

MATERIALS

☐ Osteon model

The most superficial tissue of a bone is called the **periosteum** (pehr-ee-AH-stee-um; Fig. 6.4). The outer layer of the periosteum is composed of dense irregular collagenous connective tissue richly supplied with blood vessels and nerves. The innermost layer of the periosteum contains cells called **osteoblasts** (AH-stee-oh-blasts). Osteoblasts are "builder" cells that secrete bone extracellular matrix (ECM). Also, within this layer are cells known as **osteoclasts** (AH-stee-oh-klasts). These are the "breakdown" cells, as they secrete enzymes that catalyze the digestion of bone ECM.

Deep to the periosteum, we find **osseous tissue**, which is composed of a hardened ECM and different types of bone cells. The two general forms of osseous tissue are compact bone and spongy bone.

Compact bone is hard, dense bone tissue found immediately deep to the periosteum. Its hardness comes from its structure, which consists of repeating, densely packed subunits called **osteons** (AHS-tee-ahnz). Osteons have several features, including the following, which are illustrated in Figure 6.4:

1. **Central canal.** Running down the center of each osteon is a **central canal.** Each central canal contains blood vessels and nerves and is lined with a connective tissue membrane called **endosteum** (en-DAH-stee-um). Like the periosteum, the endosteum has an inner layer composed of osteoblasts, which secrete bone ECM, and osteoclasts, which degrade bone.

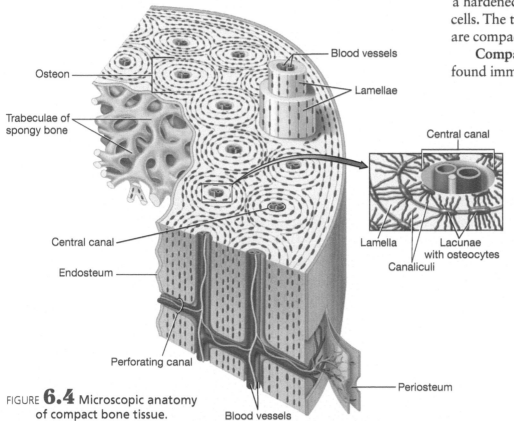

FIGURE **6.4** Microscopic anatomy of compact bone tissue.

Labels: Osteon; Trabeculae of spongy bone; Central canal; Endosteum; Perforating canal; Blood vessels; Blood vessels; Lamellae; Central canal; Lamella; Lacunae with osteocytes; Canaliculi; Periosteum

2. **Lamellae.** Surrounding the central canal are rings of bone ECM called **lamellae** (lah-MELL-ee). The lamellae give compact bone a great deal of strength, much like a tree's rings.

3. **Lacunae.** Situated between the lamellae are small cavities called **lacunae** (lah-KOO-nee). Lacunae contain mature osteoblasts called **osteocytes**, which monitor and maintain the bone ECM. Neighboring lacunae and osteocytes are connected to each other by tiny canals called **canaliculi** (kan-ah-LIK-yoo-lee).

4. **Perforating canals.** The **perforating canals** run perpendicular to the lamellae and carry blood vessels deep into the bone from the periosteum. Like the central canals, perforating canals are lined by endosteum.

Spongy bone is found on the inside of a bone deep to compact bone, and, as its name implies, it somewhat resembles a sponge. It consists of a latticework-type structure with tiny bone spicules called **trabeculae** (trah-BEK-yoo-lee; Fig. 6.5) that are lined with endosteum. The latticework-structure of spongy bone allows it to house another important tissue, the **bone marrow**. The two types of bone marrow are **red bone marrow**, which produces blood cells, and **yellow bone marrow**, which is composed primarily of adipose tissue.

As you can see in Figure 6.5, trabeculae are composed of lamellae but are not organized into osteons. For this reason, spongy bone lacks the hardness of compact bone. Between the lamellae are osteocytes in lacunae, and they are connected by canaliculi, as in compact bone.

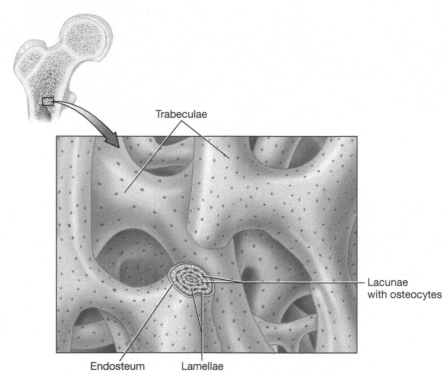

FIGURE **6.5** Microscopic anatomy of spongy bone tissue.

Procedure **1** Model Inventory for Compact and Spongy Bone

Identify the following structures of compact and spongy bone on models and diagrams, using this unit and your textbook for reference. As you examine the anatomical models and diagrams, record the name of the model and the structures you were able to identify on the model inventory in Table 6.1.

Compact Bone Structures
1. Periosteum
2. Osteons
   a. Central canal
   b. Endosteum
   c. Lamellae
   d. Lacunae
   e. Osteocyte
   f. Canaliculi
   g. Perforating canal

Spongy Bone Structures
1. Trabeculae
2. Endosteum
3. Red bone marrow
4. Yellow bone marrow
5. Osteocytes
6. Lacunae

TABLE **6.1** Model Inventory for Osseous Tissue

Model/Diagram	Bone Structures Identified

6

Exercise 6-2

Bone Markings and Bone Shapes

MATERIALS
- ❑ Disarticulated bones
- ❑ Articulated skeleton
- ❑ Long bone, sectioned
- ❑ X-rays, if available

When you examine a skeleton closely, you'll notice that few bones have smooth, flat surfaces. Instead, most bones contain depressions, openings, and projections. These features, collectively known as **bone markings**, perform numerous functions: *depressions* provide pathways along which blood vessels and nerves travel, or allow two bones to come together to form a joint; *openings* house and protect structures such as blood vessels and special sensory organs; and *projections* provide points of attachment for ligaments and tendons. The major types of bone markings are defined in Table 6.2.

TABLE **6.2** Bone Markings

Bone Marking	Description
Depressions	
Facet	Shallow indented surface where two bones meet to form a joint
Fossa	Deeper indented surface in a bone; usually allows a rounded surface of another bone to fit inside of it
Fovea	Shallow pit; often the site for the attachment of a ligament
Groove	Long, typically shallow depression that usually allows a nerve or blood vessel to travel along the bone's surface
Sulcus	Another name for a groove
Openings	
Canal	Passageway through a bone
Fissure	Slit within a bone or between bones
Foramen	Hole in a bone through which a structure such as a nerve or blood vessel passes
Meatus	Another name for a canal
Projections	
Condyle	Round end of a bone that fits into a fossa or facet of another bone at a joint
Crest	Ridge along a bone; generally a site of muscle attachment
Epicondyle	Small projection that is usually proximal to a condyle; generally the site of muscle attachment
Head	Rounded end of the bone that fits into a fossa to form a joint
Line	Ridge along a bone where a muscle attaches
Process	Any bony projection; generally the site of muscle attachment
Protuberance	An outgrowth from a bone due to repetitive pull from a muscle
Trochanter	Large bony projection to which muscles attach; only examples are in the femur (thigh bone)
Tubercle	Small rounded projection where muscles attach
Tuberosity	A larger, more prominent tubercle

Another thing you might notice about the skeleton is the variety of ways that the bones are shaped. As shown in Figure 6.6, there are four general shapes of bones:

1. **Long bones** are longer than they are wide and include the bones of the upper and lower limbs, excluding the ankle and wrist bones.

2. **Short bones** are about as long as they are wide. The bones of the wrist and the ankle are short bones.

3. **Flat bones** are shaped exactly as they're named—they are flat. Flat bones include the ribs, the sternum, certain skull bones, and the hip bones.

4. **Irregular bones** are those whose shape doesn't fit into any of the other classes. Irregular bones include the vertebrae, the sacrum, and certain bones of the skull, such as the sphenoid bone.

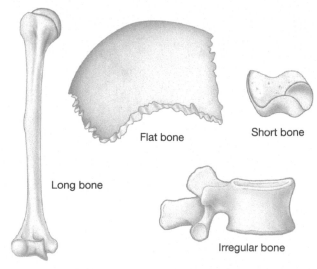

Flat bone

Short bone

Long bone

Irregular bone

FIGURE **6.6** The four shapes of bones.

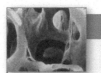

Bone Shape, not Length

Note that long bones are not named for their length but rather for their shape. Many long bones actually are quite short in length, including the bones of the fingers and the toes. Be careful when identifying bone shapes to look at the overall shape of the bone rather than its size.

Procedure **1** Identifying Bone Shapes

Obtain a set of disarticulated bones, and classify each bone according to its shape. Identify examples of long bones, short bones, flat bones, and irregular bones, and record them in Table 6.3. You may wish to use Exercises 6-3 and 6-4 for reference to help you identify the disarticulated bones.

6

TABLE **6.3** Examples of Bone Shapes

Bone	Shape

Long Bone Anatomy

All long bones share common structures and parts, as illustrated in **Figure 6.7** with the example of the femur. The **diaphysis** (dy-AEH-fih-sis) is the shaft of the long bone. As you can see in the figure, it consists of a thick collar of compact bone surrounding a hollow area called the **medullary cavity**. This collar of compact bone makes long bones quite strong. The medullary cavity has sparse trabeculae and generally is filled with yellow bone marrow in adult bones.

The ends of a long bone are called the **proximal** and **distal epiphyses** (eh-PIF-ih-seez). Each epiphysis contains a shell of compact bone surrounding the inner spongy bone. The spongy bone within the epiphyses contains either red or yellow bone marrow. The end of each epiphysis is covered with **articular cartilage** (generally composed of hyaline cartilage), which allows two bones in a joint to move around one another with minimal friction.

At certain epiphysis-diaphysis junctions, you will note a thin, calcified line called the **epiphyseal line** (eh-PIF-ih-seel). This structure is the remnant of the **epiphyseal plate**, a band of hyaline cartilage from which long bones grow in length in children and young adults (Fig. 6.8). When longitudinal growth ceases, the chondrocytes of the epiphyseal plate die and are replaced by calcified bone tissue.

6

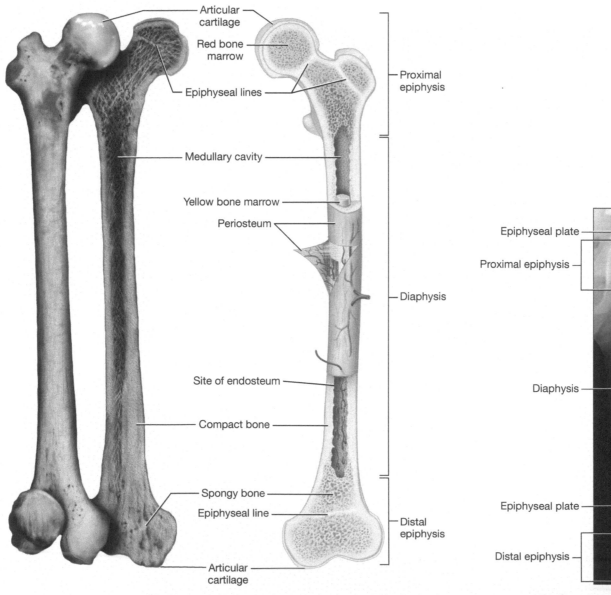

FIGURE **6.7** Adult long bone: the femur.

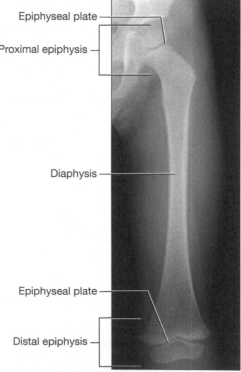

FIGURE **6.8** X-ray of a child's femur.

Procedure 2 Identifying Structures of Long Bones

Identify the following structures of long bones on specimens and X-rays (if available). Check off each structure as you identify it.

- ❑ Diaphysis
- ❑ Compact bone
- ❑ Medullary cavity
- ❑ Epiphysis
- ❑ Red bone marrow

- ❑ Yellow bone marrow
- ❑ Articular cartilage
- ❑ Epiphyseal line
- ❑ Epiphyseal plate
  (may be visible only on X-ray)

Procedure 3 Time to Draw

In the space provided, draw one of your sectioned long bone models or diagrams. In addition, write down the definition of each structure that you label.

6

Exercise 6-3

The Skull

MATERIALS

❏ Skulls, whole and sectioned
❏ Fetal skull

The skull is composed of two classes of bones—the cranial bones and the facial bones (Figs. 6.9–6.16). The **cranial bones** encase the brain and together form the **calvaria** (kal-VEHR-ee-uh; also known as the "skullcap" or *cranial vault*), which consists of several of the cranial bones joined at immovable joints called **sutures** (SOO-tchurz). These bones also form the **cranial base**, which is made up of the **anterior, middle,** and **posterior cranial fossae** (visible in Figure 6.12), indentations that support the brain. The eight cranial bones include the following:

1. **Frontal bone.** The **frontal bone** forms the anterior portion of the cranium, the superior part of the orbit, which houses the eyeball (visible in Figure 6.9), and the anterior cranial fossa. In Figure 6.9, you can see the two small holes called the **supraorbital foramina**, which are located superior to the orbits. Internally, the frontal bone contains the hollow **frontal sinuses**, which are part of the **paranasal sinuses**, a group of bony cavities surrounding the nasal cavity (see Figure 6.17, p. 137). Air from the nasal cavity enters the paranasal sinuses via small openings in the bones, and in the sinuses the air gets filtered, warmed, and humidified.

2. **Parietal bones.** The paired **parietal bones** (puh-RY-ih-tuhl) form the superior and part of the lateral walls of the cranium (best seen in Figures 6.14 and 6.15). Notice in the figures that they articulate with one another and many other cranial bones at joints called *sutures*: They meet one another at the **sagittal suture**, they meet the frontal bone at the **coronal suture**, they meet the occipital bone at the **lambdoid suture** (LAM-doyd), and they meet the temporal bones at the **squamous sutures** (seen in Figure 6.10).

3. **Temporal bones.** The paired **temporal bones** form the lateral walls of the cranium (best seen in Figure 6.10). Each temporal bone has a complex shape with several important markings:

    a. The anterior **mandibular fossa** forms the *temporomandibular joint*—or the jaw joint—with the mandible.

    b. On the bone's lateral surface is the **external acoustic (auditory) meatus**, an opening to the *external auditory canal*, which leads to the middle and inner ear.

    c. Posterior to the external acoustic meatus is the large **mastoid process**, a site of muscle attachments.

    d. Medial to the external acoustic meatus is another site of muscle attachments, the needlelike **styloid process**.

    e. Medial to the styloid process is the **jugular foramen**, through which the internal jugular vein exits the skull.

4. **Occipital bone.** The posterior cranial bone is the **occipital bone** (awk-SIP-ih-tuhl; best seen in Figures 6.11 and 6.14). Its most conspicuous feature is found in its base—a large hole called the **foramen magnum** through which the spinal cord passes. Anterior and lateral to the foramen magnum are the two **occipital condyles**, which form a joint with the first cervical vertebra. The occipital bone's posterior surface has two prominent horizontal ridges: the **superior** and **inferior nuchal lines** (NOO-kuhl). In the middle of the superior nuchal line, we find a bump called the **external occipital protuberance**.

5. **Sphenoid bone.** The butterfly-shaped **sphenoid bone** (SFEE-noyd) is posterior to the frontal bone on the interior part of the skull. Centrally, it consists of the **body**, in which we find the second paranasal sinus, called the **sphenoid sinus**, and a saddlelike formation called the **sella turcica** (SELL-uh TUR-sih-kuh) that houses the pituitary gland (see Figures 6.12 and 6.13). Extending from the body are three sets of "wings":

    a. The small **lesser wings** are the superior and most anterior set that is best seen in Figure 6.11.

    b. The larger **greater wings** are the inferior and posterior set of wings.

    c. The inferior set of wings consists of the narrow **pterygoid processes** (TEHR-ih-goyd), which form part of the posterior walls of the nasal and oral cavities (see Figures 6.11 and 6.13).

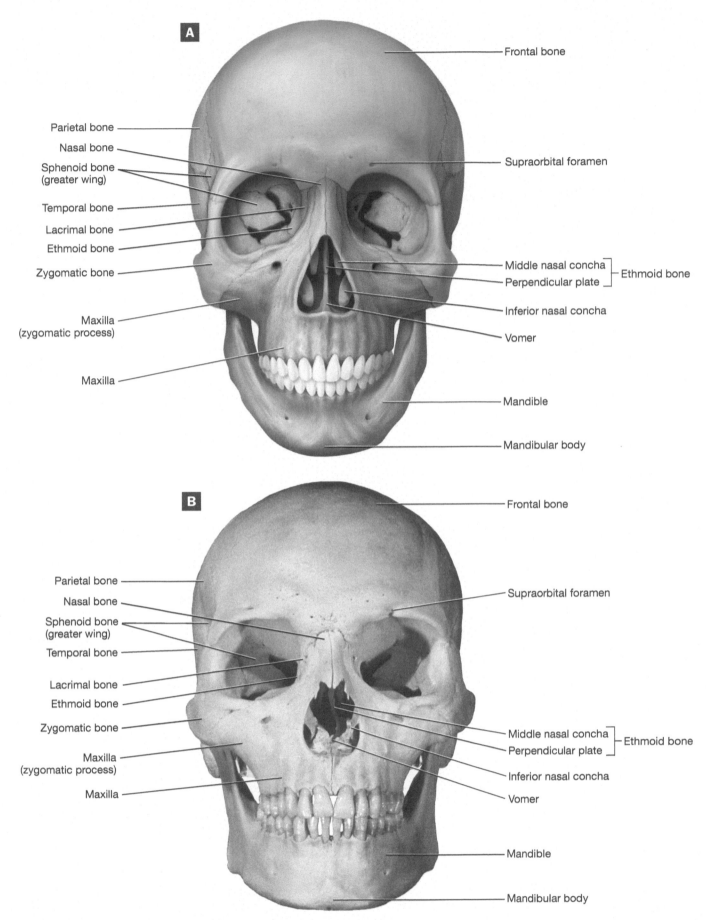

Parietal bone

Nasal bone

Sphenoid bone
(greater wing)

Temporal bone

Lacrimal bone

Ethmoid bone

Zygomatic bone

Maxilla
(zygomatic process)

Maxilla

Frontal bone

Supraorbital foramen

Middle nasal concha ⎤
                    ⎥ Ethmoid bone
Perpendicular plate ⎦

Inferior nasal concha

Vomer

Mandible

Mandibular body

A

Parietal bone

Nasal bone

Sphenoid bone
(greater wing)

Temporal bone

Lacrimal bone

Ethmoid bone

Zygomatic bone

Maxilla
(zygomatic process)

Maxilla

Frontal bone

Supraorbital foramen

Middle nasal concha ⎤
                    ⎥ Ethmoid bone
Perpendicular plate ⎦

Inferior nasal concha

Vomer

Mandible

Mandibular body

B

FIGURE **6.9** Anterior view of the skull: (**A**) illustration; (**B**) photograph.

6

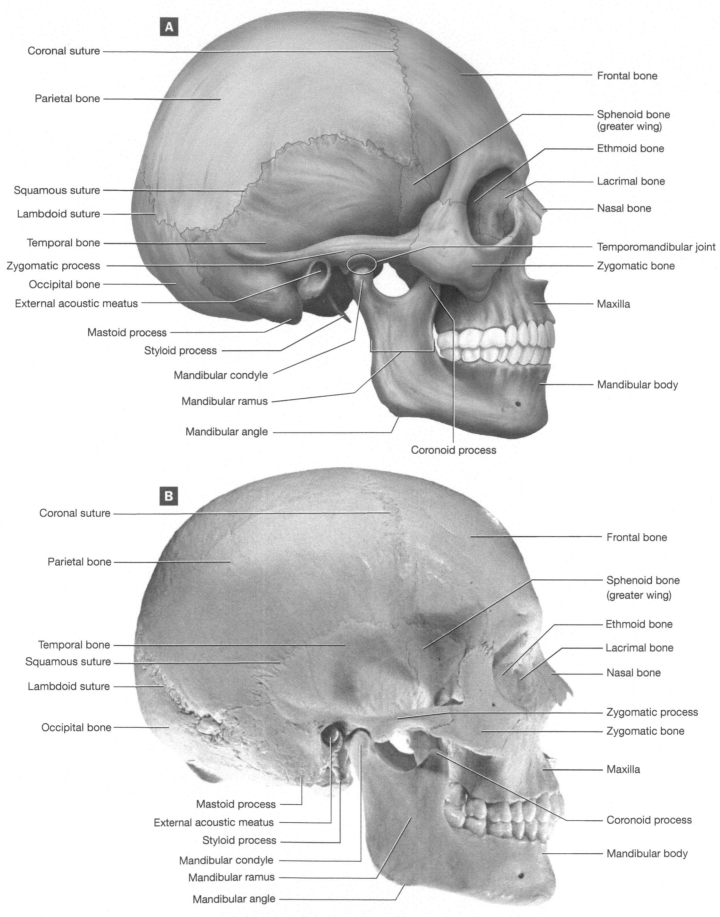

Coronal suture

Parietal bone

Squamous suture

Lambdoid suture

Temporal bone

Zygomatic process

Occipital bone

External acoustic meatus

Mastoid process

Styloid process

Mandibular condyle

Mandibular ramus

Mandibular angle

Frontal bone

Sphenoid bone
(greater wing)

Ethmoid bone

Lacrimal bone

Nasal bone

Temporomandibular joint

Zygomatic bone

Maxilla

Mandibular body

Coronoid process

Coronal suture

Parietal bone

Temporal bone

Squamous suture

Lambdoid suture

Occipital bone

Mastoid process

External acoustic meatus

Styloid process

Mandibular condyle

Mandibular ramus

Mandibular angle

Frontal bone

Sphenoid bone
(greater wing)

Ethmoid bone

Lacrimal bone

Nasal bone

Zygomatic process

Zygomatic bone

Maxilla

Coronoid process

Mandibular body

FIGURE **6.10** Lateral view of the skull: (**A**) illustration; (**B**) photograph.

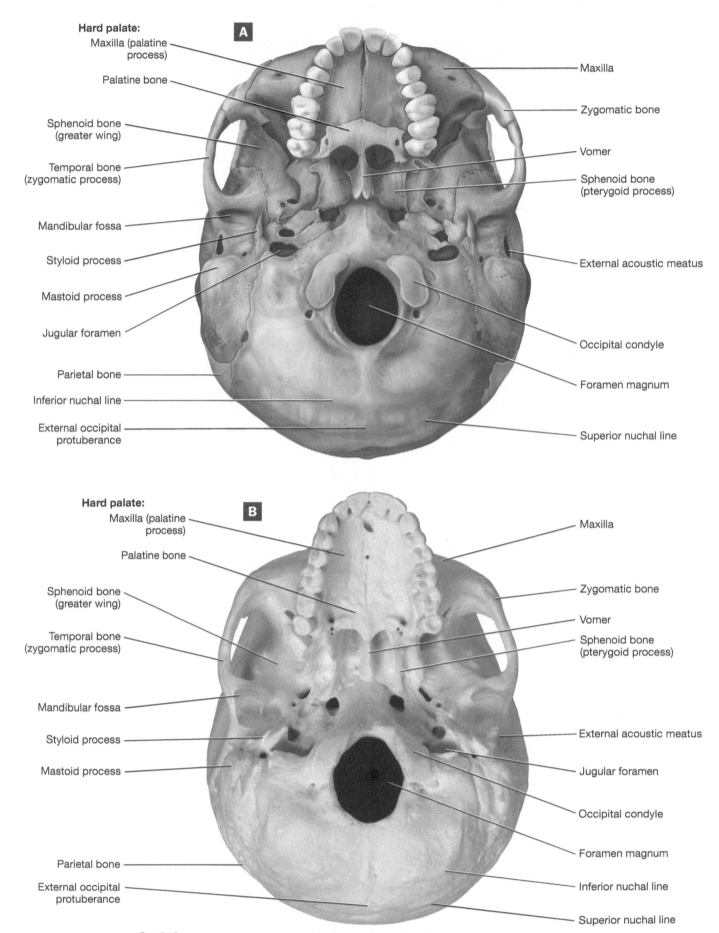

Hard palate:
Maxilla (palatine process)
Palatine bone
Sphenoid bone (greater wing)
Temporal bone (zygomatic process)
Mandibular fossa
Styloid process
Mastoid process
Jugular foramen
Parietal bone
Inferior nuchal line
External occipital protuberance

A

Maxilla
Zygomatic bone
Vomer
Sphenoid bone (pterygoid process)
External acoustic meatus
Occipital condyle
Foramen magnum
Superior nuchal line

Hard palate:
Maxilla (palatine process)
Palatine bone
Sphenoid bone (greater wing)
Temporal bone (zygomatic process)
Mandibular fossa
Styloid process
Mastoid process
Parietal bone
External occipital protuberance

B

Maxilla
Zygomatic bone
Vomer
Sphenoid bone (pterygoid process)
External acoustic meatus
Jugular foramen
Occipital condyle
Foramen magnum
Inferior nuchal line
Superior nuchal line

FIGURE **6.11** Inferior view of the skull (mandible removed): (**A**) illustration; (**B**) photograph.

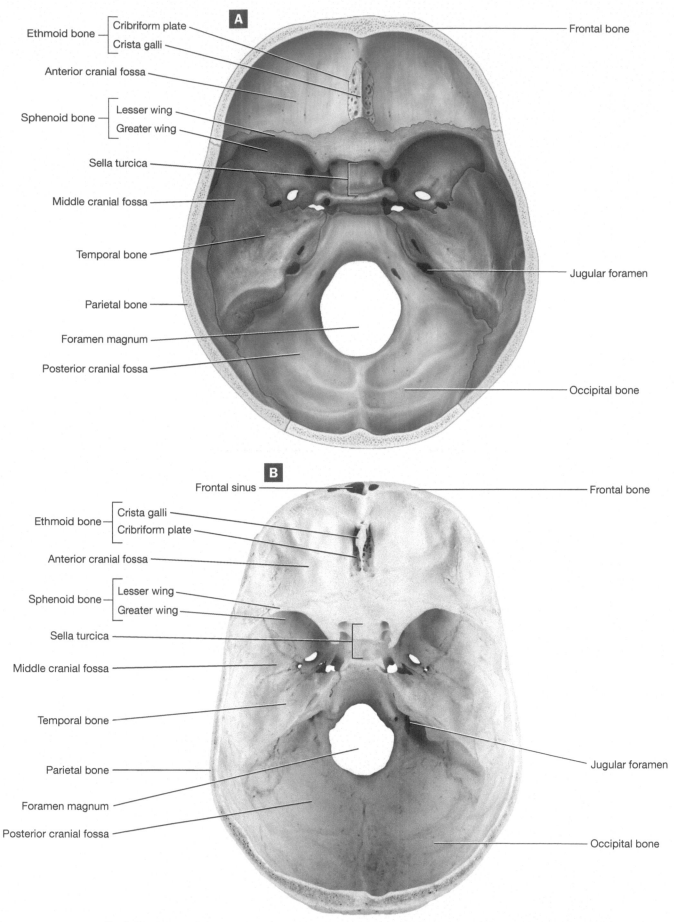

FIGURE **6.12** Interior view of the cranial base (calvaria removed): (**A**) illustration; (**B**) photograph.

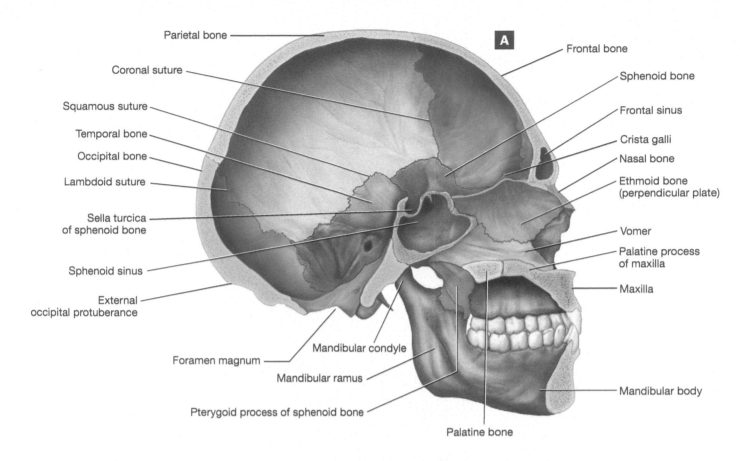

Parietal bone

Coronal suture

Squamous suture

Temporal bone

Occipital bone

Lambdoid suture

Sella turcica
of sphenoid bone

Sphenoid sinus

External
occipital protuberance

Foramen magnum

Mandibular condyle

Mandibular ramus

Pterygoid process of sphenoid bone

Palatine bone

Frontal bone

Sphenoid bone

Frontal sinus

Crista galli

Nasal bone

Ethmoid bone
(perpendicular plate)

Vomer

Palatine process
of maxilla

Maxilla

Mandibular body

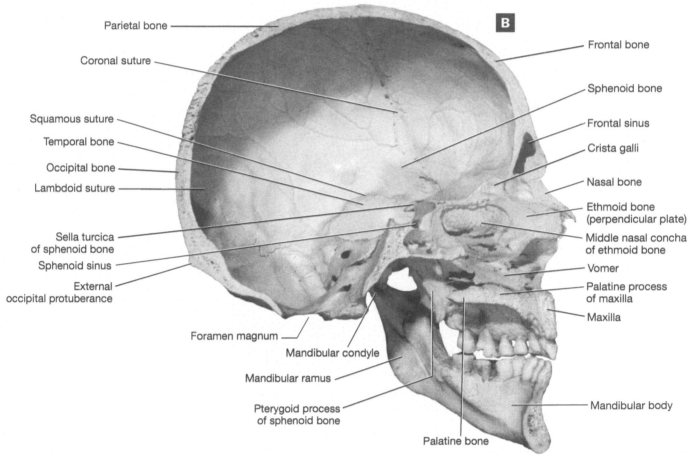

Parietal bone

Coronal suture

Squamous suture

Temporal bone

Occipital bone

Lambdoid suture

Sella turcica
of sphenoid bone

Sphenoid sinus

External
occipital protuberance

Foramen magnum

Mandibular condyle

Mandibular ramus

Pterygoid process
of sphenoid bone

Palatine bone

Frontal bone

Sphenoid bone

Frontal sinus

Crista galli

Nasal bone

Ethmoid bone
(perpendicular plate)

Middle nasal concha
of ethmoid bone

Vomer

Palatine process
of maxilla

Maxilla

Mandibular body

FIGURE **6.13** Midsagittal section of the skull: (**A**) illustration; (**B**) photograph.

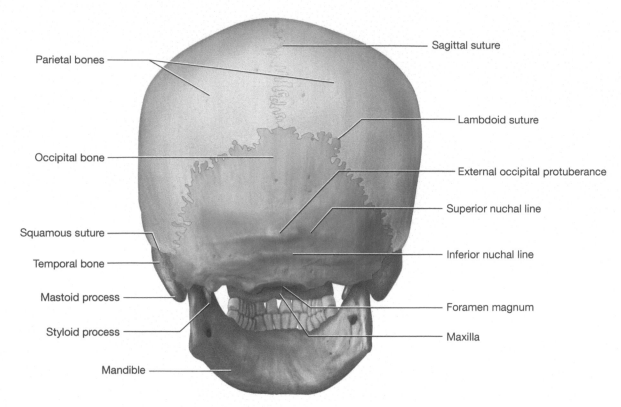

Parietal bones

Occipital bone

Squamous suture

Temporal bone

Mastoid process

Styloid process

Mandible

Sagittal suture

Lambdoid suture

External occipital protuberance

Superior nuchal line

Inferior nuchal line

Foramen magnum

Maxilla

FIGURE **6.14** Posterior view of the skull.

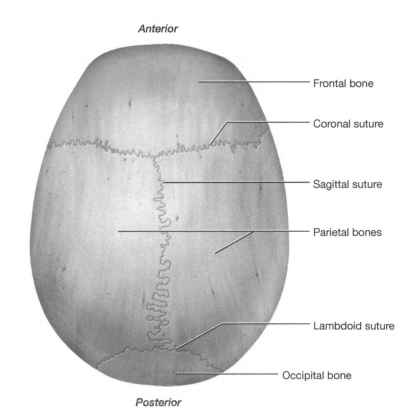

Anterior

Frontal bone

Coronal suture

Sagittal suture

Parietal bones

Lambdoid suture

Occipital bone

Posterior

FIGURE **6.15** Superior view of the calvaria.

6. **Ethmoid bone.** The complex **ethmoid bone** is the deepest cranial bone and, as such, the most difficult to see from standard views of the skull. For this reason, it is shown in a separate illustration, Figure 6.16. Located anterior to the sphenoid bone and posterior to the nasal bones of the face, it contains several features, including the following:

a. Its superior surface, called the **cribriform plate**, forms the roof of the nasal cavity and part of the base of the cranial cavity. It has a superior projection, called the **crista galli** (KRIS-tah GAL-ee), to which certain membranes around the brain attach.

b. The **lateral masses** of the ethmoid bone form part of the orbit and the walls of the nasal cavity. Internally, the lateral masses contain numerous cavities called the **ethmoid sinuses,** which are the third set of paranasal sinuses (Fig. 6.17). Extending medially from the lateral masses are two projections into the nasal cavity—the **superior nasal conchae** and **middle nasal conchae** (KAHN-kee). Note that the superior nasal conchae are quite small and are usually difficult to see.

c. The middle portion of the ethmoid bone, called the **perpendicular plate**, forms the superior part of the bony nasal septum.

The 14 **facial bones** form the framework for the face, provide openings for ventilation and eating, and form cavities for the sense organs. Several of the facial bones are located deeper in the skull, and you will want to refer to several different figures (noted with each bone) to best locate them and appreciate their structures.

1. **Mandible.** The **mandible**, or the lower jaw bone, consists of a central **body** and two "arms" called the **mandibular rami** (RAY-mee; see Figures 6.10 and 6.13). The mandibular rami turn superiorly at the **mandibular angle**. Their superior ends have two processes: an anterior **coronoid process** and a posterior **mandibular condyle**. The mandibular condyle fits into the mandibular fossa of the temporal bone to form the **temporomandibular joint**, or **TMJ**.

2. **Maxillae.** The two fused **maxillae** (mak-SILL-ee) are the upper jaw bones (Figs. 6.9–6.11). They form the inferior wall of the orbit, part of the lateral wall of the nasal cavity, and part of the superior wall of the oral cavity. In Figure 6.11, you can see that the maxillae also form the anterior portion of the *hard palate* via their **palatine processes**. Within their walls are cavities called the **maxillary sinuses**, the fourth and final set of paranasal sinuses (Fig. 6.17).

3. **Lacrimal bones.** The tiny **lacrimal bones** (LAK-rih-muhl) are located in the medial part of the orbit, where they form part of the structure that drains tears produced by the lacrimal gland of the eye. They are best seen in Figure 6.10.

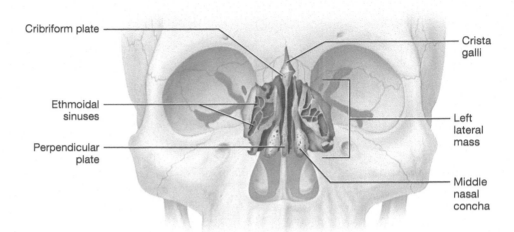

FIGURE **6.16** Ethmoid bone.

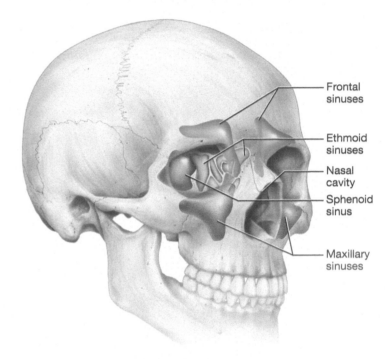

FIGURE **6.17** Paranasal sinuses.

4. **Nasal bones.** The two **nasal bones** form the anterior framework of the bridge of the nose.

5. **Vomer.** The single **vomer** (VOH-muhr) forms the inferior portion of the bony nasal septum (best seen in Figure 6.13).

6. **Inferior nasal conchae.** The small **inferior nasal conchae** form part of the lateral walls of the nasal cavity. As their name implies, they are located inferior to the middle nasal conchae of the ethmoid bone (Fig. 6.9).

7. **Palatine bones.** The two **palatine bones** (PAL-uh-ty'n) form the posterior part of the hard palate (seen in Figure 6.11) and the posterolateral walls of the nasal cavity.

8. **Zygomatic bones.** The two **zygomatic bones** form the bulk of the cheek and a significant portion of the "cheekbone" or *zygomatic arch* (Figs. 6.9 and 6.10).

The orbit and the nasal cavity are complicated structures with contributions from several bones. The **orbit** is formed by parts of seven bones: the frontal bone, maxilla, sphenoid bone, ethmoid bone, lacrimal bone, zygomatic bone, and a tiny piece of the palatine bone (Fig. 6.9). The **nasal cavity** is formed by parts of six bones: the ethmoid bone, maxilla, palatine bone, inferior nasal concha, sphenoid bone, and vomer (Figs. 6.9 and 6.13).

The features we have been examining are found in an adult skull. However, as you can see in Figure 6.18, a fetal skull contains notable differences from an adult skull. In adults, the sutures are fused, but in the fetus, the sutures have not yet fused and are instead joined by fibrous membranes. This can be seen with the **frontal suture**, which is where the two fetal frontal bones fuse. Where several sutures meet, we find large, membrane-covered areas called the **fontanels** (fawn-tuh-NELZ), known to many as "soft spots." The two main fontanels are the **anterior fontanel**, where the sagittal and coronal sutures meet, and the **posterior fontanel**, where the sagittal and lambdoid sutures meet.

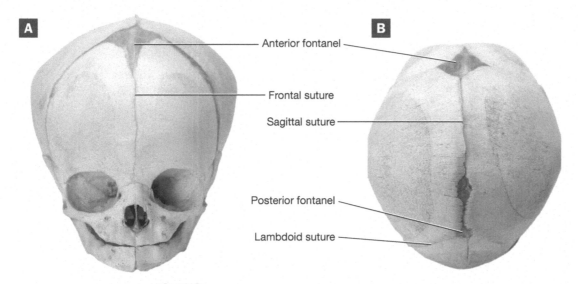

FIGURE **6.18** Fetal skull: (**A**) anterior view; (**B**) superior view.

Procedure **1** Cooperative Learning for the Skull

This exercise takes an approach called *cooperative learning*, in which you work with your lab partners to teach one another the bones and bone markings. The process may seem a bit confusing at first, but by the end of the first couple of rotations, it should move more quickly and smoothly.

1 Assemble into teams of a minimum of four students; five is optimum.

2 Distribute a skull to each member of the team, and assign each student one of the following five groups of bones and bone markings (the specific structures of each bone are listed below and on the next page):

Group 1: Calvaria, base, and structures of the frontal bone and parietal bones.

Group 2: Temporal bone and occipital bone structures.

Group 3: Ethmoid bone and sphenoid bone structures.

Group 4: Mandible and maxillary bone structures.

Group 5: Remainder of the facial bones, orbit, and anterior and posterior fontanels.

3 Spend approximately 3 minutes learning the assigned structures on your own. Explore each structure thoroughly. If there are canals and foramina, pass a pipe cleaner through the holes to see where they go.

4 Spend about 1 to 2 minutes teaching your assigned structures to your team. Then have each team member spend approximately 1 to 2 minutes teaching the team his or her assigned structures.

5 Rotate the assigned structures clockwise so each member has a new set of structures to learn and teach (the student who was assigned group 1 will take group 2, and so on).

6 Repeat steps 3 and 4, then rotate the assigned structures again. Continue this process (it begins to speed up significantly at this point), repeating until each team member has studied and taught each group of structures once.

By the end of this activity, each team member will have learned and presented each group of structures.

The following is a list of bones and bone markings of the skull that will be covered in this exercise:

Group 1 Structures: Cranium, Frontal Bone, Parietal Bones

1. Calvaria
2. Base of cranial cavity
   a. Anterior cranial fossa
   b. Middle cranial fossa
   c. Posterior cranial fossa
3. Frontal bone
   a. Frontal sinuses
4. Parietal bones
   a. Sagittal suture
   b. Coronal suture
   c. Lambdoid suture
   d. Squamous suture

Group 2 Structures: Temporal Bones, Occipital Bone

1. Temporal bones
   a. External acoustic (auditory) meatus
   b. Styloid process
   c. Mastoid process
2. Occipital bone
   a. Foramen magnum
   b. Occipital condyles

Group 3 Structures: Sphenoid Bone, Ethmoid Bone

1. Sphenoid bone
   a. Body
   b. Sphenoid sinus
   c. Sella turcica
   d. Greater and lesser wings
   e. Pterygoid processes
2. Ethmoid bone
   a. Cribriform plate
   b. Crista galli
   c. Ethmoid sinuses
   d. Superior and middle nasal conchae
   e. Perpendicular plate

Group 4 Structures: Mandible, Maxillae

1. Mandible
   a. Mandibular body
   b. Mandibular ramus
   c. Coronoid process
   d. Mandibular condyle
2. Maxillae
   a. Palatine processes
   b. Maxillary sinuses

Group 5 Structures: Other Facial Bones

1. Lacrimal bones
2. Nasal bones
3. Vomer
4. Inferior nasal conchae
5. Palatine bones
6. Zygomatic bones and zygomatic arch
7. Anterior fontanel
8. Posterior fontanel

Your instructor may wish to omit certain structures included above or add structures not included in these lists. List any additional structures below:

Exercise 6-4

Remainder of the Skeleton

MATERIALS

❑ Vertebral column, articulated
❑ Disarticulated bones
❑ Skeleton, articulated

Another key component of the axial skeleton is the **vertebral column**, which consists of 24 unfused vertebrae, the five fused vertebrae of the sacrum (SAY-krum), and the coccyx (KAHX-iks). If you view the vertebral column from the lateral side, as in Figure 6.19, you can see that the vertebral column has four curvatures: the concave **cervical** and **lumbar curvatures** and the convex **thoracic** and **sacral curvatures**. The cervical and lumbar curvatures are particularly important to our ability to walk upright.

The 24 vertebrae consist of 7 cervical vertebrae, 12 thoracic vertebrae, and 5 lumbar vertebrae (remember this as "breakfast at 7, lunch at 12, dinner at 5"). Nearly all vertebrae share certain general features, which are visible in the typical cervical vertebra shown in Figure 6.20A. These features include a posterior **spinous process**, two lateral **transverse processes**, a central **vertebral foramen**, and an anterior **vertebral body**. In between each vertebral body is a fibrocartilage pad called an **intervertebral disc** that absorbs shock as the vertebral column moves. Extending from the vertebral body to the transverse processes on both sides are two short extensions known as **pedicles**, which enclose the vertebral foramen along with two posterior extensions, the **laminae**.

The basic properties of each region of the vertebral column are as follows:

1. The seven **cervical vertebrae** are located in the neck (Fig. 6.20A–C). All cervical vertebrae have holes in their transverse processes, called **transverse foramina**, that permit the passage of blood vessels. In addition, the spinous processes of cervical vertebrae are often forked. Two cervical vertebrae are named differently than the others because of their unique features:

   a. **Atlas** (C1): The **atlas** is the first cervical vertebra, and it articulates with the occipital condyles of the occipital bone. It is easily identified because it has a large vertebral foramen, no body, and no spinous process (Fig. 6.20B).

   b. **Axis** (C2): The **axis** is the second cervical vertebra. It is also easily identified by a superior projection called the **dens** (or the **odontoid process**; Fig. 6.20C). The dens fits up inside the atlas to form the *atlantoaxial joint*, which allows rotation of the head.

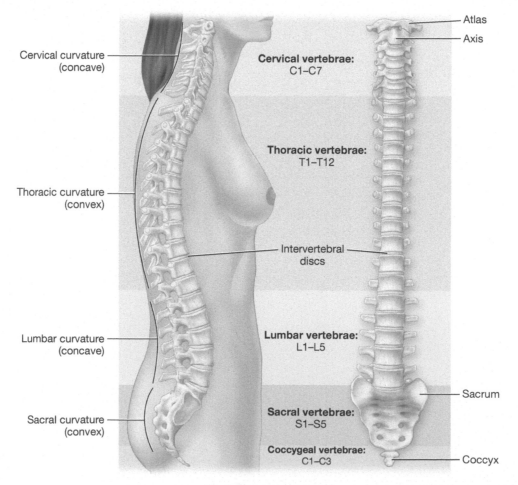

FIGURE **6.19** Vertebral column.

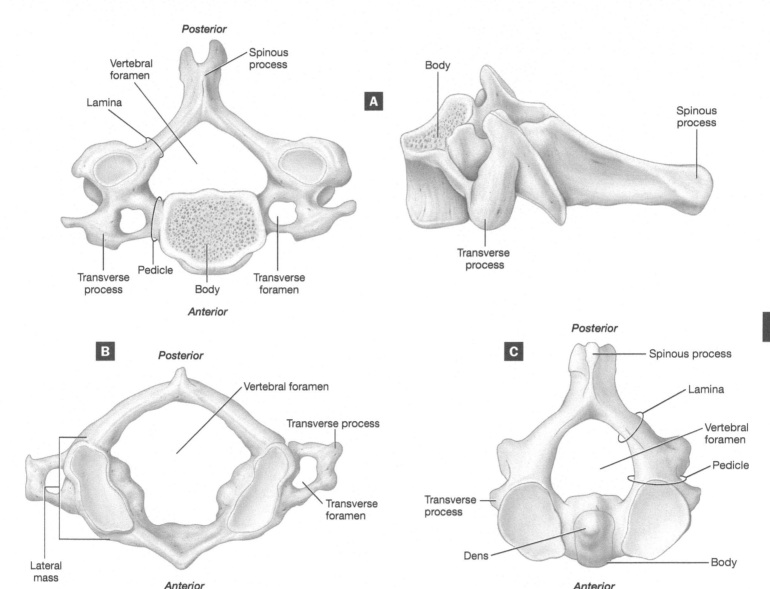

FIGURE **6.20** Cervical vertebrae: (**A**) typical cervical vertebra; (**B**) the atlas (C1); (**C**) the axis (C2).

2. The 12 **thoracic vertebrae** each articulate with a pair of ribs, and they share the following common features (Fig. 6.21A):

   a. The spinous processes are thin and point inferiorly.

   b. Most have two sets of facets that articulate with the ribs: the *costal facets* on their bodies and the *transverse costal facets* on their transverse processes.

   c. All have approximately triangular vertebral foramina.

   d. If you look at a thoracic vertebra from the posterolateral side, it looks like a giraffe. Try it—it really does!

3. The five large **lumbar vertebrae** share the following common features (Fig. 6.21B):

   a. All have a large, blocklike body.

   b. The spinous processes are thick and point posteriorly.

   c. If you look at a lumbar vertebra from the posterolateral side, it looks like a moose.

4. The **sacrum** consists of five fused vertebrae (Figs. 6.22A and 6.22B). Spinal nerves pass through holes called **anterior** and **posterior sacral foramina** that flank both sides of the sacral bodies. The lateral surfaces of the sacrum, known as the *auricular surfaces*, articulate with the hip bones to form the **sacroiliac joints** (say-kroh-ILL-ee-ak).

5. The **coccyx** consists of three to five (average: four) small, fused vertebrae that articulate superiorly with the sacrum.

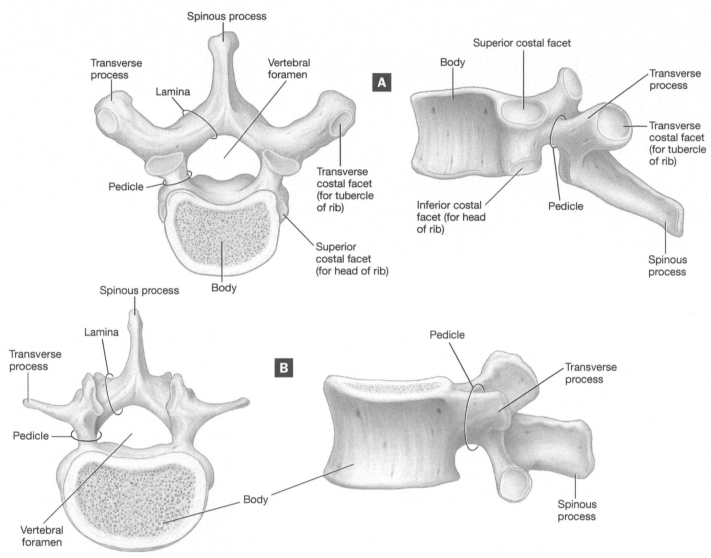

FIGURE **6.21** Vertebrae: (**A**) thoracic vertebra; (**B**) lumbar vertebra.

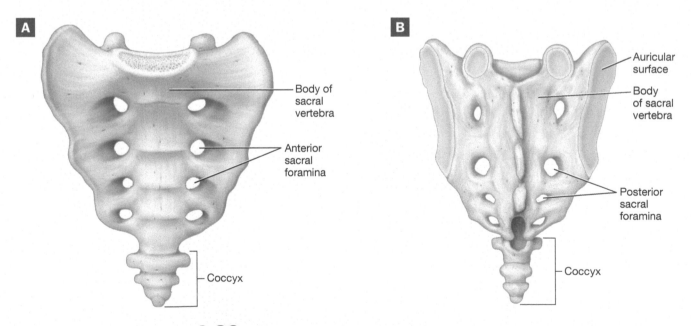

FIGURE **6.22** Sacrum and coccyx: (**A**) anterior view; (**B**) posterior view.

How to Tell the Different Vertebrae Apart

It's not uncommon to find yourself in a lab practical having to identify a single vertebra as being cervical, thoracic, or lumbar. Fortunately, they're really easy to tell apart if you remember a few key things.

ℹ️ Only cervical vertebrae have transverse foramina. So if you see holes in the transverse processes of the vertebra, you know you're holding a cervical vertebra. From there, you might need to know if you're holding the atlas (C1) or the axis (C2):

- Only the atlas lacks a vertebral body, which gives it a huge vertebral foramen.
- Only the axis has a superior projection—the dens. If you get confused as to which is the axis and which is the atlas, here's a simple visual analogy: Think about the dens as being a projection you could stick something like a DVD on top of, and the DVD would be able to spin on its axis.

ℹ️ Only thoracic vertebrae have costal facets. So if you find yourself with a vertebra that lacks transverse foramina but has indentations (facets) on its body where it articulates with ribs, you've got a thoracic vertebra.

ℹ️ Lumbar vertebrae are easy to identify by their large, blocky shape, but they can also be identified by process of elimination. If there are no transverse foramina and no costal facets, you're left with only lumbar vertebrae. If the costal facets on your bones aren't very well-defined and you're stuck between thoracic and lumbar, you can also use the giraffe/moose trick. The thoracic vertebra really does look like a giraffe and the lumbar like a "lumbering moose."

6

The remainder of the axial skeleton consists of the bones of the thoracic cavity (the sternum and the ribs; Fig. 6.23) and the hyoid bone in the neck. The **sternum** is the central bone of the thorax. It is divided into three parts: the superior **manubrium** (mah-NOO-bree-um), the middle **body**, and the inferior **xiphoid process** (ZY-foyd). It is the body of the sternum that you compress during the chest compressions of cardiopulmonary resuscitation (CPR) and the xiphoid process that you must take care not to break.

The twelve pairs of **ribs** enclose the thoracic cavity and protect its vital organs (note that men and women have the same number of ribs). The spaces between the ribs are known as **intercostal spaces,** and they contain muscles, blood vessels, nerves, and lymphatic vessels. Ribs articulate anteriorly with the sternum via **costal cartilage.** They are classified according to the structure of their costal cartilage: Ribs 1–7 are considered **true ribs,** or **vertebrosternal ribs,** because they attach directly to the sternum by their own costal cartilage. Ribs 8–12, on the other hand, are classified as **false ribs** because they lack this direct attachment to the sternum. Notice in Figure 6.23 that ribs 8–10, the **vertebrochondral ribs,** have an indirect attachment to the sternum, as their cartilage attaches to the costal cartilage of the true ribs. This attachment site of the false ribs forms a ridge of cartilage at the inferior border of the ribcage that is known as the **costal margin.** Ribs 11–12 have no attachment to the sternum at all, so they are often referred to as **floating ribs** or **vertebral ribs.**

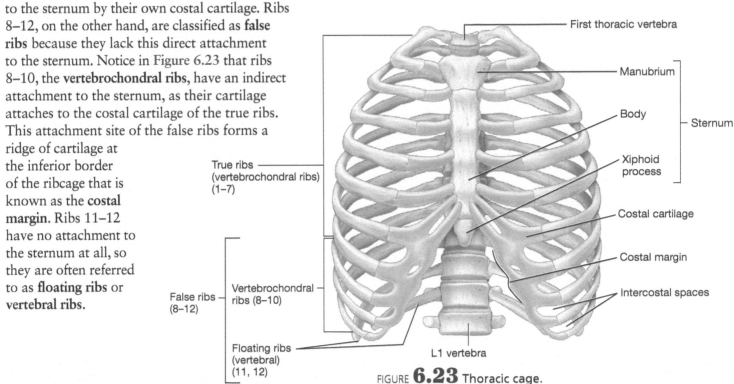

FIGURE **6.23** Thoracic cage.

The **hyoid bone**, shown in Figure 6.24, is located in the superior neck, where it is occasionally classified as a skull bone. Note, however, that it does not articulate with any skull bone, or with any other bone, for that matter. It is held in place in the superior neck by muscles and ligaments, and helps to form part of the framework for the larynx (LEHR-inks; voice box). It also serves as an attachment site for the muscles of the tongue and aids in swallowing. When a person is choked manually, the hyoid bone is often broken.

Let's move on to the appendicular skeleton, starting with the pectoral girdle. The **pectoral girdle** consists of the two bones that frame the shoulder: the scapula and the clavicle (Fig. 6.25). The **scapula** (SKAP-yoo-lah), shown in Figure 6.26, is a roughly triangular-shaped bone. From an anterior view (Fig. 6.26A), you can also get a good view of its two superior projections: the anterior **coracoid process** (KOHR-ah-koyd) and the posterior **acromion** (ah-KROH-mee-ahn). The acromion forms a joint with the lateral portion of the clavicle called the **acromioclavicular** (ah-KROH-mee-oh-klah-VIK-yoo-lur; **AC**) joint. On its lateral surface (Fig. 6.26B) is a shallow depression called the **glenoid cavity** that forms the shoulder joint with the humerus. When we turn to the scapula's posterior surface (Fig. 6.26C), we find a prominent ridge called the **spine** that you can feel under your skin as your "shoulder blade."

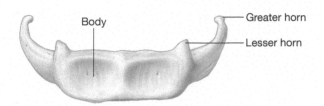

FIGURE **6.24** Hyoid bone.

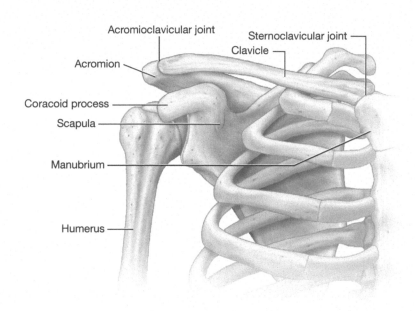

FIGURE **6.25** Pectoral girdle.

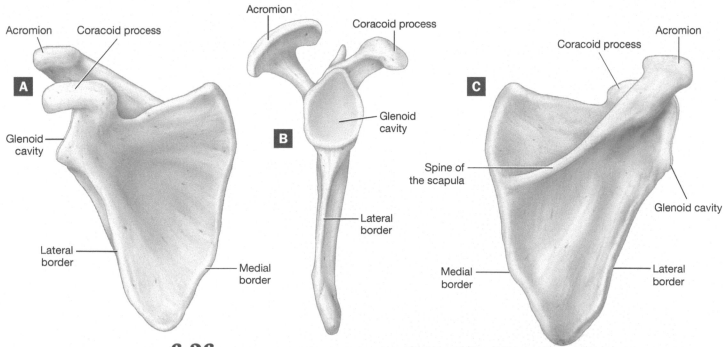

FIGURE **6.26** Right scapula: (**A**) anterior view; (**B**) lateral view; (**C**) posterior view.

The **clavicle** is a small S-shaped bone that spans between the acromion of the scapula at its lateral **acromial end** and the manubrium of the sternum at its medial **sternal end**. Note its S-shape is best appreciated from the superior view shown in Figure 6.27, as the clavicle appears nearly straight from an anterior view. Functionally, the clavicle acts somewhat like a brace, holding the upper limb in place away from the body.

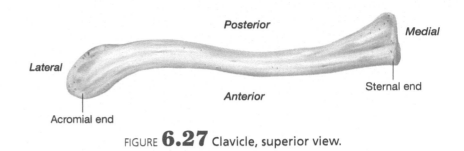

FIGURE **6.27** Clavicle, superior view.

The **upper limb** consists of the arm, the forearm, the wrist, and the hand. The only bone within the arm is the **humerus** (HYOO-mur-us), a long bone that forms the shoulder joint with the scapula at its proximal end and the elbow joint with the ulna and radius at its distal end. The humerus has many features, including the following (Fig. 6.28):

▮ At its proximal end is a rounded **head** that fits into the glenoid cavity. Just lateral to the head is the **greater tubercle**, separated from the smaller **lesser tubercle** by the **intertubercular sulcus** (in-ter-too-BUR-kyoo-lur SUL-kuss).

▮ The diaphysis of the humerus features a projection called the **deltoid tuberosity**, where the deltoid muscle attaches.

▮ At the humerus' distal end, we find two condyles: the medial **trochlea** (TROH-klee-uh), which is shaped like a spool of thread, and the lateral **capitulum** (kah-PIT-yoo-lum), which is ball-shaped. Just proximal to the trochlea are indentations in the humerus where the ulna and radius articulate: the anterior **coronoid** and **radial fossae** and the posterior **olecranon fossa** (oh-LEK-rah-nahn).

6

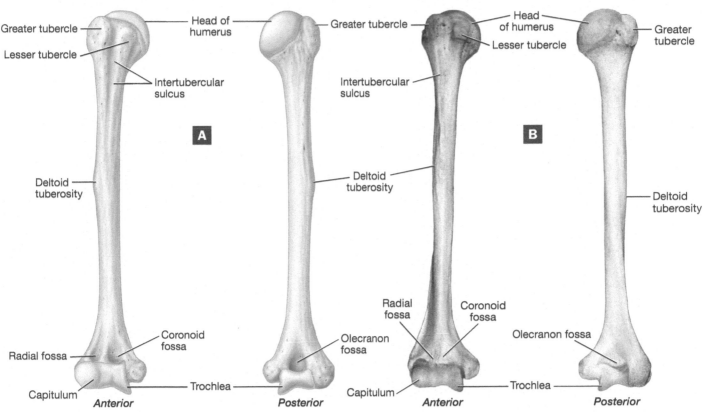

FIGURE **6.28** Right humerus: **(A)** illustration; **(B)** photograph.

The two forearm bones, shown in Figure 6.29, are the lateral **radius** (RAY-dee-us) and medial **ulna**. If you have a hard time remembering which is which, stand in anatomical position and take your radial pulse; the pulse is on the lateral side of the forearm, just like the radius. The ulna is wide proximally where it articulates with the humerus and thin distally where it articulates with the bones of the wrist. Its proximal end has two processes—the large, posterior **olecranon process** and the smaller, anterior **coronoid process**—separated by a deep curve called the **trochlear notch.**

As its name implies, the trochlear notch fits around the trochlea of the humerus to form the elbow joint. The olecranon process is the actual "elbow bone," which you can feel on your posterior arm. The trochlear notch and the ulna's two processes form a "U" shape when the ulna is held on its side. This makes the ulna easy to differentiate from the radius (think "U" for "ulna").

The distal end of the ulna is known as the **ulnar head**. It articulates with one of the wrist (or *carpal*) bones known as the *lunate*. On its medial side, it has a small projection called the **styloid process** that is palpable through the skin.

The radius' shape is opposite to that of the ulna: It is skinny proximally and wide distally. Proximally, it consists of a round, flattened **radial head** that articulates with the capitulum of the humerus to help form the elbow joint. Distal to the radial head, we find the **radial neck** and a projection called the **radial tuberosity**. At its most distal end, the radius articulates with several carpal bones. It also features a lateral **styloid process**, just as we found on the ulna.

Notice in Figure 6.29 that the radius and ulna also articulate with one another. On the ulna, just distal to the coronoid process, we find the **radial notch**, which is where the radius and ulna articulate at the **proximal radioulnar joint**. At the distal end of the radius is an indentation called the **ulnar notch** where the ulna fits into the radius. This joint is known as at the **distal radioulnar joint**. These joints are stabilized by the *interosseous membrane*, a band of dense regular collagenous connective tissue that joins the radius and ulna along their length.

The wrist is composed of eight short bones called **carpals** (KAR-pulz), labeled individually in Figure 6.30. The carpals articulate proximally with the radius and the ulna, and distally with the **metacarpals**, which are the five long bones in the hand. The metacarpals articulate distally with the fingers, which are formed from 14 long bones called **phalanges** (fuh-LAN-jeez). The second through fifth digits have three phalanges each (the *proximal*, *intermediate* [or *middle*], and *distal phalanges*); the thumb has only two (a *proximal* and a *distal phalanx*).

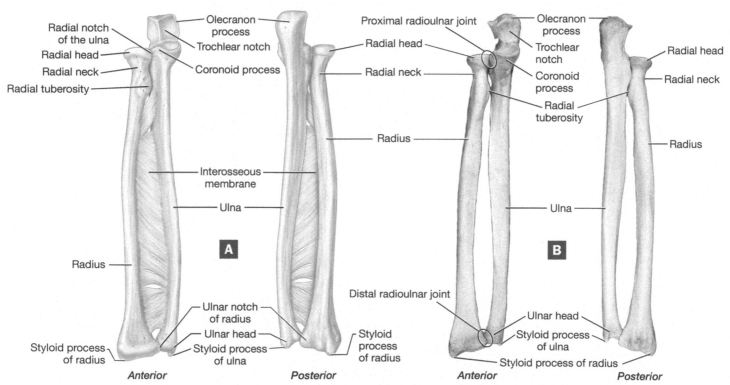

FIGURE **6.29** Right radius and ulna: (**A**) illustration; (**B**) photograph.

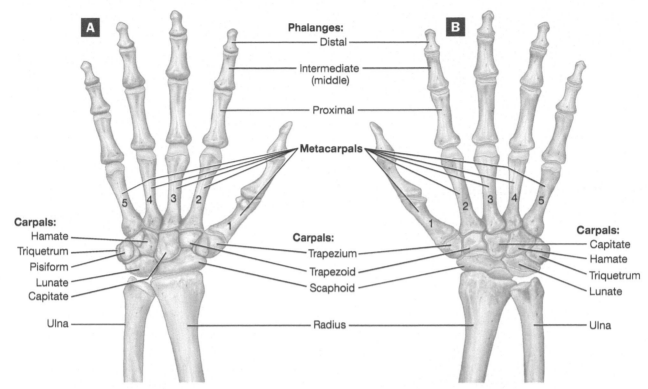

FIGURE **6.30** Right wrist and hand: **(A)** anterior view; **(B)** posterior view.

The **pelvis** connects the lower limbs to the trunk, supports the pelvic organs, and transmits the weight of the trunk to the legs. It is formed by the sacrum and the **pelvic girdle**, which itself is composed of two **coxal bones** (Fig. 6.31). Three fused bones—the *ilium*, the *ischium*, and the *pubis*—make up each coxal bone, or **hemipelvis** (Fig. 6.32). You can see in the lateral view the place where all three bones come together to form a deep socket. This socket, called the **acetabulum** (aeh-seh-TAB-yoo-lum), forms the hip joint with the femur. Notice also that where the ischium and pubis meet there is a large hole called the **obturator foramen** (AHB-too-ray-tur). In a living person, this hole is covered with a membrane and allows only small blood vessels and nerves to pass through.

Let's look more closely at the three bones of the pelvic girdle and their features.

1. **Ilium.** The **ilium** (ILL-ee-um) is the largest of the three bones. Its main portion is called the **body,** and its superior "wing" is called the **ala** (AY-luh). The ridge of the ala, called the **iliac crest,** is where you rest your hands when your hands are on your hips. At the anterior end of the crest, we find a projection called the **anterior superior iliac spine,** and at its posterior end is the smaller **posterior superior iliac spine.** The posterior ilium also features a notch called the **greater sciatic notch** (sy-AEH-tik), which allows the greater sciatic nerve to pass from the pelvis to the thigh. On its medial surface, it articulates with the sacrum, forming the sacroiliac joint.

2. **Ischium.** The **ischium** (ISS-kee-um) makes up the posteroinferior pelvis. It contains three features on its posterior side: the superior **ischial spine,** the middle **lesser sciatic notch,** and the thick, inferior **ischial tuberosity.** The ischial tuberosities are the "butt bones": the bones that bear your weight when you sit down. On the ischium's anterior side, we find the **ischial ramus,** which articulates with the pubis.

3. **Pubis.** The pelvis' anterior portion is formed by the **pubis** (PYOO-bis), or the **pubic bone.** The pubis consists of a **body** and two extensions called the **superior** and **inferior rami.** The bodies of the two pubic bones meet at a fibrocartilage pad called the **pubic symphysis** (SIM-fih-sis; see Figure 6.31).

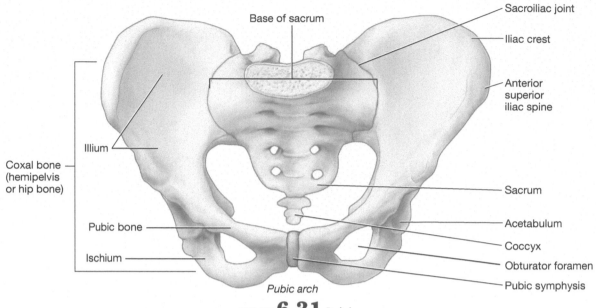

FIGURE **6.31** Pelvis.

Base of sacrum

Sacroiliac joint

Iliac crest

Anterior superior iliac spine

Illium

Coxal bone (hemipelvis or hip bone)

Sacrum

Pubic bone

Acetabulum

Coccyx

Ischium

Obturator foramen

Pubic symphysis

Pubic arch

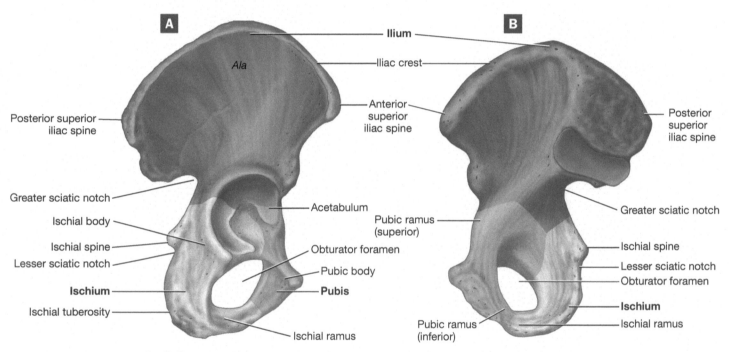

FIGURE **6.32** Hemipelvis: (**A**) right coxal bone, lateral view; (**B**) right coxal bone, medial view.

The lower limb consists of the thigh, the **patella** (puh-TEL-uh; kneecap), the leg, the ankle, and the foot. The thigh contains only a single bone, the large, heavy **femur** (Fig. 6.33). Proximally the femur articulates with the acetabulum at its rounded **head**. Just distal to the femoral head is the **neck** of the femur, the weakest part of the femur and the most common location for it to fracture (when the femoral neck fractures, it is usually called a "broken hip," even though the hip itself doesn't fracture). Where the femoral neck meets the femoral shaft, we find two large prominences—the anterolateral **greater trochanter** (TROH-kan-tur) and the posteromedial **lesser trochanter**. Between the two trochanters on the anterior side is a ridge called the **intertrochanteric line** (in-ter-troh-kan-TEHR-ik); on the posterior side in the same location is another ridge known as the **intertrochanteric crest**.

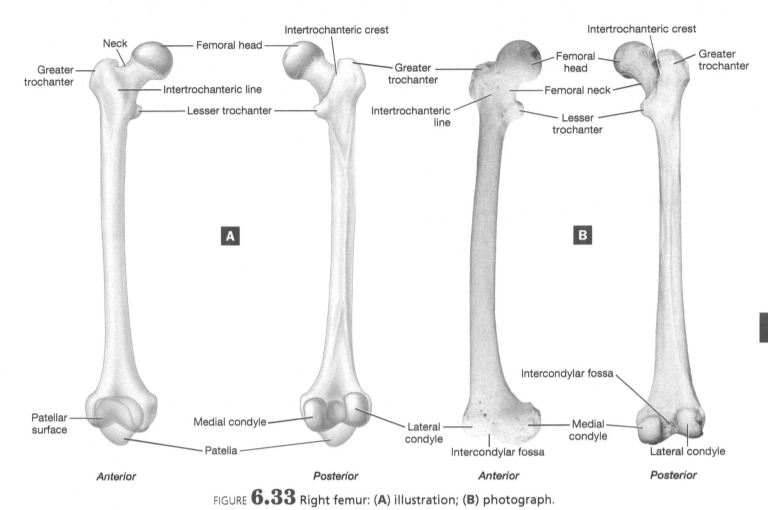

Anterior   Posterior   Anterior   Posterior

FIGURE **6.33** Right femur: **(A)** illustration; **(B)** photograph.

Distally the femur expands into the large **medial** and **lateral condyles,** which form the knee joint with the largest bone of the leg, the tibia. The **tibia** (TIB-ee-ah) is the medial and larger of the two leg bones (Fig. 6.34). It is flattened proximally at its articular surface, where its **medial** and **lateral condyles** fit together with those of the femur. Just distal to the condyles on the tibia's anterior surface is a rough projection called the **tibial tuberosity,** which is where the patellar ligament inserts. Along the anterior tibial diaphysis is a sharpened ridge known as the **anterior crest,** which we commonly call the "shin." At the tibia's distal epiphysis, it articulates with a tarsal bone called the **talus** (TAY-luss), with which it forms the ankle joint. At its terminal end is a projection, the **medial malleolus** (mal-ee-OH-lus), which you may think of as your "medial ankle bone."

The other leg bone is the thin, lateral **fibula** (FIB-yoo-lah). The proximal fibular end is called the **head,** and its distal end is the **lateral malleolus** (which you may think of as your "lateral ankle bone"). The head has no articulation with the femur, but the lateral malleolus articulates with the talus to help form the ankle joint.

Notice in Figure 6.34 that the tibia and fibula articulate with one another just as the radius and ulna do. At the proximal end, we find the **proximal tibiofibular joint,** and at the distal end we find the **distal tibiofibular joint.** The two bones are also held together by a stabilizing interosseous membrane.

The ankle is composed of seven short bones called **tarsals** (TAHR-sulz). You have already been introduced to the talus, the tarsal bone that forms the ankle joint with the tibia and fibula. Distal to the talus is the large **calcaneus** (kal-KAYN-ee-us), or heel bone. The other tarsals are labeled individually in Figure 6.35. The tarsals articulate with the five long bones in the foot called **metatarsals** (met-uh-TAHR-sulz). Like the bones of the fingers, the bones of the toes also consist of 14 **phalanges.** The second through fifth digits have three phalanges each (the *proximal, intermediate* [or *middle*], and *distal phalanges*), and the big toe, or *hallux,* has only two (a *proximal* and a *distal phalanx*).

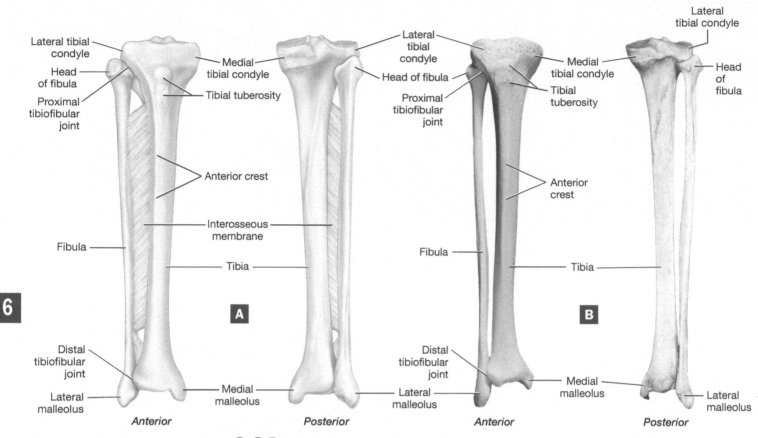

FIGURE **6.34** Right tibia and fibula: (**A**) illustration; (**B**) photograph.

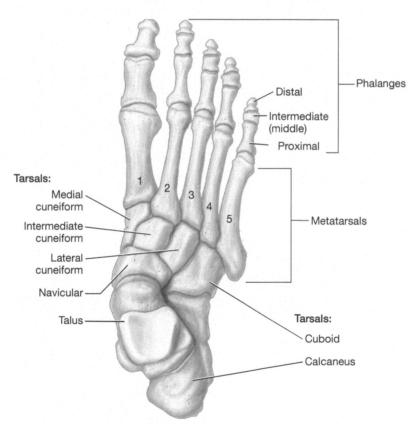

FIGURE **6.35** Right foot and ankle, superior view.

Procedure 1 Cooperative Learning for the Skeleton

We will follow essentially the same procedure here as we did in Exercise 6-3, but with different groups of bones and bone markings. Note that this cooperative learning exercise includes the structures from Exercise 6-2.

1 Assemble into teams with a minimum of four students; five is optimum.

2 Distribute a bone or set of bones to each member of the team, and assign each student one of the following five groups of bones and bone markings (the specific structures of each bone are listed below and on the next page):

Group 1: Cervical, thoracic, and lumbar vertebrae and vertebral markings.

Group 2: Ribs and rib markings; sternum structures; the hyoid bone; scapula structures; the clavicle.

Group 3: Humerus structures; radius and ulna markings; carpals, metacarpals, and phalanges.

Group 4: Ilium, ischium, and pubis structures.

Group 5: Femur, tibia, and fibula structures; tarsals, metatarsals, and phalanges.

3 Spend approximately 3 minutes learning the assigned structures on your own.

4 Spend about 1 to 2 minutes teaching your assigned structures to your team. Then have each team member spend approximately 1 to 2 minutes teaching the team his or her assigned structures.

5 Rotate the assigned structures clockwise so each member has a new set of structures to learn and teach (the student who was assigned group 1 will take group 2, and so on).

6 Repeat steps 3 and 4, then rotate the assigned structures again. Continue this process (it begins to speed up significantly at this point) until each team member has taught each group of structures once.

The following is a list of bone and bone markings of the axial and appendicular skeletons covered in this exercise:

Group 1 Structures: Vertebral Column

1. Vertebral column
   a. Cervical curvature
   b. Lumbar curvature
   c. Thoracic curvature
   d. Sacral curvature
2. Vertebrae
   a. Spinous process
   b. Transverse processes
   c. Vertebral foramen
   d. Vertebral body
   e. Intervertebral disc
3. Cervical vertebrae
   a. Transverse foramina
   b. Atlas
   c. Axis
   d. Dens
4. Thoracic vertebrae
5. Lumbar vertebrae
6. Sacrum
7. Coccyx

Group 2 Structures: Remaining Bones of the Axial Skeleton

1. Sternum
   a. Manubrium
   b. Body
   c. Xiphoid process
2. Ribs
   a. Intercostal spaces
   b. Costal cartilage
   c. True (vertebrosternal) ribs
   d. False ribs
      (1) Vertebrochondral ribs
      (2) Floating ribs (vertebral ribs)
3. Hyoid bone
4. Scapula
   a. Coracoid process
   b. Acromion
   c. Glenoid cavity
   d. Spine
5. Clavicle

Group 3 Structures: Upper Limb

1. Humerus
   a. Head
   b. Greater tubercle
   c. Lesser tubercle
   d. Trochlea
   e. Capitulum
   f. Olecranon fossa
2. Ulna
   a. Olecranon process
   b. Coronoid process
   c. Trochlear notch
   d. Ulnar head
   e. Styloid process
3. Radius
   a. Radial head
   b. Radial neck
   c. Radial tuberosity
   d. Styloid process
4. Carpals
5. Metacarpals
6. Phalanges

Group 4 Structures: Pelvic Girdle

1. Pelvis
   a. Sacroiliac joints
   b. Coxal bone (hemipelvis)
   c. Acetabulum
   d. Obturator foramen
2. Ilium
   a. Ala
   b. Iliac crest
   c. Anterior superior iliac spine
   d. Posterior superior iliac spine
   e. Greater sciatic notch
3. Ischium
   a. Ischial spine
   b. Lesser sciatic notch
   c. Ischial tuberosity
4. Pubis
   a. Body
   b. Superior and inferior rami
   c. Pubic symphysis

Group 5 Structures: Lower Limb

1. Femur
   a. Head
   b. Neck
   c. Greater trochanter
   d. Lesser trochanter
   e. Intertrochanteric line and crest
   f. Medial and lateral condyles
2. Patella
3. Tibia
   a. Tibial tuberosity
   b. Anterior crest
   c. Medial malleolus
4. Fibula
   a. Head
   b. Lateral malleolus
5. Tarsals
   a. Calcaneus
   b. Talus
6. Metatarsals
7. Phalanges

Your instructor may wish to omit certain structures included above or add structures not included in these lists. List any additional structures below:

Procedure 2 Building a Skeleton

1 Obtain a set of disarticulated bones (real bones are best).

2 Assemble the bones into a full skeleton. (If you have an articulated vertebral column and rib cage, go ahead and use them.)

3 Be certain to keep your skeleton in anatomical position. Figure 6.36 gives an overall "big picture" view of the skeleton that you may use for reference.

4 Assemble the bones into a full skeleton.

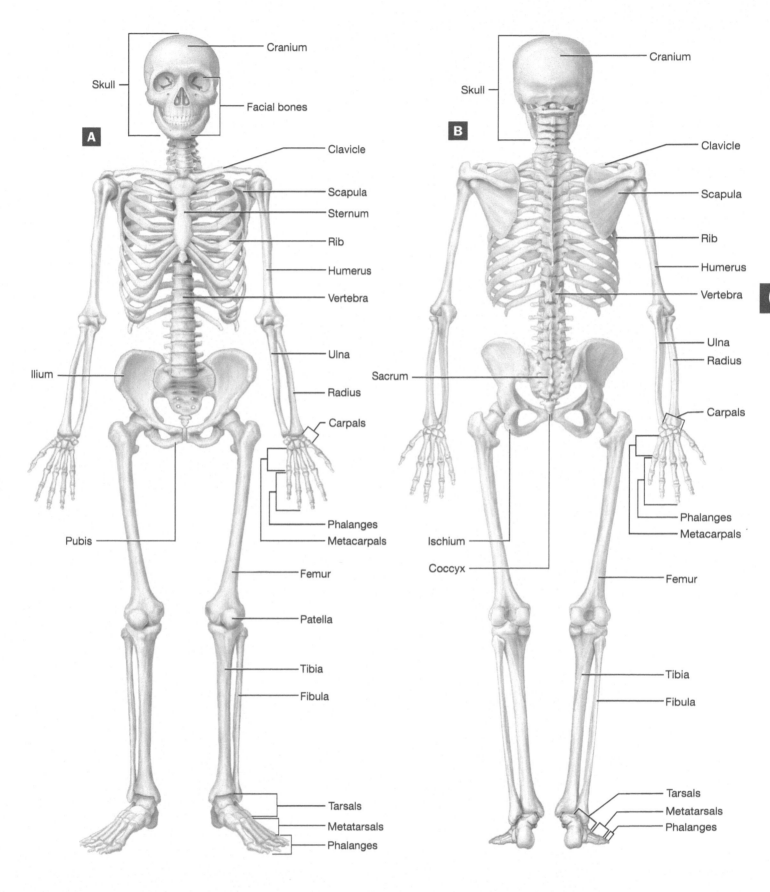

FIGURE **6.36** Articulated skeleton: (**A**) anterior view; (**B**) posterior view.

Procedure **3** Identifying Bones Blindly

Work with a lab partner to try to identify bones blindly, based only on feeling their shapes and bone markings.

1 Obtain a set of disarticulated bones and place them in a box.

2 Close your eyes, reach into the box, and grab a bone randomly. Attempt to identify the bone only by its feel.

3 If necessary, have your lab partner give you hints to help you find out the bone's identity.

Procedure **4** Time to Draw

Remember from your previous drawing exercises that drawing even the crudest picture of an anatomical structure or physiological pathway engages multiple parts of the brain and greatly enhances memory. So, with that in mind: (1) draw a skull and a skeleton in the spaces provided, and (2) label and color your drawings.

These drawings are not meant to be works of art, so don't worry if you aren't even remotely skilled as an artist. (However, if they turn out well, feel free to take them out of your lab book and put them on your refrigerator.)

1 Skull: anterior view

2 Skull: lateral view

3 Skeleton: anterior view

Exercise 6-5

Articulations

MATERIALS
☐ Skeleton, articulated
☐ Skull

Let's now turn to the other structures of the skeletal system: joints, or **articulations**, where two bones come together. There are three structural classes of joints, which include the following (Fig. 6.37):

1. **Fibrous joints. Fibrous joints** consist of two bones joined by short collagen fibers. Most fibrous joints allow no motion. Examples of fibrous joints include sutures in the skull and the joint between a tooth and its bony socket in the mandible or maxilla.

2. **Cartilaginous joints. Cartilaginous joints** consist of bones united by cartilage rather than fibrous connective tissue. Examples include the pubic symphysis, the costochondral joints (the locations where the ribs meet their cartilages), and the epiphyseal plate. Most cartilaginous joints allow some motion; however, the epiphyseal plate is an immovable cartilaginous joint.

3. **Synovial joints. Synovial joints** (sih-NOH-vee-uhl) are freely movable joints. They have a true joint cavity and consist of two bones with articular ends covered with **articular,** or hyaline, **cartilage.** The joint is surrounded by a **joint capsule** composed of dense connective tissue. Internally, the capsule is lined by a **synovial membrane** that secretes a watery fluid called **synovial fluid,** which is similar in composition to blood plasma without the proteins. The fluid bathes the joint to permit nearly frictionless motion. The bones in a synovial joint are held together by **ligaments** that reinforce the joint. Certain synovial joints feature structures known as **menisci** (men-ISS-kee; singular, *meniscus*) or *articular discs*, fibrocartilage pads that improve the fit of two bones to prevent dislocation. Synovial joints typically are surrounded by tendons that move the bones involved in the joint. The tendons may be wrapped in a sheath of connective tissue known as a **tendon sheath,** in which they can slide with a minimum of friction. Fluid-filled sacs called **bursae** (BURR-see) are often located between tendons and joints, and this also reduces friction (note that a tendon sheath is essentially an elongated bursa). Examples of synovial joints include the knee, elbow, hip, and shoulder joints.

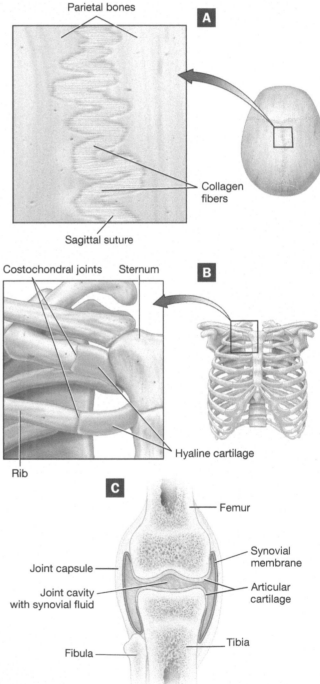

FIGURE **6.37** Structural classes of joints: **(A)** fibrous joint; **(B)** cartilaginous joint; **(C)** synovial joint.

Procedure 1 Classifying Joints by Structure

Classify each joint listed in Table 6.4 by its structure. Then examine and manipulate the joint to determine the amount of motion allowed at the joint (freely movable, slightly movable, or immovable).

TABLE **6.4** Structural Classification of Joints

Joint	Structural Classification	Amount of Motion
Intervertebral joint		
Shoulder (glenohumeral) joint		
Coronal suture		
Interphalangeal joint		

Procedure 2 Identifying Structures of Synovial Joints

Identify the following structures on fresh specimens, such as pigs' feet. If fresh specimens are not available, use anatomical models instead. Check off each structure as you identify it.

- ❏ Joint cavity
- ❏ Articular cartilage
- ❏ Joint capsule
- ❏ Synovial membrane

- ❏ Ligaments
- ❏ Articular discs (menisci)
- ❏ Tendon with tendon sheath
- ❏ Bursae

Knee Joint

The **knee joint**, illustrated in Figure 6.38, is made up of the medial and lateral femoral condyles and the tibial surface. It is stabilized by a joint capsule and numerous ligaments, including the following:

1. **Anterior cruciate ligament (ACL).** The **anterior cruciate ligament** (KROO-shee-iht), or **ACL**, extends from the anterior tibial surface to the posterior part of the lateral femoral condyle. Its function is to prevent hyperextension of the knee.

2. **Posterior cruciate ligament (PCL).** The **posterior cruciate ligament**, or **PCL**, extends from the posterior tibial surface to the anterior part of the medial femoral condyle. It crosses under the ACL, and the two together form an "X." The PCL prevents posterior displacement of the tibia on the femur (i.e., it stops the tibia from sliding backward on the femur).

3. **Tibial collateral ligament.** The **tibial collateral ligament**, also known as the *medial collateral ligament (MCL)*, extends from the medial tibia to the medial femur. It resists stresses that pull the tibia laterally on the femur.

4. **Fibular collateral ligament.** The **fibular collateral ligament**, also known as the *lateral collateral ligament (LCL)*, extends from the lateral fibula to the lateral femur. It resists stresses that pull the tibia medially on the femur.

Other important supportive structures of the knee joint are menisci and bursae. The **medial** and **lateral menisci** are half-moon-shaped articular discs located on the outer edges of the tibial surface that improve the fit of the tibia and femur. The **bursae** around the knee joint are particularly concentrated around the patella and *patellar ligament*, as this apparatus generates a great deal of friction over the femur during motion (see Figure 6.38B). For this reason, these bursae can become inflamed as a result of repetitive activities or simply "overdoing" it.

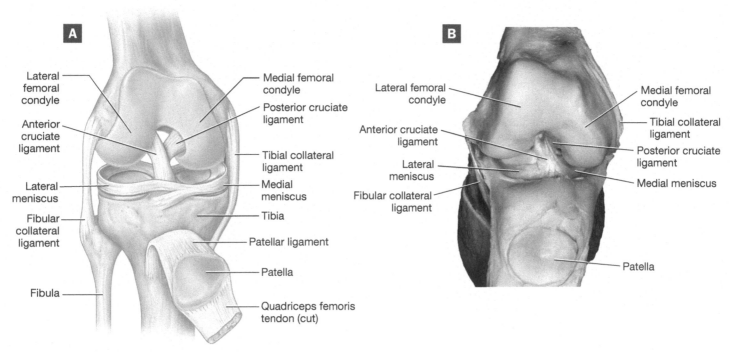

FIGURE **6.38** Right knee joint: (**A**) anterior view; (**B**) midsagittal section.

Procedure **3** Identifying Structures of the Knee Joint

Identify the following structures of the knee joint on anatomical models or fresh specimens. Check off each structure as you identify it.

- ❏ Medial and lateral femoral condyles
- ❏ Joint capsule
- ❏ Anterior cruciate ligament (ACL)
- ❏ Posterior cruciate ligament (PCL)
- ❏ Tibial collateral ligament
- ❏ Fibular collateral ligament
- ❏ Medial meniscus
- ❏ Lateral meniscus
- ❏ Bursae
- ❏ Patellar ligament

Exercise 6-6

Motions of Synovial and Cartilaginous Joints

Each time you move your body in a seemingly routine fashion (such as walking or climbing stairs), you are producing motion at a tremendous number of joints. Many possible motions can occur at synovial and cartilaginous joints. These motions are illustrated in Figure 6.39.

In the following procedure, you will perform two common movements: walking up the stairs and doing jumping jacks. You will determine which joints you are moving with each action, and then which motions are occurring at each joint.

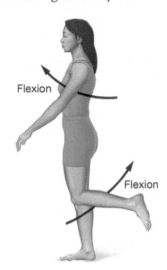

A Angular movements: extension and flexion at the shoulder and knee

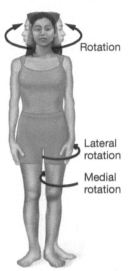

B Rotation of the head, neck, and lower limb

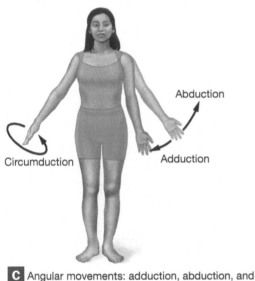

C Angular movements: adduction, abduction, and circumduction of the upper limb at the shoulder

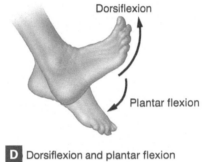

D Dorsiflexion and plantar flexion

E Pronation (P) and supination (S)

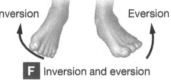

F Inversion and eversion

G Opposition and reposition

H Protraction and retraction

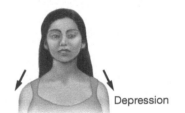

I Elevation and depression

FIGURE **6.39** Motions of synovial and cartilaginous joints.

Procedure 1 Identifying Joint Motions of Common Movements

Team up with a partner, and have your partner perform the following actions. Watch carefully as the actions are performed, and list the joints in motion.

Ask your instructor whether you should use the technical name or the common name for each joint (e.g., glenohumeral versus shoulder joint). Some of the joints, such as the hip joint and knee joint, will be obvious. Others, such as the radioulnar joint, the fingers and toes, and the intervertebral joints, are less obvious and easily overlooked.

After you have listed the joints being used, determine which motions are occurring at each joint. Keep in mind the type and the range of motion of each joint as you answer each question.

1 Walking up stairs

Joints moving: Motions occurring:

2 Doing jumping jacks

Joints moving: Motions occurring:

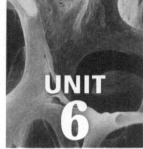

1 Label the following parts of compact bone on Figure 6.40.

❑ Blood vessels

❑ Central canal

❑ Lacunae

❑ Lamellae

❑ Osteon

❑ Perforating canal

❑ Trabeculae of spongy bone

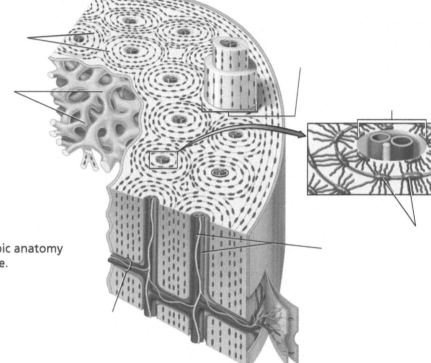

FIGURE **6.40** Microscopic anatomy of compact bone tissue.

6

2 *True/False:* Mark the following statements as true (T) or false (F). If the statement is false, correct it to make it a true statement.

_____ The periosteum contains osteoblasts and osteocytes.

_____ Osteoclasts secrete bone matrix.

_____ Compact bone is the hard, outer bone.

_____ The epiphyseal line is located between the epiphysis and the diaphysis.

_____ Spongy bone houses red and yellow bone marrow.

_____ The shaft of a long bone is called the epiphysis.

3 Long bones are

a. named for their length.

b. about as long as they are wide.

c. irregular in shape.

d. longer than they are wide.

4 Label the following bones in Figure 6.41.

❑ Clavicle
❑ Femur
❑ Fibula
❑ Humerus
❑ Ilium
❑ Metacarpals
❑ Pubis
❑ Radius
❑ Sternum
❑ Tarsals
❑ Tibia
❑ Ulna

6

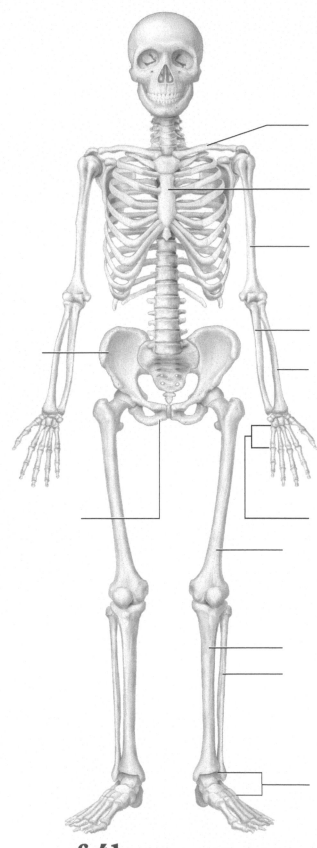

FIGURE **6.41** Anterior view of the skeleton.

UNIT 6 QUIZ
(continued)

5 Label the following bones of the skull in **Figure 6.42**.

❑ Ethmoid bone   ❑ Occipital bone
❑ Frontal bone   ❑ Parietal bone
❑ Mandible   ❑ Sphenoid bone
❑ Maxilla   ❑ Temporal bone
❑ Nasal bones   ❑ Zygomatic bone

A

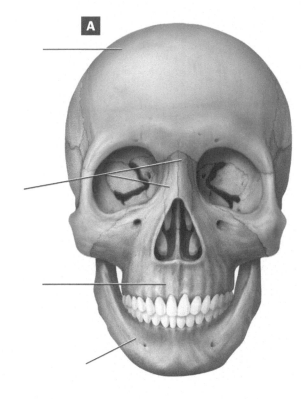

B

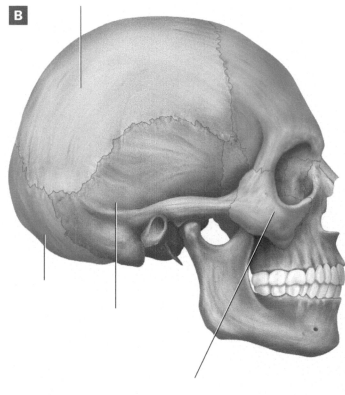

6

C

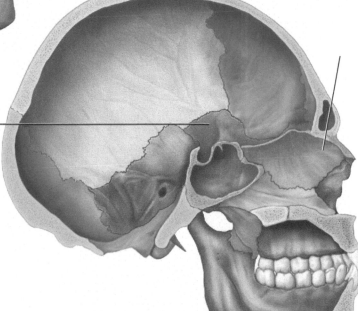

FIGURE **6.42** Skull: (**A**) anterior view; (**B**) lateral view; (**C**) midsagittal section.

6 Label the following parts of the upper limb in **Figure 6.43**.

- ❑ Capitulum
- ❑ Coronoid process
- ❑ Deltoid tuberosity
- ❑ Olecranon
- ❑ Trochlea
- ❑ Trochlear notch

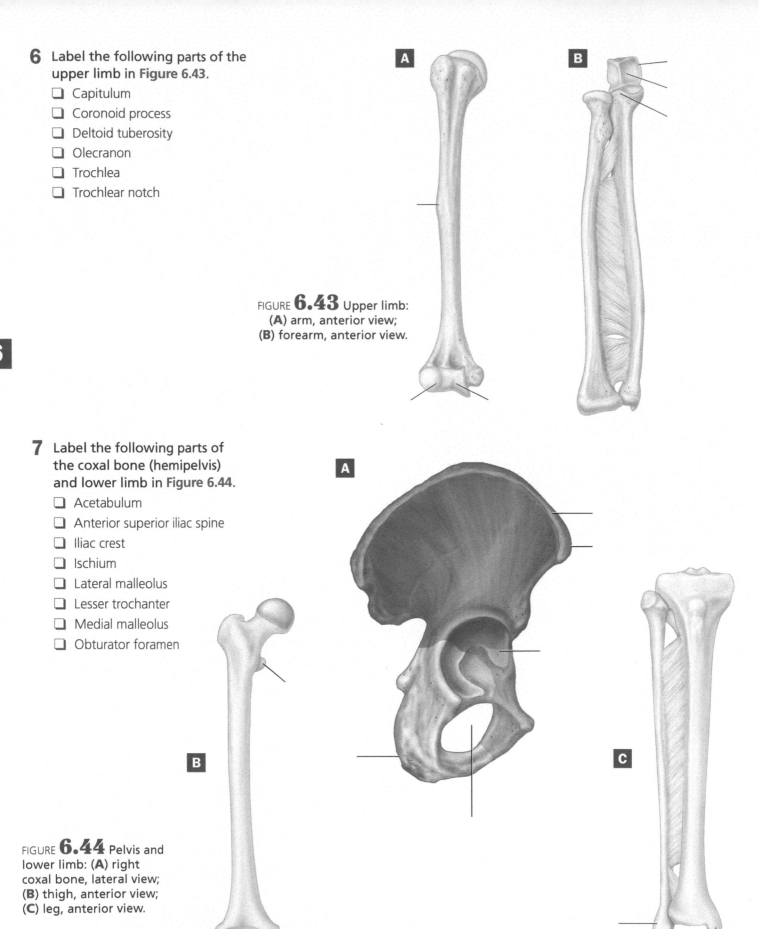

FIGURE **6.43** Upper limb: **(A)** arm, anterior view; **(B)** forearm, anterior view.

7 Label the following parts of the coxal bone (hemipelvis) and lower limb in **Figure 6.44**.

- ❑ Acetabulum
- ❑ Anterior superior iliac spine
- ❑ Iliac crest
- ❑ Ischium
- ❑ Lateral malleolus
- ❑ Lesser trochanter
- ❑ Medial malleolus
- ❑ Obturator foramen

FIGURE **6.44** Pelvis and lower limb: **(A)** right coxal bone, lateral view; **(B)** thigh, anterior view; **(C)** leg, anterior view.

UNIT 6 QUIZ
(continued)

8   The epiphyseal plate is

    a. the structure from which long bones grow in length.

    b. a remnant of the structure from which long bones grow in length.

    c. composed of osseous tissue.

    d. found lining the surface of the epiphysis.

9   A bone tumor disrupts the normal structure of osteons, replacing the organized rings with disorganized, irregular masses of bone. How will this affect the ability of a bone to perform its functions?

_____

_____

_____

_____

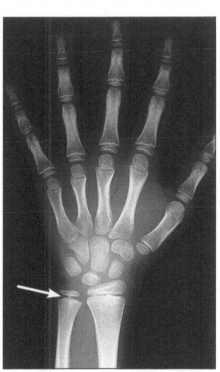

FIGURE **6.45** X-ray from six-year-old child.

10   You are presented the following X-ray from a six-year-old child (Fig. 6.45).

    a Identify the bones in the X-ray.

_____

_____

_____

    b Your colleague thinks there may be a fracture present at the arrow in **Figure 6.45**. What do you think? Is this a fracture or a normal anatomical feature? Explain.

_____

_____

_____

_____

11   A gymnast lands badly after a vault and "twists" her ankle. She now has significant pain and swelling over her lateral ankle area, and her coach is worried that she might have an ankle fracture. Which bone is involved in her injury? What is the specific part of the bone that is injured?

_____

_____

_____

_____

12 Most fibrous joints allow _____ between the articulating bones.

   a. no motion

   b. some motion

13 Synovial joints are surrounded by a/an _____ and filled with _____.

   a. articular disc; serous fluid

   b. joint capsule; synovial fluid

   c. articular cartilage; serous fluid

   d. articular disc; synovial fluid

14 Label the following parts of the knee joint in **Figure 6.46**.

   ❑ Anterior cruciate ligament

   ❑ Fibular collateral ligament

   ❑ Lateral meniscus

   ❑ Medial meniscus

   ❑ Posterior cruciate ligament

   ❑ Tibial collateral ligament

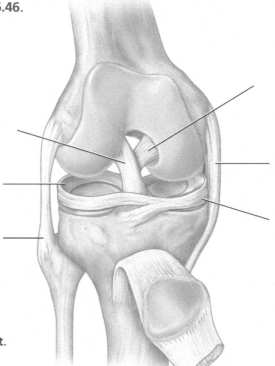

FIGURE **6.46** Knee joint.

15 Which of the following correctly describes abduction?

   a. Abduction decreases the angle between two bones.

   b. Abduction moves a body part toward the midline of the body.

   c. Abduction moves a bone around its own axis.

   d. Abduction moves a body part away from the midline of the body.

16 How would the function of synovial joints be changed if they lacked joint cavities and the bones were united instead? Explain.

_____

_____

_____

_____

Muscle Tissue and the Muscular System

When you have completed this unit, you should be able to:

1 Describe the microscopic anatomy of skeletal muscle fibers.

2 Identify structures of skeletal muscle.

3 Identify muscles of the upper and lower limbs, trunk, head, and neck.

4 List the muscles required to perform common movements.

167

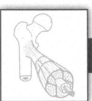

PRE-LAB EXERCISES

Complete the following exercises prior to coming to lab, using your lab manual and textbook for reference.

Pre-Lab Exercise **7-1**

 Key Terms

You should be familiar with the following terms before coming to lab. For the skeletal muscles, describe the muscles' location and appearance.

Term	Definition
Skeletal Muscle Structures	
Epimysium	
Fascicle	
Perimysium	
Muscle fiber	
Endomysium	
Structures of the Skeletal Muscle Fiber	
Sarcolemma	
T-tubule	
Sarcoplasmic reticulum	
Myofibril	
Myofilament	
Sarcomere	

7

A band

I band

General Terms

Origin

Insertion

Muscle action

Skeletal Muscles of the Head, Neck, and Thorax

Sternocleidomastoid m.

Trapezius m.

Erector spinae m.

Intercostal muscles

Diaphragm m.

Rectus abdominis m.

Skeletal Muscles That Move the Upper Limb

Deltoid m.

Latissimus dorsi m.

Pectoralis major m.

Biceps brachii m.

Triceps brachii m. _____

Skeletal Muscles That Move the Lower Limb

Gluteal muscle group

Sartorius m.

Quadriceps femoris muscle group

Hamstrings muscle group

Gastrocnemius m.

Pre-Lab Exercise **7-2**

Muscle Fiber Microanatomy

Color the microscopic anatomy of the skeletal muscle fiber in Figure 7.1, and label it with the following terms from Exercise 7-1 (p. 176). Use Exercise 7-1 in this unit and your text for reference.

❏ Myofibril
❏ Sarcolemma
❏ Sarcomere
 ☐ A band
 ☐ I band
 ☐ Z disc

❏ Transverse tubules
❏ Sarcoplasmic reticulum
❏ Terminal cisternae
❏ Triad

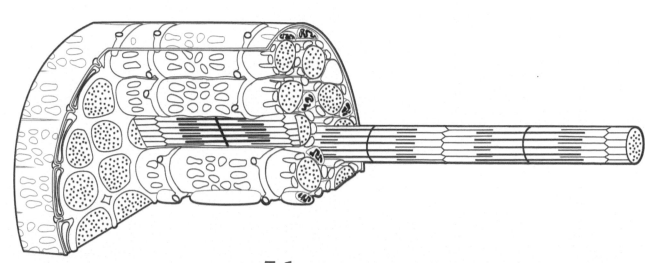

FIGURE **7.1** Skeletal muscle fiber.

Pre-Lab Exercise **7-3**

Skeletal Muscle Anatomy: Muscles That Move the Face and Head

Color the muscles in Figure 7.2, and label them with the following terms from Exercise 7-2 (p. 179). Use Exercise 7-2 in this unit and your text for reference.

❑ Orbicularis oculi m.

❑ Zygomaticus major and minor mm.

❑ Orbicularis oris m.

❑ Temporalis m.

❑ Masseter m.

❑ Sternocleidomastoid m.

❑ Trapezius m.

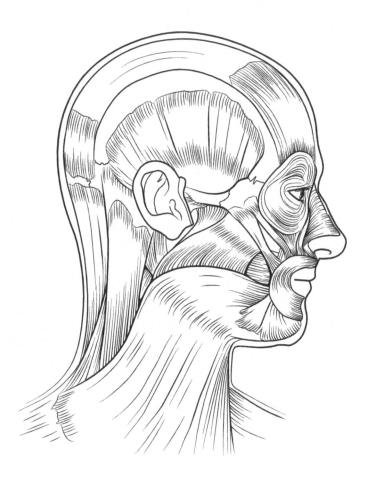

FIGURE **7.2** Human musculature, lateral view of the head and neck.

Pre-Lab Exercise **7-4**

Skeletal Muscle Anatomy: Muscles of the Rest of the Body, Anterior Side

Color the muscles in **Figure 7.3,** and label them with the following terms from Exercise 7-2 (p. 187). Use Exercise 7-2 in this unit and your text for reference.

Muscles That Move the Trunk
❏ Rectus abdominis m.
❏ External oblique m.

Muscles That Move the Arm
❏ Deltoid m.
❏ Pectoralis major m.

Muscles That Move the Forearm
❏ Biceps brachii m.
❏ Brachialis m.
❏ Brachioradialis m.

Muscles That Move the Thigh and Leg
❏ Sartorius m.
❏ Rectus femoris m.
❏ Vastus lateralis m.
❏ Vastus medialis m.

Muscle That Moves the Foot
❏ Tibialis anterior m.

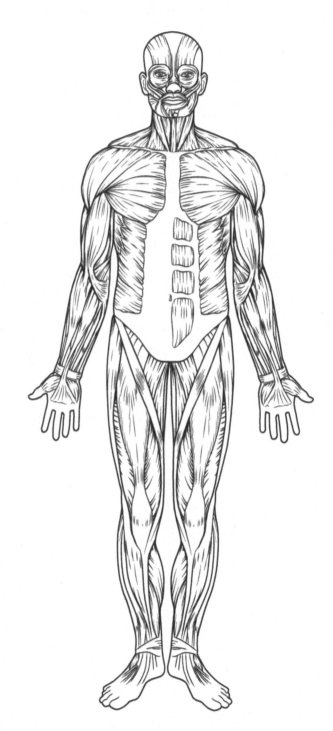

FIGURE **7.3** Human musculature, anterior view.

Skeletal Muscle Anatomy: Muscles of the Rest of the Body, Posterior Side

Color the muscles in Figure 7.4, and label them with the following terms from Exercise 7-2 (p. 188). Use Exercise 7-2 in this unit and your text for reference.

Muscles That Move the Scapula and Arm

❏ Trapezius m.
❏ Latissimus dorsi m.

Muscle That Moves the Forearm

❏ Triceps brachii m.

Muscles That Move the Thigh and Leg

❏ Gluteus maximus m.
❏ Semitendinosus m.
❏ Semimembranosus m.
❏ Biceps femoris m.

Muscles That Move the Foot

❏ Gastrocnemius m.
❏ Soleus m.

7

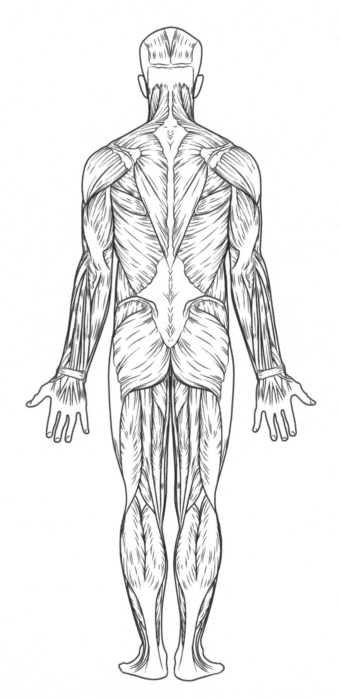

FIGURE **7.4** Human musculature, posterior view.

EXERCISES

Colored SEM of striated muscle fiber.

The human body has nearly 700 **skeletal muscles** that make up the **muscular system**. The skeletal muscles range dramatically in size and shape, from the large trapezius muscle to the tiny corrugator supercilii muscle. Luckily for you, we will be learning only about 40 muscles rather than the full 700. The exercises in this unit help you become familiar with the basic gross and microscopic anatomy of skeletal muscles, the main muscle groups, and their component muscles.

Exercise 7-1

Skeletal Muscle Anatomy and Muscle Tissue

MATERIALS
- ❏ Skeletal muscle models
- ❏ Skeletal muscle fiber model
- ❏ Skeletal muscle tissue slide
- ❏ Light microscope with oil-immersion objective
- ❏ Oil
- ❏ Colored pencils

Recall from the tissues unit (p. 92) that skeletal muscle tissue is composed of skeletal muscle cells, also called **muscle fibers**. Individual skeletal muscle fibers are surrounded by their extracellular matrix, known as the **endomysium** (en-doh-MY-see-um). (Note that the endomysium blends with surrounding connective tissue and, for this reason, is often referred to as connective tissue itself.) Muscle fibers are arranged into groups called **fascicles** (FASS-ih-kullz; Fig. 7.5). Each fascicle is surrounded by a connective tissue sheath called the **perimysium** (pehr-ih-MY-see-um). The muscle as a whole is covered by another connective tissue sheath called the **epimysium** (ep-ih-MY-see-um). The epimysium and perimysium merge near the end of a muscle to form *tendons* and *aponeuroses*, which connect the muscle to bones or soft tissue. The most superficial connective tissue around a skeletal muscle is thick **fascia** (FASH-uh), which separates individual muscles and binds them into groups.

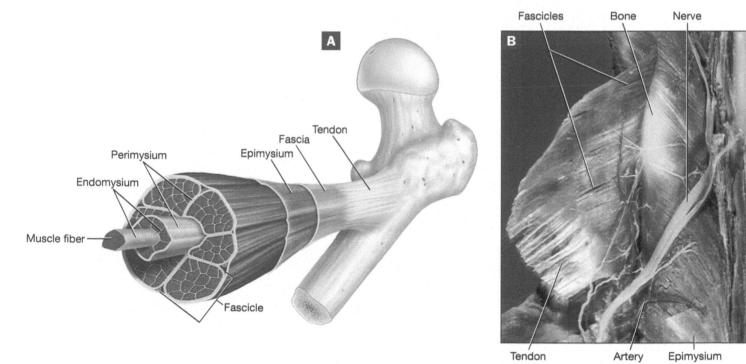

FIGURE **7.5** Overview of skeletal muscle structure: (**A**) skeletal muscle structural levels of organization; (**B**) dissected muscle.

As you can see in **Figure 7.6**, the long, multinucleated, cylindrical skeletal muscle fiber is quite different in appearance from the cells you are used to seeing. Due to these variations, we change the terms to describe some of a muscle fiber's structures. For example, the muscle fiber's plasma membrane is known as the **sarcolemma** (sar-koh-LEM-uh; Figs. 7.6 and 7.7), and its cytoplasm is called **sarcoplasm** (SAR-koh-plazm).

Small, cylindrical organelles called **myofibrils** (my-oh-FY-brillz) make up about 80 percent of the skeletal muscle fiber's sarcoplasm. Other organelles, such as mitochondria, are packed in between the myofibrils. Notice in the figure that myofibrils are wrapped by the weblike **sarcoplasmic reticulum** (sar-koh-PLAZ-mik reh-TIK-yoo-lum; **SR**), which is a modified smooth endoplasmic reticulum that stores calcium ions. Myofibrils are also wrapped by inward extensions of the sarcolemma known as **transverse tubules** or **T-tubules**. On either side of a T-tubule, the SR swells to form **terminal cisternae** (sis-TER-nee); a T-tubule and two terminal cisternae form a structure called a **triad**. These structures are important for coordinating electrical stimulation of the muscle fiber with release of calcium ions from the SR and initiation of a muscle contraction.

Myofibrils are composed of protein subunits called **myofilaments**. The two types of myofilaments involved in contraction are: (1) **thick filaments**, composed of the contractile protein **myosin** (MY-oh-sin), and (2) **thin filaments**, composed of the contractile protein **actin** and the regulatory proteins **troponin** (TROH-poh-nin) and **tropomyosin** (trohp-oh-MY-oh-sin). The arrangement of myofilaments within the myofibrils is what gives skeletal muscle its characteristic **striated** appearance. The dark regions of the striations, called **A bands**, are dark because this is where we find overlapping thick and thin filaments or thick filaments alone. The light regions, called **I bands**, appear light because they contain only thin filaments. Bisecting the I band is a dark line called the **Z disc**. The terms of this alphabet soup are used to describe the fundamental unit of contraction: the **sarcomere** (SAR-koh-meer). A sarcomere, defined as the space from one Z disc to the next Z disc, consists of a full A band and two half-I bands.

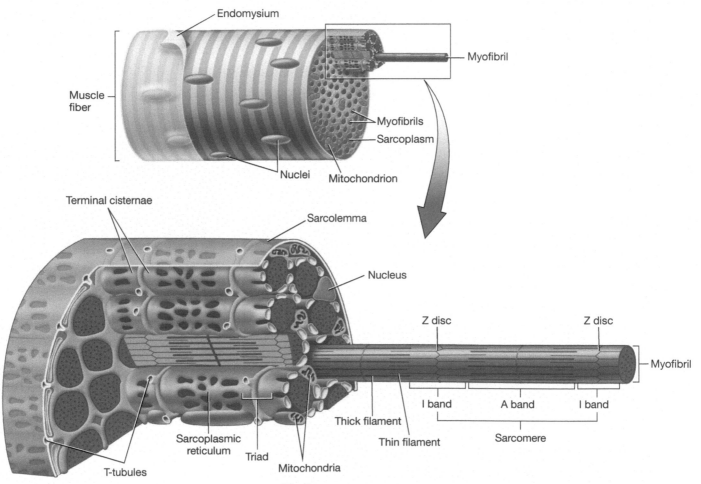

FIGURE **7.6** Skeletal muscle fiber.

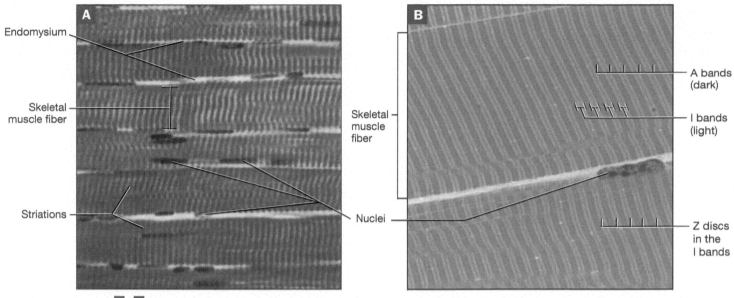

FIGURE **7.7** Skeletal muscle tissue, photomicrograph: (**A**) 40× objective; (**B**) 100× objective oil immersion.

 Procedure 1 Model Inventory for Skeletal Muscle and Skeletal Muscle Fibers

Identify the following structures of skeletal muscle and skeletal muscle fibers on models and diagrams, using this unit and your textbook for reference. As you examine the anatomical models and diagrams, record the name of the model and the structures you were able to identify on the model inventory in Table 7.1.

Skeletal Muscle Anatomy

1. Muscle fiber
2. Endomysium
3. Fascicles
4. Connective tissue coverings
   a. Epimysium
   b. Fascia

Muscle Fiber Microanatomy

1. Sarcolemma
2. Myofibrils
3. Sarcoplasmic reticulum
4. T-tubules
5. Myofilaments
   a. Thick filaments
   b. Thin filaments
6. Sarcomere
   a. A band
   b. I band
   c. Z disc

TABLE **7.1** Model Inventory for Skeletal Muscle Anatomy and Microanatomy

Model/Diagram	Structures Identified

Procedure 2 Microscopy of Skeletal Muscle Tissue

View a prepared slide of skeletal muscle tissue.

1 First, examine the slide with the regular high-power objective of your light microscope.

2 Draw and color what you see, and label your drawing with the terms indicated.

3 Then switch to an oil-immersion lens (your instructor may have one set up as a demonstration), and identify the structures of the sarcomere.

High-Power (40×) Objective

1. Striations
   a. A band
   b. I band
2. Sarcolemma
3. Nuclei
4. Endomysium

Oil-Immersion Lens

1. Sarcomere
2. A band
3. I band
4. Z disc

7

MATERIALS

☐ Muscle models: upper limb, lower limb, trunk, head, neck

In this exercise, we divide skeletal muscles into muscle groups that have similar functions. For example, we classify the latissimus dorsi muscle with the muscles of the upper limb, rather than the muscles of the back or trunk, because it moves the upper limb. We use the following groupings of skeletal muscles in this unit: muscles that move the head, neck, and face; muscles that move the trunk; muscles that move the arm; muscles that move the forearm and fingers; muscles that move the thigh and leg; and muscles that move the foot and toes.

Muscles That Move the Head, Neck, and Face

We can subdivide the muscles that move the head, neck, and face into the muscles of facial expression, the muscles of mastication, and the muscles that move the head and neck. The muscles of facial expression control the various facial expressions humans are capable of making, and they all insert into skin or other muscles (Figs. 7.8 and 7.9). One of the largest muscles of facial expression is the *epicranius muscle* (eh-pih-KRAY-nee-uhs), which elevates the eyebrows and skin of the forehead. The muscle is bipartite, meaning it has two parts. Those parts, the **frontalis** (frun-TAL-iss) and **occipitalis** (awk-sip-ih-TAL-iss) **muscles**, are connected by a sheet of connective tissue called the *epicranial aponeurosis* (ap-oh-noo-ROH-sis). Just inferior to the frontalis muscle are the two circular **orbicularis oculi muscles** (ohr-bik-yoo-LEHR-iss AWK-yoo-lye). Their circular structure enables them to close and squint the eyes.

Other muscles of facial expression move the lips and skin around the mouth:

▮ The most obvious muscle around the mouth is the circular **orbicularis oris muscle**, which purses the lips.

▮ Under the skin of the cheeks, we find the **buccinator muscle** (BUK-sin-ay-tur). (Remember that "buccal" means "cheek.") It pulls the cheeks inward to produce sucking and whistling movements.

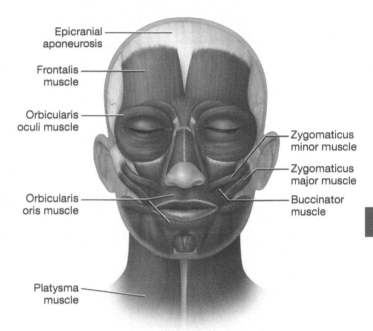

FIGURE **7.8** Facial musculature, anterior view.

7

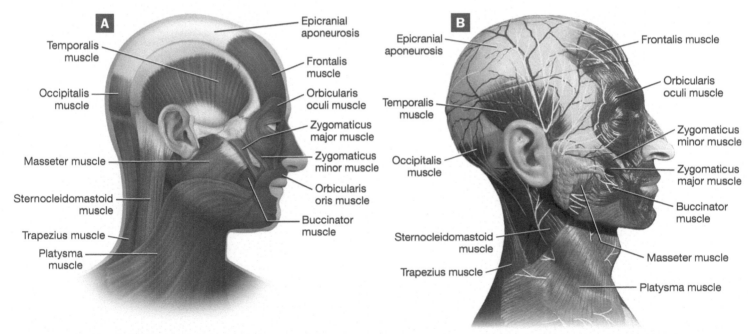

FIGURE **7.9** Muscles of the head and neck, lateral view: **(A)** illustration; **(B)** anatomical model photo.

- Covering the superficial surface of the neck is the broad, flat **platysma muscle** (plah-TIZ-muh). It depresses the mandible (opens the jaw) and tightens the skin of the neck.

- Muscles involved in smiling include the **zygomaticus major** and **minor muscles**, which pull the corners of the mouth superiorly and laterally.

The **muscles of mastication**, another group of muscles found in the head, are involved in *mastication*, or chewing (Fig. 7.9). They include the thick **masseter muscle** (MASS-uh-tur), located over the lateral mandible, and the fan-shaped **temporalis muscle** (tem-pur-AL-iss), located over the temporal bone.

Several muscles move the head, including the diamond-shaped **trapezius muscle** (trah-PEE-zee-uhs) on the back, which holds the head upright and hyperextends the head and the neck (best seen in Figure 7.14A, p. 181). In the neck, we find the strap-like **sternocleidomastoid muscle** (stern-oh-kly-doh-MASS-toy'd). When the two muscles contract together, they flex the head and neck. Acting individually, each muscle rotates the head toward the opposite side and flexes the head and neck laterally.

Muscles That Move the Trunk

The muscles that move the trunk include the muscles of the thorax, muscles of the abdominal wall, and postural muscles of the back.

1. The muscles of the thorax are generally involved in the muscle movements that produce ventilation: the **external** and **internal intercostal muscles** are located between the ribs and are involved in both inspiration and expiration, respectively (Fig. 7.10); the dome-shaped **diaphragm muscle** (DY-uh-fram) produces the movements necessary for inspiration; and the small **pectoralis minor muscle** draws the scapula anteriorly and the rib cage superiorly during forced inspiration and expiration (Fig. 7.11).

2. The abdominal muscles move the trunk and increase intra-abdominal pressure (Figs. 7.12 and 7.13). The **rectus abdominis muscle** is the central and superficial muscle that flexes the trunk. It is covered by a thick layer of connective tissue called the *rectus sheath*. On the lateral torso, we find the **external** and **internal oblique muscles**, which rotate and laterally flex the trunk. Deep to all three muscles, the **transversus abdominis muscle** squeezes the abdominal contents like a belt to increase intra-abdominal pressure.

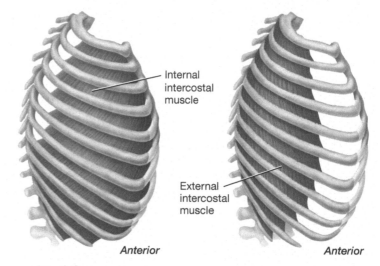

FIGURE **7.10** Internal and external intercostal muscles, lateral view.

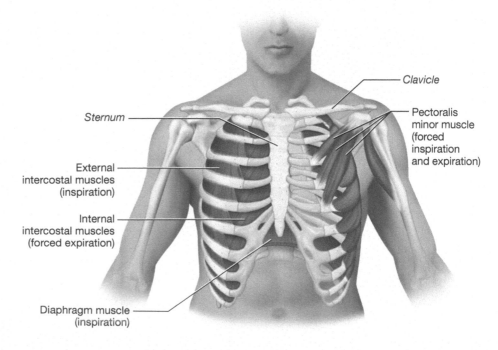

FIGURE **7.11** Deep muscles of the anterior thorax.

3. The postural muscles of the back work to ensure that we remain upright. Most of these postural muscles also extend the back. One of the main postural muscles is a group known collectively as the **erector spinae muscle** (eh-REK-tohr SPY-nee). The three components of the erector spinae muscle include the lateral **iliocostalis** (ill-ee-oh-kawst-AL-iss), the middle **longissimus** (lawn-JISS-ih-muss), and the medial **spinalis muscles** (spy-NAL-iss; Fig. 7.14). See Figure 7.14B for the other major postural muscles.

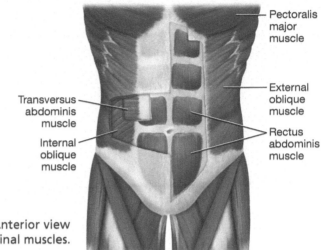

FIGURE **7.12** Anterior view of the abdominal muscles.

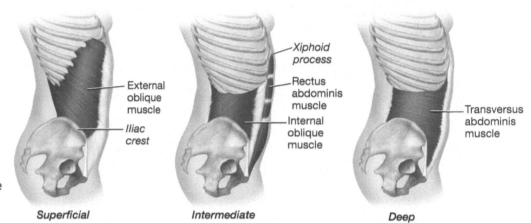

FIGURE **7.13** Lateral view of the individual abdominal muscles.

Superficial   *Intermediate*   *Deep*

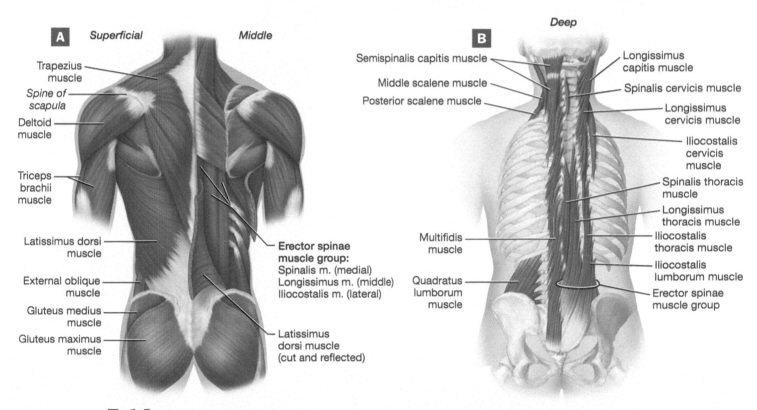

FIGURE **7.14** Muscles of the trunk, posterior view: (**A**) superficial and middle dissection; (**B**) deep dissection.

Muscles That Move the Arm

The shoulder joint has the greatest range of motion of any joint in the body. For this reason, there are many muscles acting on it, both to move the arm and to stabilize the joint. Some of these muscles include the following:

▋ The posterior **latissimus dorsi muscle** (lah-TISS-ih-muss DOHR-sye) is the prime muscle of arm extension, and it also adducts and rotates the arm (Figs. 7.14 and 7.15).

▋ The **pectoralis major muscle**, or "chest muscle," is the prime muscle of arm flexion, and it also adducts the arm (Fig. 7.16).

▋ The **deltoid muscle**, or "shoulder muscle," is the prime abductor of the arm, and assists in both flexion and extension.

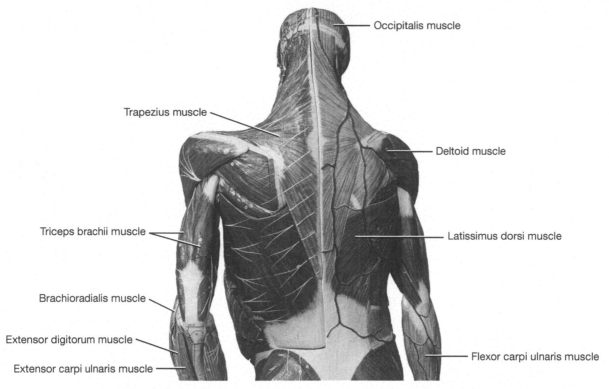

FIGURE **7.15** Anatomical model photo of muscles of the trunk, posterior view.

Muscles That Move the Forearm, Hand, and Fingers

The four main muscles that move the forearm are the anterior **biceps brachii** (BY-seps BRAY-kee-aye), the anterior and deep **brachialis** (bray-kee-AL-iss), the anterior and lateral **brachioradialis** (bray-kee-oh-ray-dee-AL-iss), and the posterior **triceps brachii muscles** (Figs. 7.16 and 7.17). The first three muscles flex the forearm, while the triceps brachii extends the forearm.

The remainder of the muscles of the upper limb act on the hand at the wrist, the hand itself (the *intrinsic hand muscles*), and/or the digits. Many of these muscles are labeled in Figures 7.16 and 7.17 for your reference.

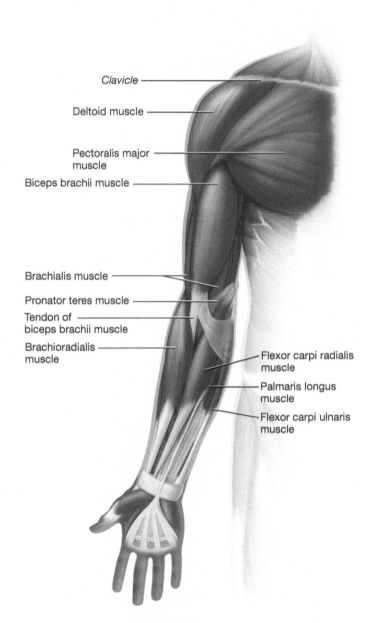

FIGURE **7.16** Anterior view of the muscles of the right upper limb.

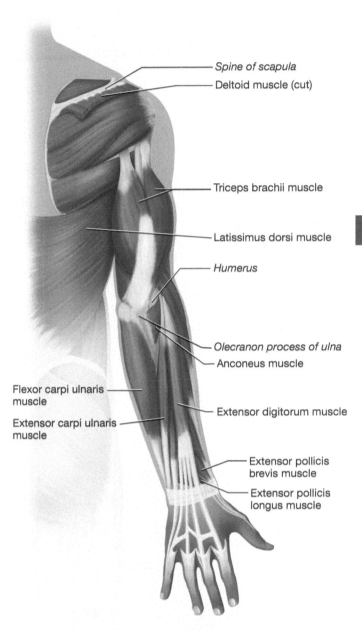

FIGURE **7.17** Posterior view of the muscles of the right upper limb.

Muscles That Move the Thigh and Leg

We can divide the muscles that move only the thigh into three groups based on their location: on the posterior pelvis (the *gluteal muscles*), on the medial thigh (the *adductor muscle group*), or in the deep anterior thigh and pelvis (the *iliopsoas muscle*).

- There are three gluteal muscles. The first and largest is the superficial **gluteus maximus muscle,** which is the prime extensor of the thigh (Fig. 7.18). The **gluteus medius muscle** (MEE-dee-uhs), located deep to the gluteus maximus muscle, is a major abductor of the thigh. The small, deep *gluteus minimus muscle* assists the gluteus medius muscle in its actions. (Other deep muscles of the gluteal region are illustrated in Figure 7.18 for your reference.)

- As implied by their names, the muscles in the *adductor group*, found on the medial femur, are the prime adductors of the thigh (Fig. 7.19). This group includes the superior **adductor brevis,** the inferior **adductor longus,** and the deep, broad **adductor magnus muscles.**

- The **iliopsoas muscle** (ill-ee-oh-SOH-uhs) is the prime flexor of the thigh. It is composed of two separate muscles: the *iliacus muscle* (ill-ee-AK-uhs) and the *psoas major muscle* (SOH-us; Fig. 7.20).

The remainder of the thigh muscles may move both the thigh and leg, or just the leg. One of the most prominent groups of thigh muscles is located in the anterior thigh and is known as the **quadriceps femoris group.** It includes the **rectus femoris, vastus medialis, vastus intermedius,** and **vastus lateralis muscles.** These four muscles converge to form the common quadriceps femoris or patellar tendon, which envelops the patella. Distal to the patella, the tendon becomes the *patellar ligament,* which inserts into the tibial tuberosity. Note that the vastus intermedius muscle is deep to the rectus femoris muscle and so is not visible in Figure 7.20. The rectus femoris muscle both flexes the thigh and extends the leg, but the vastus group doesn't cross the hip joint and so only extends the leg.

Other prominent thigh muscles include the strap-like **sartorius muscle** (sar-TOHR-ee-uhs), which crosses from lateral to medial across the anterior thigh, and the medial **gracilis muscle** (gruh-SILL-iss). Both of these muscles flex the thigh and extend the leg. The sartorius muscle also laterally rotates the thigh (it's what pulls your leg and thigh into the position you assume when you're looking at something on the bottom of your shoe). The posterior thigh muscles include the three muscles of the **hamstrings group:** the lateral **biceps femoris** and the medial **semitendinosus** (sem-aye-ten-din-OH-suhs) and **semimembranosus** (sem-aye-mem-brah-NOH-suhs) **muscles** (Fig. 7.21). The superficial semitendinosus muscle is named for its long, narrow tendon. These three muscles both extend the thigh and flex the leg.

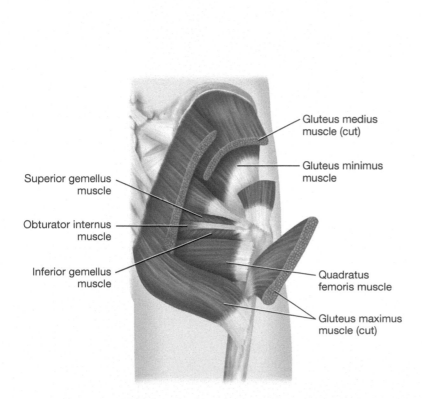

FIGURE **7.18** Deeper muscles of the gluteal region.

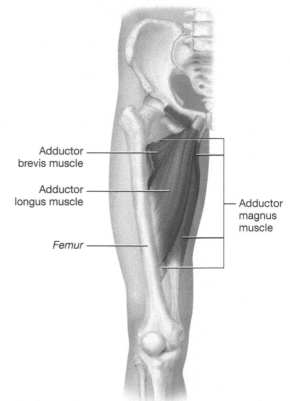

FIGURE **7.19** Muscles of the adductor muscle group.

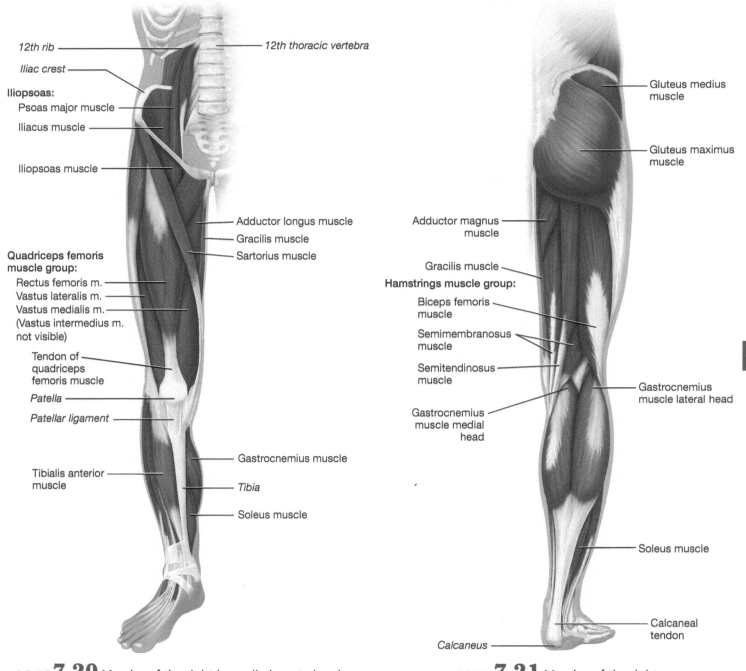

FIGURE **7.20** Muscles of the right lower limb, anterior view.

FIGURE **7.21** Muscles of the right lower limb, posterior view.

Muscles That Move the Foot and Toes

The most obvious muscle of the posterior leg is the large **gastrocnemius** muscle (gas-trawk-NEE-mee-uhs), also known as the "calf muscle," a two-headed muscle that originates from the distal femur (Fig. 7.22). Just deep to the gastrocnemius muscle is the **soleus** muscle (SOHL-ee-uhs). Together, these two muscles insert into the posterior calcaneus via a common tendon called the **calcaneal tendon** (kal-KAY-nee-uhl), more commonly known as the *Achilles tendon*. The gastrocnemius and soleus muscles, also known together as the *triceps surae muscle*, are the prime plantarflexors of the foot. The gastrocnemius muscle also contributes to flexion of the leg, as it crosses the knee joint.

On the lateral leg, we find two muscles, the **fibularis longus** (fib-yoo-LEHR-iss) and **fibularis brevis muscles**, the main everters of the foot. Anteriorly on the leg, we find extensors such as the **tibialis anterior** (tib-ee-AL-is), which dorsiflexes the foot. These and other leg muscles are labeled in Figure 7.22 for reference purposes.

Figures 7.23 and 7.24 provide whole-body views of many of the muscles we have just discussed.

HINTS & TIPS

Muscle Names

What's in a name? When it comes to a skeletal muscle, there's often a lot. The number of muscles that we cover in this lab may seem intimidating, but it can be easier to figure out a muscle's location or actions than you may think. This is because many muscles have very descriptive names, which you may have noticed from your reading of the preceding section. Some of the common features used in muscle names include the following:

ⓘ **Size.** Many muscle names include clues about their size. Some of these are obvious—*longus* describes a long muscle, and a pair of *major* and *minor* muscles describes a set in which one is larger and one is smaller. Others are a little less obvious, such as *brevis*, which means short (think "brief"), or *vastus*, which means wide (think "vast").

ⓘ **Location.** A great many muscles are named for their location. This can include directional terms, such as *lateralis*, *superficialis*, or *anterior*. It can also include regional terms such as *abdominis*, *cervicis* (neck), *brachii* (arm), and *costal* (rib). Finally, some of the terms that describe location use words or prefixes with which you're likely already familiar such as *sub-* (below) and *profundus* (*deep*—think "profound").

ⓘ **Appearance or shape.** Several muscles have names based on their appearance or shape. For example, the semitendinosus muscle has a very long tendon, and the trapezius muscle is shaped like a trapezoid. Similarly, the orbicularis oris and orbicularis oculi muscles are named for their circular structure, as "orb" means "sphere."

ⓘ **Points of origin.** Related to appearance is a muscle's number of points of origin, or "heads." We see this with muscles named *biceps*, *triceps*, and *quadriceps*, which mean "two-headed," "three-headed," and "four-headed," respectively.

ⓘ **Function.** Finally, the function of many muscles is revealed in their names. As you can imagine, an *extensor* has extension as a primary action, and an *adductor* has adduction as a primary action.

When you take all of this into consideration, it's much easier to figure out the location, appearance, and, occasionally, function of complicated-sounding muscles. For example, if you are tasked with identifying the brachioradialis muscle, just from its name you can infer that it is a muscle that attaches to the arm (brachio-) and the radius (radialis). Similarly, the tibialis anterior muscle is clearly a muscle on the anterior tibia, and the iliocostalis muscle has attachments to the ilium and the ribs. If we take a more complicated muscle name, such as the extensor carpi radialis longus muscle, we can say that it is a long muscle that originates from the radius, inserts into the carpals, and extends the hand at the wrist. See? It's not as difficult as you thought.

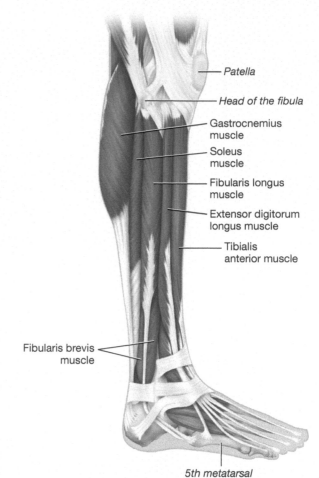

FIGURE **7.22** Lateral view of muscles of the leg.

- Patella
- Head of the fibula
- Gastrocnemius muscle
- Soleus muscle
- Fibularis longus muscle
- Extensor digitorum longus muscle
- Tibialis anterior muscle
- Fibularis brevis muscle
- 5th metatarsal

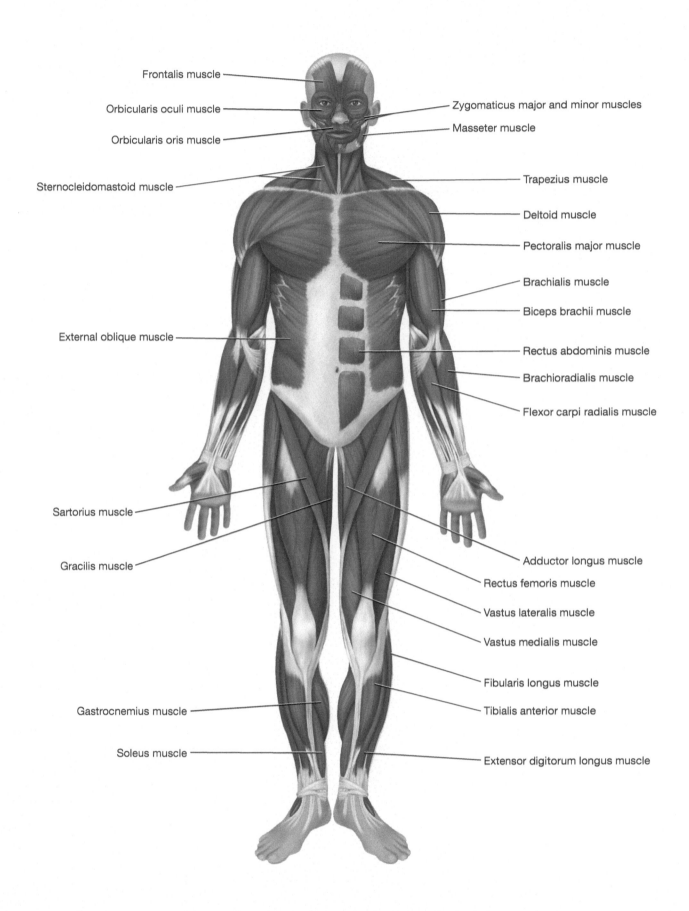

Frontalis muscle

Orbicularis oculi muscle

Orbicularis oris muscle

Sternocleidomastoid muscle

External oblique muscle

Sartorius muscle

Gracilis muscle

Gastrocnemius muscle

Soleus muscle

Zygomaticus major and minor muscles

Masseter muscle

Trapezius muscle

Deltoid muscle

Pectoralis major muscle

Brachialis muscle

Biceps brachii muscle

Rectus abdominis muscle

Brachioradialis muscle

Flexor carpi radialis muscle

Adductor longus muscle

Rectus femoris muscle

Vastus lateralis muscle

Vastus medialis muscle

Fibularis longus muscle

Tibialis anterior muscle

Extensor digitorum longus muscle

FIGURE **7.23** Muscles of the body, anterior view.

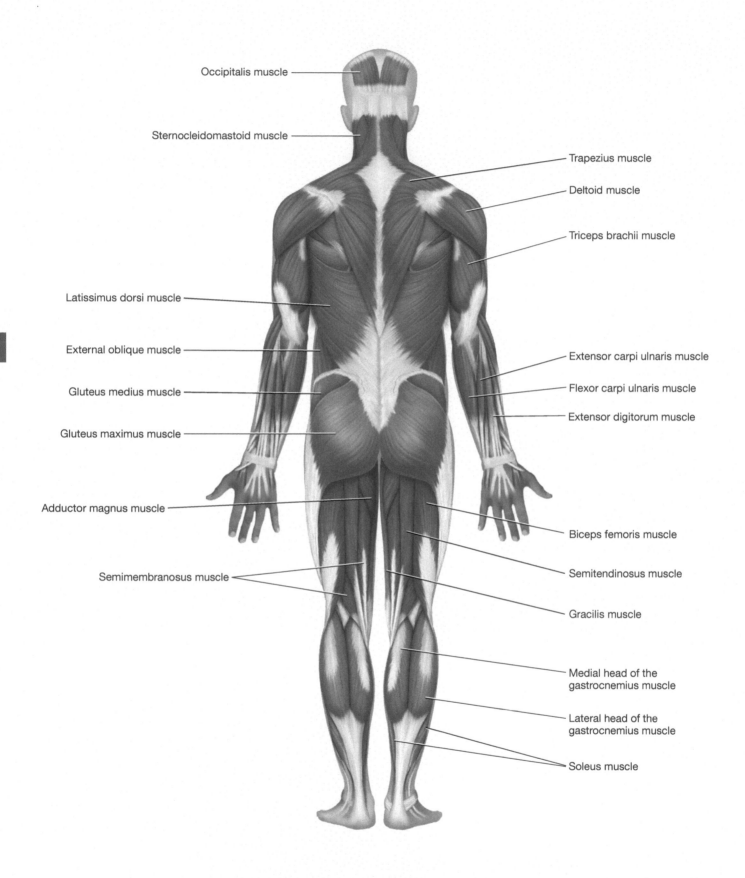

Occipitalis muscle

Sternocleidomastoid muscle

Trapezius muscle

Deltoid muscle

Triceps brachii muscle

Latissimus dorsi muscle

External oblique muscle

Gluteus medius muscle

Gluteus maximus muscle

Extensor carpi ulnaris muscle

Flexor carpi ulnaris muscle

Extensor digitorum muscle

Adductor magnus muscle

Biceps femoris muscle

Semitendinosus muscle

Gracilis muscle

Semimembranosus muscle

Medial head of the gastrocnemius muscle

Lateral head of the gastrocnemius muscle

Soleus muscle

FIGURE **7.24** Muscles of the body, posterior view.

Procedure 1 Model Inventory for the Skeletal Muscles

Identify the following muscles on models and diagrams, using this unit and your textbook for reference. As you examine the anatomical models and diagrams, record the name of the model and the structures you were able to identify on the model inventory in Table 7.2.

Muscles That Move the Head, Neck, and Face
1. Epicranius m. (frontalis and occipitalis mm.)
2. Orbicularis oculi m.
3. Orbicularis oris m.
4. Buccinator m.
5. Platysma m.
6. Zygomaticus major and minor mm.
7. Masseter m.
8. Temporalis m.
9. Trapezius m.
10. Sternocleidomastoid m.

Muscles That Move the Trunk
1. External intercostal mm.
2. Internal intercostal mm.
3. Diaphragm m.
4. Pectoralis minor m.
5. Rectus abdominis m.
6. External oblique m.
7. Internal oblique m.
8. Transversus abdominis m.
9. Erector spinae muscle group

Muscles That Move the Arm
1. Latissimus dorsi m.
2. Pectoralis major m.
3. Deltoid m.

Muscles That Move the Forearm
1. Biceps brachii m.
2. Brachialis m.
3. Brachioradialis m.
4. Triceps brachii m.

Muscles That Move the Thigh and Leg
1. Gluteus maximus m.
2. Gluteus medius m.
3. Adductor muscle group
4. Iliopsoas m.
5. Quadriceps femoris muscle group
   a. Rectus femoris m.
   b. Vastus medialis m.
   c. Vastus intermedius m.
   d. Vastus lateralis m.
6. Sartorius m.
7. Gracilis m.
8. Hamstring muscle group
   a. Biceps femoris m.
   b. Semitendinosus m.
   c. Semimembranosus m.

Muscles That Move the Foot
1. Gastrocnemius m.
2. Soleus m.
3. Tibialis anterior m.

Your instructor may wish to omit certain muscles included above or add muscles not included in these lists. List any additional structures below:

7

TABLE **7.2** Model Inventory for Skeletal Muscles

Model/Diagram	Structures Identified

Exercise 7-3

Muscle Origins and Insertions

MATERIALS
❑ Small skeleton
❑ Modeling clay

You can best understand a muscle's actions by first understanding its origin and insertion. A muscle begins at its **origin**, generally the more stationary part, and attaches to its **insertion**, generally the part the muscle moves. For example, you can see in Figure 7.25 that the biceps brachii muscle originates on the scapula and crosses the elbow joint, where it inserts into the proximal radius. Notice also that the triceps brachii muscle originates from the humerus and the inferior scapula and crosses the elbow joint to insert into the posterior ulna.

After you determine a muscle's origin and insertion, figuring out its actions becomes easy. Let's examine the biceps brachii and triceps brachii muscles again. The biceps brachii muscle inserts into the radius, so we know it will move the forearm. Given how it crosses the anterior elbow joint, we can conclude it will cause forearm flexion at the elbow joint. The triceps brachii muscle inserts into the ulna, so we know it also will move the forearm. Given how it crosses the posterior elbow joint, we can conclude it will cause forearm extension at the elbow joint. Now wasn't that easy?

So you don't need to actually memorize most muscle actions—all you need to do is look at a muscle's origin and insertion and use some basic logic to figure out its actions. We practice doing this in the following procedure.

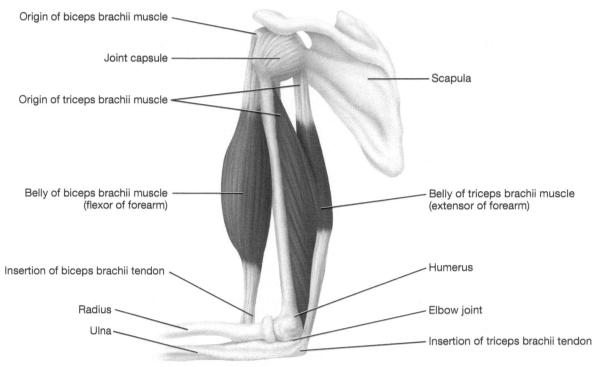

Origin of biceps brachii muscle

Joint capsule

Origin of triceps brachii muscle

Scapula

Belly of biceps brachii muscle
(flexor of forearm)

Belly of triceps brachii muscle
(extensor of forearm)

Insertion of biceps brachii tendon

Humerus

Radius

Elbow joint

Ulna

Insertion of triceps brachii tendon

FIGURE **7.25** Origin and insertion of the biceps brachii and triceps brachii muscles, posterior view.

Procedure 1 Build Muscles

In this procedure, you will use small skeletons and modeling clay to build specific muscle groups. This may sound easy, but there is a catch: You must determine the actions of the muscle by looking only at the origin and insertion of each muscle you build.

1 Obtain a small skeleton and four colors of modeling clay.

2 Build the indicated muscles, using a different color of clay for each muscle. As you build, pay careful attention to the origin and insertion of each muscle.

3 Determine the primary actions for each muscle you have built by looking *only* at the origin and insertion. Record this information in Tables 7.3 through 7.5.

TABLE **7.3** Muscles and Actions for Group 1

Muscle	Actions
Biceps femoris m.	
Rectus femoris m.	
Gluteus medius m.	
Adductor m. group	

TABLE **7.4** Muscles and Actions for Group 2

Muscle	Actions
Biceps brachii m.	
Triceps brachii m.	
Deltoid m.	
Pectoralis major m.	

TABLE **7.5** Muscles and Actions for Group 3

Muscle	Actions
Gastrocnemius m.	
Tibialis anterior m.	
Rectus abdominis m.	
Trapezius m.	

1 Label the following terms on **Figure 7.26**.
- ❏ Epimysium
- ❏ Fascicle
- ❏ Muscle fiber
- ❏ Perimysium
- ❏ Tendon

2 Label the following terms on **Figure 7.27**.
- ❏ A band
- ❏ I band
- ❏ Myofibril
- ❏ Sarcolemma
- ❏ Sarcomere
- ❏ Sarcoplasmic reticulum
- ❏ T-tubule
- ❏ Z disc

7

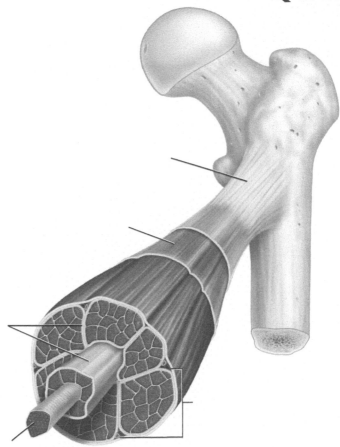

FIGURE **7.26** Basic skeletal muscle structural levels of organization.

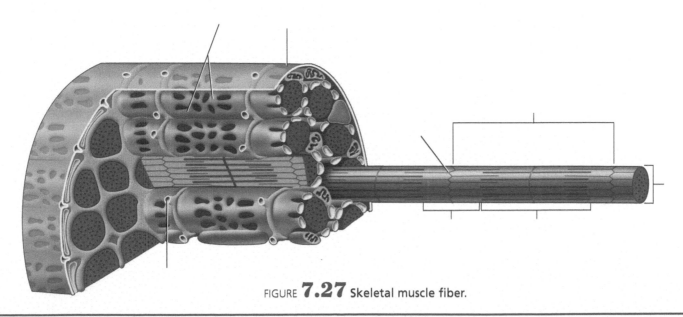

FIGURE **7.27** Skeletal muscle fiber.

3 The striations in skeletal muscle fibers are attributable to

    a. light and dark pigments found in the sarcoplasm.

    b. the arrangement of thick and thin filaments in the myofibril.

    c. overlapping Z discs.

    d. overlapping adjacent skeletal muscle fibers.

4 Label the following muscles on **Figure 7.28**.

    ❑ Frontalis m.

    ❑ Orbicularis oculi m.

    ❑ Orbicularis oris m.

    ❑ Platysma m.

    ❑ Zygomaticus mm.

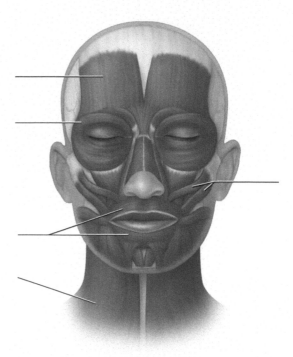

FIGURE **7.28** Facial musculature, anterior view.

5 Which of the following muscles extends the forearm at the elbow joint?

    a. Brachioradialis m.

    b. Triceps brachii m.

    c. Biceps brachii m.

    d. Brachialis m.

6 Which of the following muscles extends the thigh and flexes the leg?

    a. Sartorius m.

    b. Iliopsoas m.

    c. Semimembranosus m.

    d. Rectus femoris m.

7 Which of the following muscles is the prime flexor of the arm at the shoulder joint?

    a. Pectoralis major m.

    b. Latissimus dorsi m.

    c. Pectoralis minor m.

    d. Deltoid m.

UNIT 7 QUIZ
(continued)

8 Label the following muscles on Figure 7.29.

- ❏ Biceps brachii m.
- ❏ Biceps femoris m.
- ❏ Deltoid m.
- ❏ External oblique m.
- ❏ Gastrocnemius m.

- ❏ Gluteus maximus m.
- ❏ Latissimus dorsi m.
- ❏ Pectoralis major m.
- ❏ Rectus abdominis m.
- ❏ Rectus femoris m.

- ❏ Sartorius m.
- ❏ Semimembranosus m.
- ❏ Tibialis anterior m.
- ❏ Trapezius m.
- ❏ Triceps brachii m.

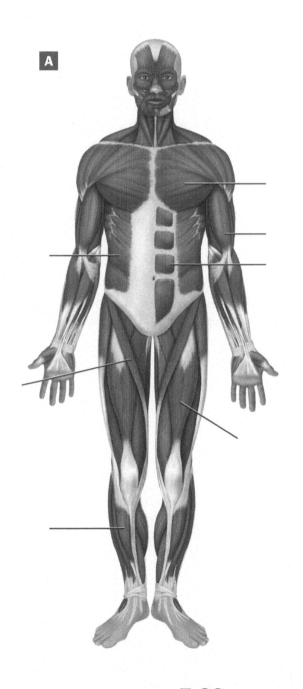
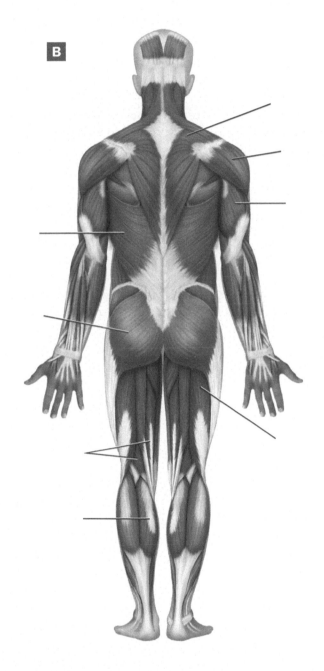

FIGURE **7.29** Muscles of the body: (**A**) anterior view; (**B**) posterior view.

9 Muscle strain, or "pulling a muscle," may result from overuse injuries or from trauma. Typically muscle strain causes pain around the muscle with movement and with pressure. Predict which muscle or muscles may be strained if a patient complains of pain in each of the following locations:

a Medial thigh

b Anterior arm

c Posterior neck

d Lateral abdomen (the "side")

10 Predict the location, appearance, and function of the flexor carpi radialis longus muscle by looking only at its name.

11 The serratus anterior muscle normally originates from the first nine ribs and inserts into the medial border of the scapula. Predict its action from this information.

12 When a person is in respiratory distress, the origin and insertion of the serratus anterior muscle may actually switch, and the part of the muscle attached to the medial border of the scapula acts as the origin. How will this change the muscle's action?

13 A stroke, caused by a clot in a blood vessel of the brain, may lead to a loss of function of certain muscles. Which motions would an individual be unable to perform if the following muscles lost function?

a Orbicularis oculi m.

b Sternocleidomastoid m.

c Gluteus maximus m.

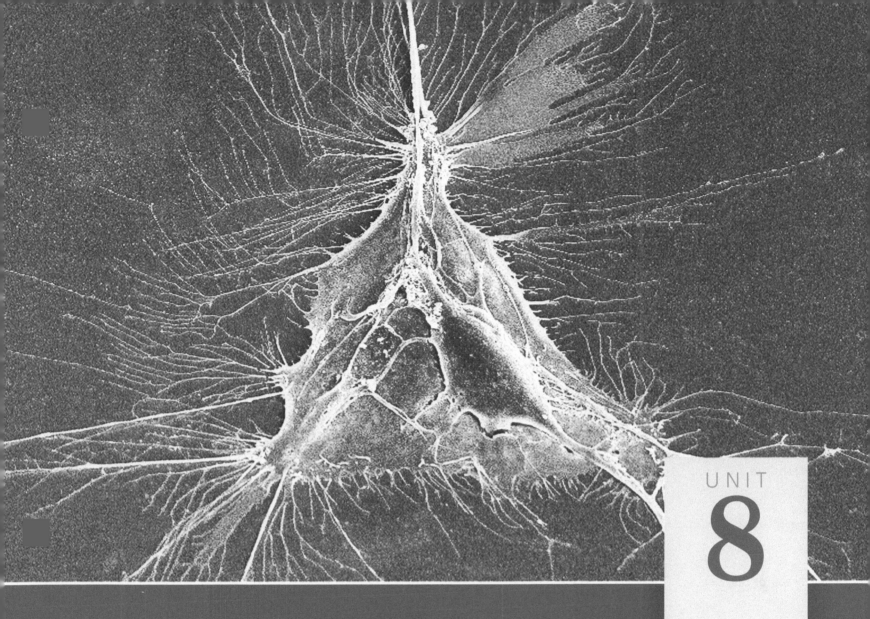

Nervous System

When you have completed this unit, you should be able to:

1 Describe the microanatomy of nervous tissue.

2 Describe and identify the gross structures of the brain.

3 Describe and identify structures of the spinal cord.

4 Identify, describe, and demonstrate functions of cranial nerves.

5 Identify spinal nerves and plexuses.

6 Describe a simple spinal reflex arc.

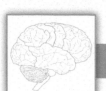

PRE-LAB EXERCISES

Complete the following exercises prior to coming to lab, using your lab manual and textbook for reference.

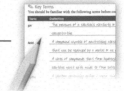

Pre-Lab Exercise **8-1**

 Key Terms

You should be familiar with the following terms before coming to lab.

Term	Definition
Parts of the Neuron	
Neuron	
Cell body	
Axon	
Dendrite	
Myelin sheath	
Node of Ranvier	
Structures of the Brain	
Cerebral hemispheres	
Cerebral cortex	
Corpus callosum	
Diencephalon	
Thalamus	
Hypothalamus	

8

Midbrain _____

Pons _____

Medulla oblongata _____

Cerebellum _____

Meninges _____

Ventricles _____

Structures of the Spinal Cord

Gray matter horns (anterior, posterior, lateral) _____

_____

Cauda equina _____

Nerves: General Terms

Cranial nerve _____

Spinal nerve _____

Nerve plexus _____

Spinal Nerve Plexuses

Cervical plexus _____

Brachial plexus _____

Lumbar plexus _____

Sacral plexus _____

8

Neuron Structure

Color the neuron in Figure 8.1, and label it with the following terms from Exercise 8-1 (p. 206). Use Exercise 8-1 in this unit and your text for reference.

❑ Axon

❑ Axon hillock

❑ Telodendria

❑ Axon terminal

❑ Dendrite(s)

❑ Cell body (soma)

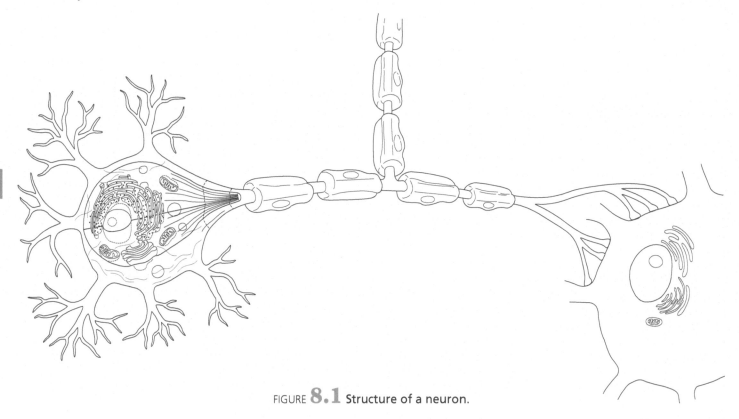

FIGURE **8.1** Structure of a neuron.

Pre-Lab Exercise **8-3**

Brain Anatomy

Color the diagrams of the brain in Figures 8.2 and 8.3, and label them with the following terms from Exercise 8-2 (pp. 211–212). Use Exercise 8-2 in this unit and your text for reference.

Cerebrum

- ❑ Corpus callosum (cerebral white matter)
- ❑ Lobes of the cerebrum
  - ❑ Frontal lobe
  - ❑ Parietal lobe
  - ❑ Occipital lobe
  - ❑ Temporal lobe
- ❑ Sulci
  - ❑ Central sulcus
  - ❑ Lateral sulcus

Diencephalon

- ❑ Thalamus
- ❑ Hypothalamus
  - ❑ Infundibulum
  - ❑ Pituitary gland
- ❑ Pineal gland

Brainstem

- ❑ Midbrain
- ❑ Pons
- ❑ Medulla oblongata

Cerebellum

Ventricles

- ❑ Fourth ventricle
- ❑ Cerebral aqueduct

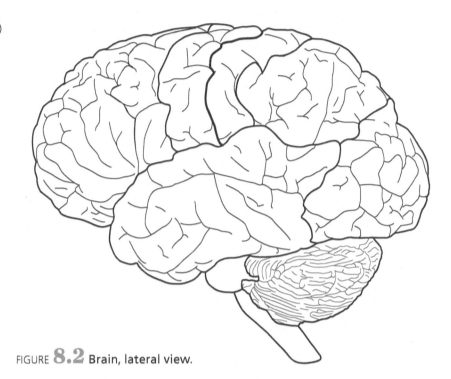

FIGURE **8.2** Brain, lateral view.

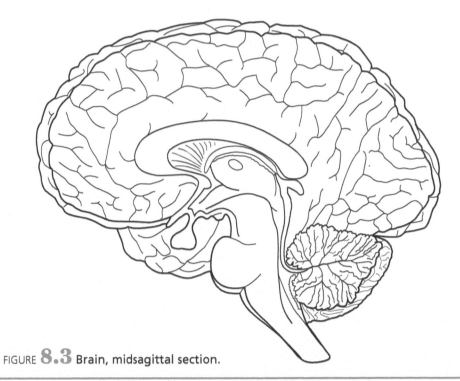

FIGURE **8.3** Brain, midsagittal section.

Pre-Lab Exercise **8-4**

Spinal Cord Anatomy

Color the diagram of the spinal cord in Figure 8.4, and label it with the following terms from Exercise 8-3 (p. 220). Use Exercise 8-3 in this unit and your text for reference.

- ❏ Meninges
  - ☐ Dura mater
  - ☐ Arachnoid mater
  - ☐ Pia mater
- ❏ Spinal gray matter
  - ☐ Anterior horn
  - ☐ Lateral horn
  - ☐ Posterior horn
- ❏ Spinal white matter
- ❏ Central canal

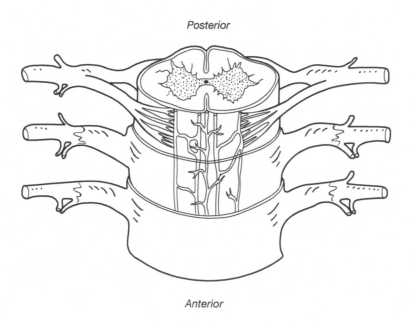

FIGURE **8.4** Spinal cord, transverse section.

Pre-Lab Exercise **8-5**

Cranial Nerve Functions

Complete Table 8.1 by listing the functions of each pair of cranial nerves and indicating whether the nerve is motor, sensory, or mixed.

TABLE **8.1** Cranial Nerve Functions

Cranial Nerve	Functions	Motor, Sensory, or Mixed
CN I: Olfactory nerve		
CN II: Optic nerve		
CN III: Oculomotor nerve		
CN IV: Trochlear nerve		
CN V: Trigeminal nerve		
CN VI: Abducens nerve		
CN VII: Facial nerve		
CN VIII: Vestibulocochlear nerve		
CN IX: Glossopharyngeal nerve		
CN X: Vagus nerve		
CN XI: Accessory nerve		
CN XII: Hypoglossal nerve		

8

Spinal Nerve Plexuses and the Anterior Rami of the Spinal Nerves

Label Figure 8.5, which shows the spinal nerve plexuses and the anterior rami of the spinal nerves, with the following terms from Exercise 8-5 (pp. 226 and 228). Use Exercise 8-5 in this unit and your text for reference. Note that this diagram is presented in color to facilitate identification of the nerves.

❑ Cervical plexus

❑ Brachial plexus
- ☐ Axillary nerve
- ☐ Radial nerve
- ☐ Musculocutaneous nerve
- ☐ Ulnar nerve
- ☐ Median nerve

❑ Thoracic (intercostal) nerves

❑ Lumbar plexus
- ☐ Femoral nerve

❑ Sacral plexus
- ☐ Sciatic nerve

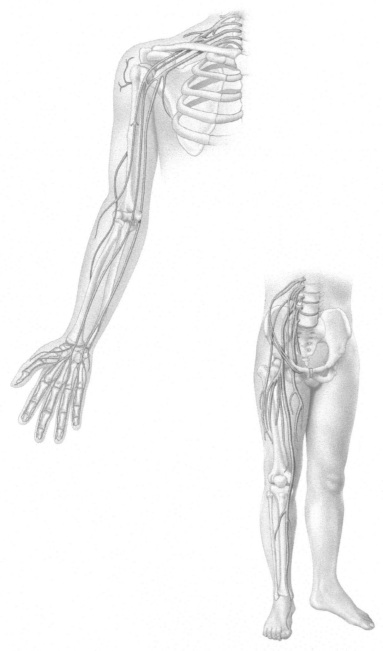

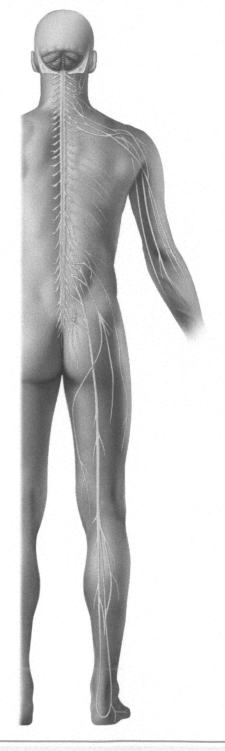

FIGURE **8.5** Nerve plexuses and anterior rami of the spinal nerves.

8

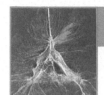

SEM of oligodendrocyte.

We now begin our study of the **nervous system**, which is one of the two systems in the body charged with maintaining homeostasis of the body as a whole (the other is the *endocrine system*). This group of organs regulates the activities of other cells by sending nerve impulses, also called **action potentials**, via cells called *neurons*. Neurons are supported by a variety of smaller cells, collectively called *neuroglial cells*, and together these two cell types make up **nervous tissue**.

Nervous tissue makes up the bulk of the organs of the two divisions of the nervous system: (1) the **central nervous system (CNS)**, which consists of the **brain** and the **spinal cord**, and (2) the **peripheral nervous system (PNS)**, which consists of organs called **peripheral nerves**, or simply *nerves* (Fig. 8.6). There are two types of nerves classified by location: (1) nerves that originate from the brain and the brainstem, called **cranial nerves**, and (2) nerves that originate from the spinal cord, called **spinal nerves**.

The following exercises will introduce you first to nervous tissue, after which you will examine the anatomy of the brain, spinal cord, and peripheral nerves. You will also look at some of the functions of the nervous system, including those of the cranial nerves and spinal reflexes.

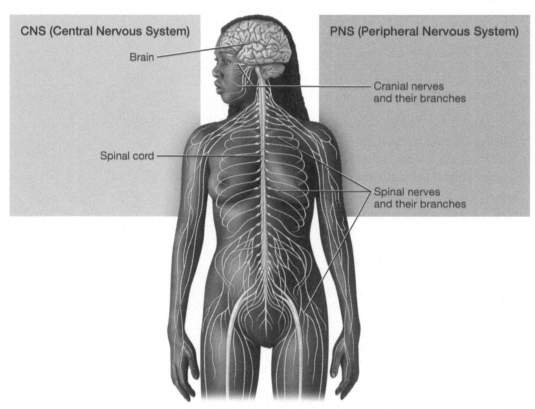

FIGURE **8.6** Anatomical organization of the nervous system.

Exercise 8-1

Nervous Tissue

MATERIALS

☐ Neuron models
☐ Modeling clay in four colors

There are two types of cells within nervous tissue: neurons and neuroglial cells (Fig. 8.7). **Neurons** (NOOR-ahnz) are large cells that transmit and generate signals in the form of nerve impulses, or **neuronal action potentials**. Although they vary widely in size and structure, most have the following three components: the cell body, one axon, and one or more dendrites (Fig. 8.8).

1. **Cell body.** The **cell body** contains the nucleus and many of the neuron's organelles, and so is where synthesis of proteins and other major cellular compounds takes place. Its cytoskeleton consists of densely packed **neurofilaments** that compartmentalize the rough endoplasmic reticulum into dark-staining structures called **Nissl bodies.** Near the axon and dendrites, neurofilaments form larger **neurofibrils,** which extend out into these processes. Cell bodies are often found in clusters.

2. **Axon.** A single **axon** exits the cell body at the **axon hillock,** at which point the neuron's plasma membrane is known as the **axolemma** (aks-oh-LEM-uh). Axons are defined by their ability to generate action potentials across their axolemmae. The action potential spreads, or *propagates,* down the entire length of the axon. Some axons have branches known as **axon collaterals** that stem off the main axon at right angles. Both the main axon and the axon collaterals end in multiple terminal branches called **telodendria** (tee-loh-DEN-dree-uh). At the end of each telodendrion is an **axon terminal,** which is sometimes called a synaptic bulb. The axon terminal contains synaptic vesicles with neurotransmitters that communicate with the axon's target cell.

3. **Dendrites.** Most neurons have one or more branching processes called **dendrites** (DEN-dryt'z) that receive signals from other neurons. They can transmit these signals to the neuron's cell body, but they are not capable of generating action potentials.

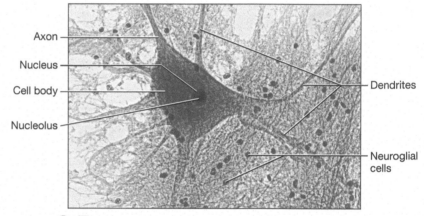

FIGURE **8.7** Nervous tissue, photomicrograph (motor neuron smear).

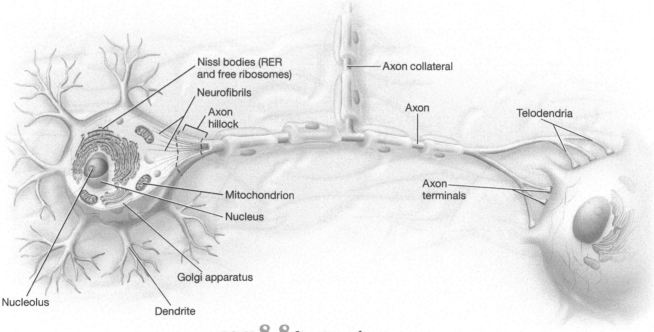

FIGURE **8.8** Structure of a neuron.

Neuroglial cells (noor-oh-GLEE-uhl) are much smaller than neurons, and they outnumber neurons about 50 to 1—no small feat considering that the nervous system contains about a trillion neurons. Following are the neuroglial cells of the central nervous system (CNS; brain and spinal cord) and the peripheral nervous system (PNS; cranial and spinal nerves). The neuroglia are shown in Figure 8.9, but note that this figure shows only neuroglial cells of the CNS.

1. **Astrocytes. Astrocytes** are the most numerous neuroglial cell type in the CNS. These star-shaped cells have many functions, including anchoring neurons and blood vessels in place with processes called **perivascular feet,** and regulating the extracellular environment of the brain. In addition, they facilitate the formation of the **blood brain barrier,** created by tight junctions in the brain capillaries. The blood brain barrier prevents many substances in the blood from entering the brain tissue.

2. **Oligodendrocytes. Oligodendrocytes** (oh-lig-oh-DEN-droh-syt'z) have long extensions that wrap around the axons of certain neurons in the CNS to form a structure called the *myelin sheath.* Note that one oligodendrocyte can myelinate segments of several axons.

3. **Microglial cells.** The small **microglial cells** (my-kroh-GLEE-uhl) are very active phagocytes that clean up debris surrounding neurons. Microglia also degrade and ingest damaged or dead neurons.

4. **Ependymal cells.** The ciliated **ependymal cells** (eh-PEN-dih-muhl) line the hollow spaces of the brain and spinal cord. They assist in forming **cerebrospinal fluid,** the liquid that bathes the brain and spinal cord, and circulate it with their cilia.

5. **Schwann cells. Schwann cells,** also known as *neurolemmocytes,* form the myelin sheath around the axons of certain neurons in the PNS. A Schwann cell can myelinate a segment of only one axon.

6. **Satellite cells. Satellite cells** surround the cell bodies of neurons in the PNS. These cells enclose and support the cell bodies and regulate the extracellular environment around them.

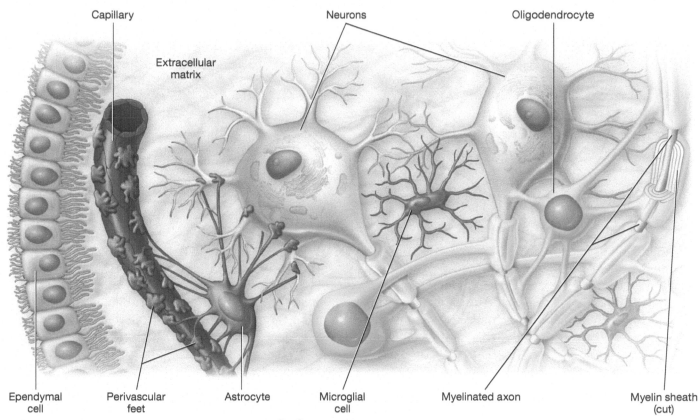

FIGURE **8.9** Neuroglial cells of the CNS.

The **myelin sheath** (MY-lin) is composed of many layers of the plasma membrane of the oligodendrocyte or Schwann cell wrapped around an axon. It performs critical functions within the nervous system, including protecting and insulating the axons and speeding up conduction of action potentials. As you can see in Figure 8.10, Schwann cells and oligodendrocytes myelinate axons differently. Schwann cells, which myelinate a segment of only a single axon, wrap clockwise around the axon (Fig. 8.10A). The outer edge of the Schwann cell, called the **neurilemma** (noor-ih-LEM-uh), contains most of its cytoplasm and the nucleus. Oligodendrocytes, on the other hand, wrap segments of multiple axons in a counterclockwise direction (Fig. 8.10B).

Notice that there are small gaps in the myelin sheath between each oligodendrocyte arm or Schwann cell where the plasma membrane of the axon is exposed. These gaps are called **myelin sheath gaps**, also known as **nodes of Ranvier** (rahn-vee-ay), and the myelin-covered segments between the nodes are **internodes**.

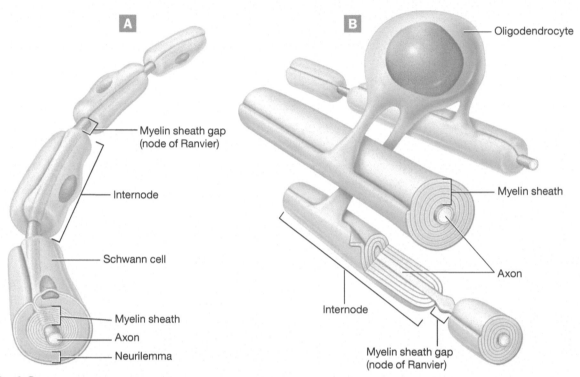

FIGURE **8.10** Myelin sheath: (**A**) Schwann cells around a PNS axon; (**B**) an oligodendrocyte around multiple CNS axons.

Procedure **1** Model Inventory for Nervous Tissue

Identify the following structures of nervous tissue on models and diagrams using this unit and your textbook for reference. As you examine the anatomical models and diagrams, record the name of the model and the structures you were able to identify on the model inventory in Table 8.2.

1. Neuron
   a. Cell body
      (1) Neurofilaments
      (2) Nissl bodies
      (3) Neurofibrils
   b. Axon
      (1) Axon hillock
      (2) Telodendria
      (3) Axon terminal
   c. Dendrite(s)

2. Neuroglial cells
   a. Oligodendrocytes
   b. Schwann cells
3. Other structures
   a. Myelin sheath
   b. Myelin sheath gap (node of Ranvier)
   c. Internode

TABLE **8.2** Model Inventory for Nervous Tissue

Model/Diagram	Structures Identified

Procedure **2** Building a Myelin Sheath

In this procedure, you will demonstrate the difference between the methods by which Schwann cells and oligodendrocytes myelinate the axons of neurons in the peripheral nervous system and the central nervous system, respectively.

1 Obtain four colors of modeling clay (blue, green, yellow, and red, if available).

2 Use the following color code to build three central nervous system (CNS) axons, one peripheral nervous system (PNS) axon, one oligodendrocyte, and two Schwann cells:

a CNS axons: blue

b PNS axon: green

c Oligodendrocyte: yellow

d Schwann cells: red

3 Build your oligodendrocyte so it reaches out to myelinate the three CNS axons.

4 Build your Schwann cells so they myelinate the single PNS axon, making sure to leave space for the myelin sheath gaps. Be sure to pay attention to the direction in which oligodendrocytes and Schwann cells myelinate their axons (clockwise or counterclockwise).

Exercise 8-2

Anatomy of the Brain

MATERIALS

- ❏ Brain models: whole and sectioned
- ❏ Ventricle models
- ❏ Brainstem models
- ❏ Dural sinus model
- ❏ Sheep brain
- ❏ Dissection equipment and trays
- ❏ Colored pencils

The anatomy of the brain may seem complex, but really the brain is simply a hollow, highly folded tube. Its many functions include maintaining the homeostasis of many physiological variables, interpreting sensation, planning and executing movement, and controlling our higher mental functions. Grossly, the brain is a whitish-gray mass of nervous tissue, connective tissue, and modified epithelium that we can divide into four regions: the *cerebral hemispheres* (collectively called the *cerebrum*), the *diencephalon*, the *cerebellum*, and the *brainstem*.

Internally, the brain contains hollow spaces called **ventricles** that are filled with a fluid similar to plasma called **cerebrospinal fluid** (seh-ree-broh-SPY-nuhl; **CSF**). As you can see in Figure 8.11, the largest of the ventricles, called the **lateral ventricles**, are located in the right and left cerebral hemispheres. Note in Figure 8.11A that the lateral ventricles resemble rams' horns when viewed from the anterior side.

The two lateral ventricles are connected by a small opening that leads to the smaller **third ventricle**, which is housed within the diencephalon. The third ventricle is continuous with the **fourth ventricle**, found in the brainstem, via a small canal called the **cerebral aqueduct**. The fourth ventricle is continuous with a canal that runs down the central spinal cord called the **central canal**.

Within each of the four ventricles, we find collections of blood vessels known as **choroid plexuses** (KOHR-oyd; look ahead to Fig. 8.17, p. 214). As blood flows through the choroid plexuses, fluid filters out into the ventricles and, at that point, is called CSF. The largest choroid plexuses are within the lateral ventricles. One of the main functions of CSF is to reduce brain weight, as the brain is buoyant in the CSF. Without CSF, your brain literally could crush itself under its own weight.

The **cerebrum** (seh-REE-brum) is the most superior portion of the brain and is responsible for the brain's cognitive functions, including learning and language, conscious interpretation of sensory information, conscious planning of movement, and personality (Fig. 8.12). The two cerebral hemispheres are separated from one another by a deep groove called the **longitudinal fissure**, and they are separated from the cerebellum by the **transverse fissure**. The cerebral surface features elevated ridges called **gyri** (JY-ree) and shallow grooves called **sulci** (SUL-kee). The cerebrum consists of five lobes: the **frontal, parietal, temporal, occipital,** and deep **insula** lobes (remember this last one by the mnemonic "the *insula* is *insula*ted"; note that the insula is not visible in the figure). Notice that the **lateral sulcus** separates the frontal and temporal lobes, and the **central sulcus** separates the frontal and parietal lobes. On either side of the central sulcus are two prominent gyri: the anterior **precentral gyrus**, which houses the primary motor cortex, and the posterior **postcentral gyrus**, which houses the primary somatosensory cortex.

The outer 2 millimeters of the cerebrum is known as the **cerebral cortex** (Fig. 8.13). Here we find unmyelinated parts of the cerebral neurons, including cell bodies, dendrites, and unmyelinated axons. As you can see in the figure, the lack of myelin gives the cerebral cortex a gray-brown color, and, for this reason, it is called **gray matter**. The cell bodies and

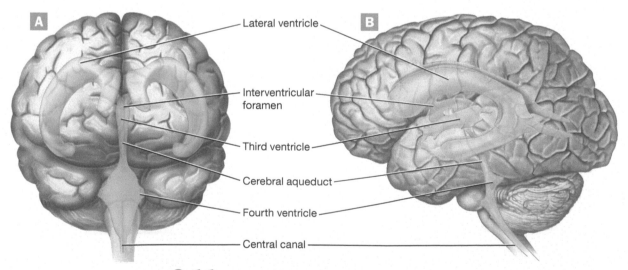

FIGURE **8.11** Ventricles: (**A**) anterior view; (**B**) left lateral view.

processes of the cerebral cortex communicate with other parts of the nervous system by bundles of myelinated axons called **white matter**. The largest tract of cerebral white matter is called the **corpus callosum** (KOHR-pus kal-OH-sum), which connects the right and left cerebral hemispheres.

Gray matter isn't confined to the cerebral cortex. Clusters of cell bodies called **nuclei** are found throughout the white matter of the cerebrum. An important group of nuclei, the **basal nuclei** (the *caudate nucleus*, *putamen*, and *globus pallidus*), monitors voluntary motor functions. The neurons of these nuclei are connected to other parts of the nervous system by various tracts of cerebral white matter.

Deep to the cerebral hemispheres in the central core of the brain, we find the **diencephalon** (dy-en-SEF-ah-lahn), which is composed of three main parts (Fig. 8.14):

1. **Thalamus.** The **thalamus** (THAL-uh-muss), which makes up 80 percent of the diencephalon, is the large, central, egg-shaped mass of gray and white matter. Functionally, the thalamus is a major integration and relay center that edits and sorts information going into the cerebrum. It essentially functions as the "gateway" into the cerebrum.

2. **Hypothalamus.** The **hypothalamus** (hy-poh-THAL-uh-muss) is the anterior and inferior component of the diencephalon. It is a deceptively small structure that contains the nuclei whose neurons help to regulate the endocrine system; monitor the sleep-wake cycle; control thirst, hunger, and body temperature; and monitor the autonomic nervous system. An endocrine organ called the *pituitary gland* (pih-TOO-ih-tehr-ee) is connected to the hypothalamus by a stalk called the *infundibulum* (in-fun-DIB-yoo-lum).

3. **Epithalamus.** The **epithalamus** (ep-ih-THAL-ih-mus) is the posterior and superior part of the diencephalon. It contains an endocrine organ called the **pineal gland** (pin-EE-uhl) that secretes the hormone **melatonin** (mel-uh-TOH-nin), which helps to regulate the sleep-wake cycle.

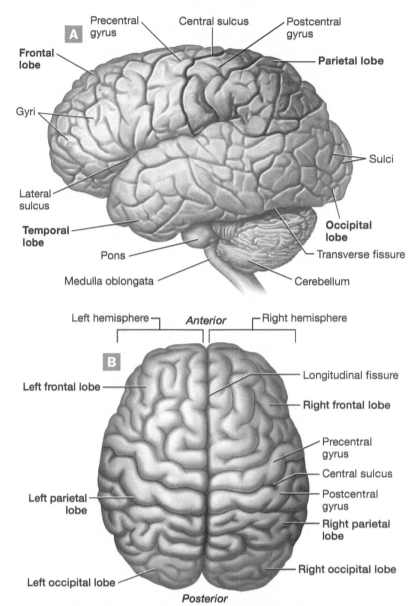

FIGURE **8.12** Brain: (**A**) lateral view; (**B**) superior view.

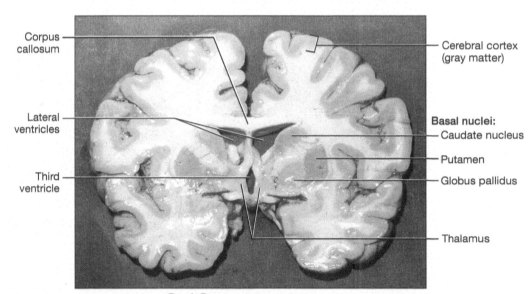

FIGURE **8.13** Frontal section through the brain.

The third major component of the brain is the large posterior **cerebellum** (sehr-eh-BELL-um). It consists of two highly folded lobes with an outer **cerebellar cortex**, composed of gray matter, surrounding an inner region of white matter. The cerebellum coordinates and plans ongoing motor activities and is critical in reducing and preventing motor error with movement.

The final division of the brain, the **brainstem**, influences the automatic functions of the body (Fig. 8.15). The most superior portion of the brainstem is the **midbrain**, which has roles in movement, sensation, and certain reflexes. Inferior to the midbrain we find the rounded **pons**, which bulges anteriorly. Nuclei of the pons are involved in controlling the rhythm for breathing and the sleep cycle. The last segment of the brainstem, the **medulla oblongata** (or simply *medulla*), is continuous inferiorly with the spinal cord. The nuclei in the medulla work with those of the pons to control ventilation and are also involved in reflexes that influence variables such as the heart rate and blood pressure.

Note in Figure 8.16 that a set of three membranes, collectively called the **meninges** (meh-NIN-jeez; singular: *meninx*), surrounds the brain. The meninges include the following:

1. **Dura mater**. The outermost meninx is the thick, leathery, double-layered **dura mater** (DOO-rah MAH-tur). The superficial *periosteal layer* is fused to the skull, and the deeper *meningeal layer* is continuous with the dura mater of the spinal cord. The two layers of the dura are fused, but in three regions the meningeal layer separates from the periosteal layer to form structures that separate regions of the brain. At these locations there are spaces between the two dural layers collectively called the **dural sinuses** (see Fig. 11.21, p. 304). All deoxygenated blood from the brain drains into the dural sinuses. From the dural sinuses, blood drains into veins exiting the head and neck. You can see the large superior sagittal sinus, which runs along the longitudinal fissure, in Figures 8.16 and 8.17.

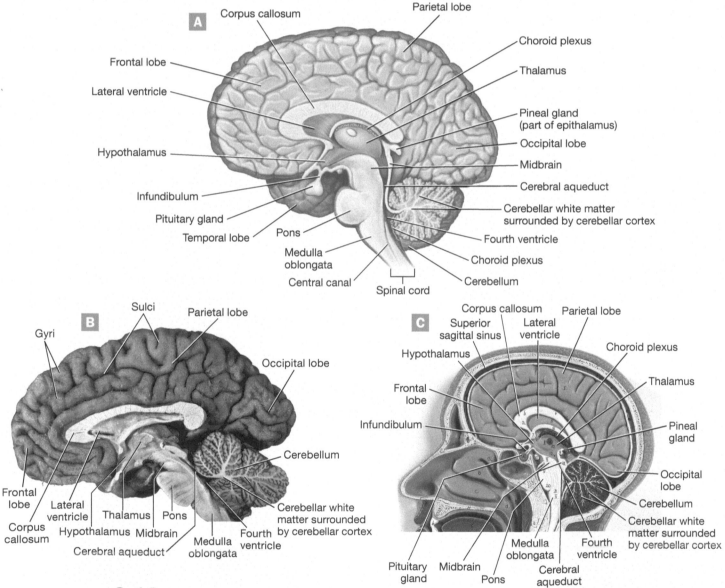

FIGURE **8.14** Midsagittal section of the brain: (**A**) illustration; (**B**) photograph; (**C**) anatomical model.

2. **Arachnoid mater.** The middle meninx, the **arachnoid mater** (ah-RAK-noyd), is separated from the dura by a potential space called the **subdural space.** Small bundles of the arachnoid mater, called the **arachnoid granulations** (or **arachnoid villi**), project into the dural sinuses and allow CSF to reenter the blood. You can see the pattern of CSF circulation from its formation by the choroid plexuses to its return to the blood in Figure 8.17.

3. **Pia mater.** The thinnest, innermost meninx is the **pia mater** (PEE-ah), which follows the contours of the sulci and gyri and is richly supplied with blood vessels. There is a space between the pia mater and the arachnoid mater, called the **subarachnoid space,** which is filled with CSF.

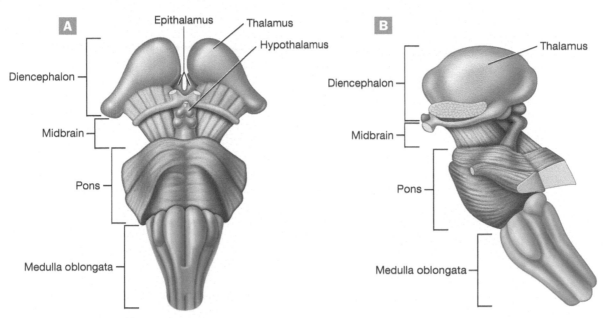

FIGURE **8.15** Diencephalon and brainstem: (**A**) anterior view; (**B**) lateral view.

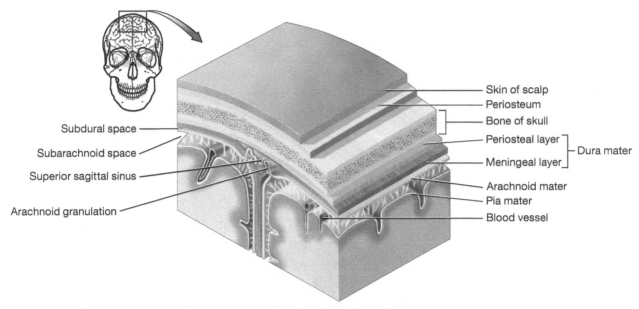

FIGURE **8.16** Brain and meninges, frontal and parasagittal section.

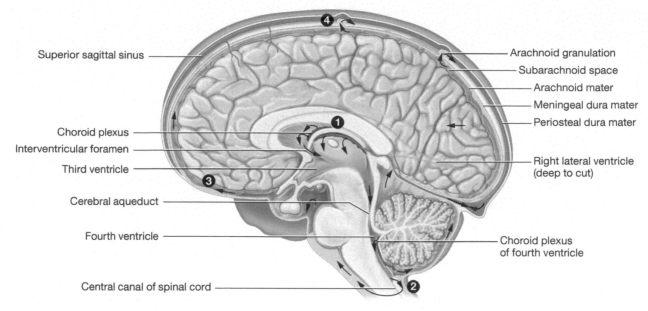

Superior sagittal sinus

Choroid plexus

Interventricular foramen

Third ventricle

Cerebral aqueduct

Fourth ventricle

Central canal of spinal cord

Arachnoid granulation

Subarachnoid space

Arachnoid mater

Meningeal dura mater

Periosteal dura mater

Right lateral ventricle
(deep to cut)

Choroid plexus
of fourth ventricle

1 CSF is produced by the choroid plexuses of each ventricle.

2 CSF flows through the ventricles and into the subarachnoid space. Some CSF flows through the central canal of the spinal cord.

3 CSF flows through the subarachnoid space.

4 CSF is absorbed into the dural venous sinuses via the arachnoid granulations.

FIGURE **8.17** Circulation of CSF through the brain and spinal cord.

Procedure **1** Model Inventory for the Brain

Identify the following structures of the brain on models and diagrams, using this unit and your textbook for reference. As you examine the anatomical models and diagrams, record the name of the model and the structures you were able to identify on the model inventory in Table 8.3. The brain's structure is fairly complex, and it's best to examine models in as many different planes of section as possible.

1. Ventricles
   a. Lateral ventricles
   b. Third ventricle
   c. Fourth ventricle
   d. Cerebral aqueduct
   e. Choroid plexuses
2. Cerebrum
   a. Cerebral hemispheres
   b. Fissures
      (1) Longitudinal fissure
      (2) Transverse fissure
   c. Lobes of the cerebrum
      (1) Frontal lobe
      (2) Parietal lobe
      (3) Temporal lobe
      (4) Occipital lobe
      (5) Insula lobe
   d. Sulci
      (1) Lateral sulcus
      (2) Central sulcus
         (a) Precentral gyrus
         (b) Postcentral gyrus
   e. Cerebral cortex (gray matter)
   f. Corpus callosum (cerebral white matter)
3. Diencephalon
   a. Thalamus
   b. Hypothalamus
   c. Epithalamus (contains pineal gland)
4. Cerebellum
5. Brainstem
   a. Midbrain
   b. Pons
   c. Medulla oblongata
6. Brain coverings
   a. Dura mater
      (1) Dural sinuses
   b. Arachnoid mater
      (1) Subdural space
      (2) Arachnoid granulations
   c. Pia mater
      (1) Subarachnoid space

8

TABLE **8.3** Model Inventory for the Brain

Model/Diagram	Structures Identified

Procedure **2** Brain Dissection

Often, structures of the brain and spinal cord are difficult to see on anatomical models. This procedure allows you to examine these structures more closely by dissecting a preserved sheep brain. Note that the process of preservation makes many structures of the brain and spinal cord much tougher than they would be in a fresh specimen.

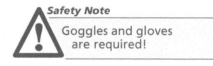

Safety Note

Goggles and gloves are required!

1 First note the thick part of the dura mater covering the longitudinal fissure. If you cut through this with scissors, you will enter a dural sinus called the superior sagittal sinus.

2 Next remove the dura mater to reveal the thin membrane on top of the brain. This is the arachnoid mater.

3 Remove an area of the arachnoid mater to see the shiny inner membrane—the pia mater—directly touching the surface of the brain. Note how the pia mater follows the convolutions of the gyri and sulci.

4 Examine the surface anatomy of both the superior and the inferior surfaces of the sheep brain (Figs. 8.18 and 8.19).

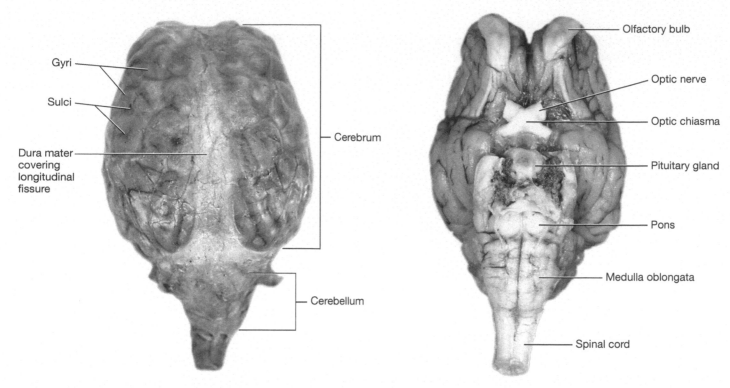

Gyri

Sulci

Dura mater covering longitudinal fissure

Cerebrum

Cerebellum

Olfactory bulb

Optic nerve

Optic chiasma

Pituitary gland

Pons

Medulla oblongata

Spinal cord

FIGURE **8.18** Superior view of the sheep brain.

FIGURE **8.19** Inferior view of the sheep brain.

Draw what you see on both surfaces in the space provided, and label your drawing with the following structures:

❏ Arachnoid mater

❏ Cerebellum

❏ Cerebrum

❏ Dura mater

❏ Gyri

❏ Longitudinal fissure

❏ Medulla oblongata

❏ Pituitary gland

❏ Pons

❏ Sulci

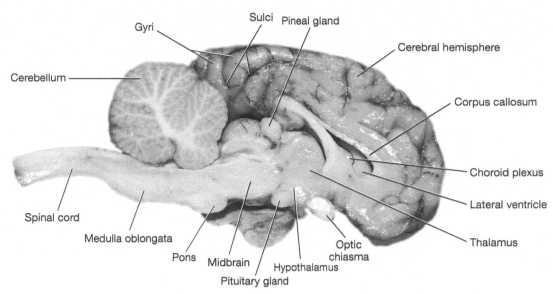

Gyri Sulci Pineal gland

Cerebellum

Cerebral hemisphere

Corpus callosum

Choroid plexus

Lateral ventricle

Thalamus

Spinal cord

Medulla oblongata

Pons Midbrain

Hypothalamus

Optic chiasma

Pituitary gland

FIGURE **8.20** Lateral view of the sheep brain, midsagittal section.

5 Separate the two cerebral hemispheres with your fingers, and identify the corpus callosum.

6 Make a cut along the brain's midsagittal plane through the corpus callosum to separate the two cerebral hemispheres.

7 Examine the brain's internal anatomy (Fig. 8.20), and stick your finger in the lateral ventricle. You will see (or feel) that it is much larger than it appears. Draw what you see in the space provided, and label your drawing with the following structures:

❑ Corpus callosum   ❑ Medulla oblongata   ❑ Thalamus
❑ Hypothalamus   ❑ Midbrain
❑ Lateral ventricle   ❑ Pons

8

8 Section one of the halves of the brain in the frontal plane, approximately along the central sulcus. Notice the outer cerebral cortex (the gray matter) and the inner white matter. From this view, you also can see the lateral and third ventricles. Again, draw what you see in the space provided, and label your drawing with the following structures:

- ❏ Cerebral cortex
- ❏ Cerebral white matter
- ❏ Lateral ventricle
- ❏ Thalamus
- ❏ Third ventricle

Exercise 8-3

Spinal Cord

MATERIALS
- ❏ Spinal cord models: whole and sectioned

The medulla oblongata passes through the foramen magnum of the occipital bone and becomes the **spinal cord** (Fig. 8.21). Along its length are two notable bulges, the **cervical** and **lumbar enlargements**. The spinal cord is wider in these areas due to the high number of nerve roots that attach at these locations going to and coming from the upper and lower limbs. Notice in Figure 8.21 that the spinal cord does not extend the entire length of the vertebral column; rather, it ends between the first and second lumbar vertebrae. At this point, it tapers to form the end of the spinal cord, called the **conus medullaris** (KOHN-us med-yoo-LEHR-us), which gives off a tuft of nerve roots called the **cauda equina** (KOW-dah eh-KWY-nah; "horse's tail"). The cauda equina fills the remainder of the vertebral column to the sacrum and exits out of the appropriate foramina to become spinal nerves.

Like the brain, the spinal cord is protected by a set of **spinal meninges**, which are continuous with the cranial meninges (Figs. 8.22 and 8.23). The cranial and spinal meninges are similar in name and structure: the dura mater is the outer meninx; the arachnoid mater is the middle meninx, and is separated from the dura by a thin potential space called the subdural space; and the pia mater is the inner meninx, and is separated from the arachnoid mater by the CSF-filled subarachnoid space. A key difference between the cranial and spinal meninges is that the spinal dura mater consists of only *one* layer rather than two like the cranial dura. This single dural layer does not attach to the vertebral column, which creates a fat-filled space between the spinal dura and the interior vertebral foramen called the **epidural space** (ep-ih-DOO-ruhl). There is no epidural space around the brain because the periosteal layer of the cranial dura is fused to the skull.

Internally, the spinal cord consists of a butterfly-shaped core of gray matter that surrounds the CSF-filled **central canal**. Notice that this arrangement is different from what we saw in the brain, where the gray matter was on the outside and the white matter on the inside. The gray matter is divided into regions, or **horns**: the **anterior horns** contain the cell bodies of motor neurons; the **posterior horns** contain the cell bodies of sensory neurons; and the **lateral horns**, found in the thoracic and lumbar regions of the spinal cord, contain the cell bodies of autonomic neurons. As you can see in Figure 8.23, it is possible to tell the anterior and posterior spinal cord apart by looking at the shapes of the anterior and posterior horns. The anterior horns are broad and flat on the ends, whereas the posterior horns are more tapered, and they extend farther out toward the edge.

Surrounding the spinal gray matter is the spinal white matter, which contains myelinated axons grouped into bundles called **tracts**. Tracts contain axons that have the same beginning and end points and the same general function. *Ascending tracts* carry sensory information from sensory neurons to the brain, and *descending tracts* carry motor information from the brain to motor neurons.

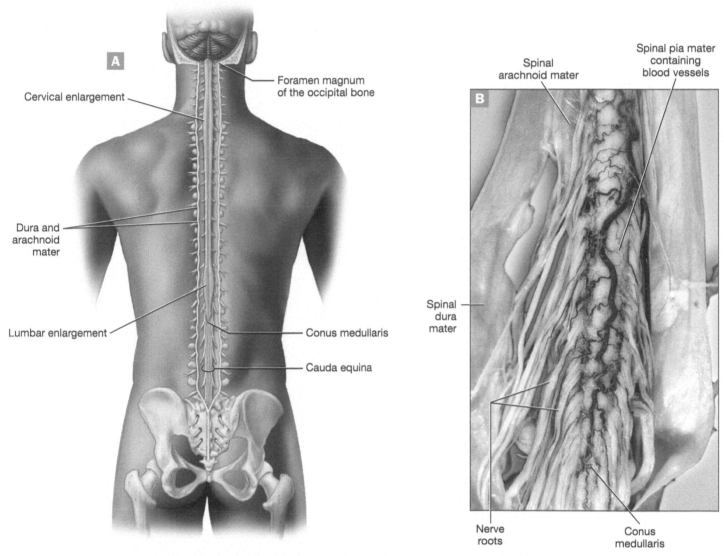

FIGURE **8.21** Spinal cord, posterior view: (**A**) illustration; (**B**) photograph.

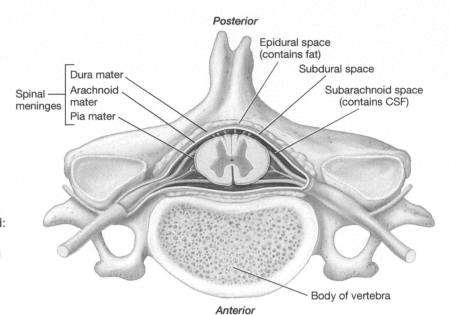

FIGURE **8.22** Spinal cord: transverse section in the vertebral column showing spinal meninges and associated spaces.

Posterior

Spinal meninges
Dura mater
Arachnoid mater
Pia mater

Epidural space (contains fat)
Subdural space
Subarachnoid space (contains CSF)

Body of vertebra

Anterior

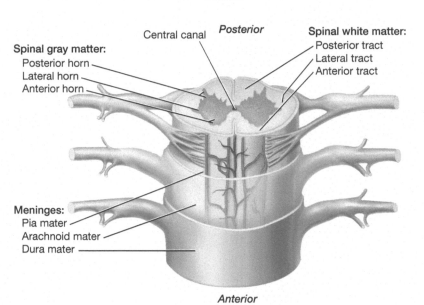

FIGURE **8.23** Internal anatomy of the spinal cord.

Central canal
Posterior

Spinal gray matter:
Posterior horn
Lateral horn
Anterior horn

Spinal white matter:
Posterior tract
Lateral tract
Anterior tract

Meninges:
Pia mater
Arachnoid mater
Dura mater

Anterior

Procedure **1** Model Inventory for the Spinal Cord

Identify the following structures of the spinal cord on models and diagrams, using this unit and your textbook for reference. As you examine the anatomical models and diagrams, record the name of the model and the structures you were able to identify on the model inventory in Table 8.4.

1. Cervical enlargement
2. Lumbar enlargement
3. Conus medullaris
4. Cauda equina
5. Meninges
   a. Dura mater
      (1) Epidural space
   b. Arachnoid mater
      (1) Subdural space
   c. Pia mater
      (1) Subarachnoid space

6. Central canal
7. Spinal gray matter
   a. Anterior horn
   b. Posterior horn
   c. Lateral horn
8. Spinal white matter

TABLE **8.4** Model Inventory for the Spinal Cord

Model/Diagram	Structures Identified

8

Exercise 8-4

Cranial Nerves

MATERIALS

❏ Penlight
❏ Snellen vision chart
❏ Tuning fork
❏ Unknown samples to smell
❏ PTC, thiourea, and sodium benzoate tasting papers

The 12 pairs of **cranial nerves** originate from or bring information to the brain (Fig. 8.24). Each nerve is given two names: (1) a sequential Roman numeral in order of its attachment to the brain, and (2) a name that describes the nerve's location or function. For example, cranial nerve III is the third cranial nerve that attaches to the brain. It is also called the oculomotor nerve because one of its functions is to provide motor axons to some of the muscles that move the eyeball.

Most cranial nerves innervate structures of the head and neck. Three cranial nerves are **sensory nerves** whose axons have purely sensory functions, four are **mixed nerves** that contain both sensory and motor axons, and five are **motor nerves** that contain primarily motor axons. Following is an overview of the main functions of each cranial nerve (note that CN = "cranial nerve"):

1. **CN I: Olfactory Nerve.** The **olfactory nerve** is a purely sensory nerve that innervates the olfactory mucosa in the superior nasal cavity, where it provides for the sense of smell.

2. **CN II: Optic Nerve.** The **optic nerve** is also a purely sensory nerve that provides for the sense of vision. Its axons emerge from the retina of the eye and meet at the **optic chiasma** (ky-AZ-mah), where the nerves partially exchange axons before diverging to form the **optic tracts.**

3. **CN III: Oculomotor Nerve.** The **oculomotor nerve** (awk-yoo-loh-MOH-tohr) is a motor cranial nerve. It innervates four of the six extrinsic eye muscles that move the eyeball, the muscle that opens the eyelid, the muscle that constricts the pupil, and the muscle that changes the shape of the lens for near vision, an adjustment called **accommodation.**

4. **CN IV: Trochlear Nerve.** The **trochlear nerve** (TROH-klee-ur) is a small motor nerve that innervates one of the six extrinsic eye muscles that moves the eyeball (the *superior oblique muscle*).

5. **CN V: Trigeminal Nerve.** The **trigeminal nerve** is a large mixed nerve named for the three branches that provide sensory innervation from the face and motor innervation to the muscles of mastication (chewing).

6. **CN VI: Abducens Nerve.** The **abducens nerve** (ab-DOO-senz) is a small motor nerve that innervates the final extrinsic eye muscle (the *lateral rectus muscle*).

7. **CN VII: Facial Nerve.** The mixed fibers of the **facial nerve** innervate many structures and provide for the following: motor to the muscles of facial expression; taste from the anterior two-thirds of the tongue; motor to the glands that produce tears (the lacrimal glands), mucus, and saliva; and sensory from part of the face and mouth.

8. **CN VIII: Vestibulocochlear Nerve.** The final sensory nerve, the **vestibulocochlear nerve** (ves-tib-yoo-loh-KOHK-lee-ur), innervates the structures of the inner ear and provides for the senses of hearing and equilibrium.

9. **CN IX: Glossopharyngeal Nerve.** The small mixed **glossopharyngeal nerve** (glah-soh-fehr-IN-jee-uhl) provides motor axons to the muscles of the pharynx (throat) involved in swallowing and sensory fibers from the posterior one-third of the tongue for taste sensation.

10. **CN X: Vagus Nerve.** The mixed **vagus nerve** (VAY-gus) is the only cranial nerve that "wanders" outside of the head and neck ("vagus" means "wanderer"). In the head and the neck, it provides some sensory innervation from the skin of the head and the pharynx, motor axons to muscles involved in speech and swallowing, and motor axons to certain salivary glands. Outside the head and the neck, it innervates most of the thoracic and abdominal viscera as the main nerve of the parasympathetic nervous system.

11. **CN XI: Accessory Nerve.** The **accessory nerve** is the only cranial nerve that has both a cranial component, originating from the brainstem, and a spinal component, originating from the spinal cord. Its motor axons innervate the muscles that move the head and the neck, such as the trapezius and sternocleidomastoid muscles.

12. **CN XII: Hypoglossal Nerve.** The **hypoglossal nerve** is a small motor nerve that innervates the muscles that move the tongue. Note that the hypoglossal nerve *moves* the tongue but does not provide any taste sensation to the tongue.

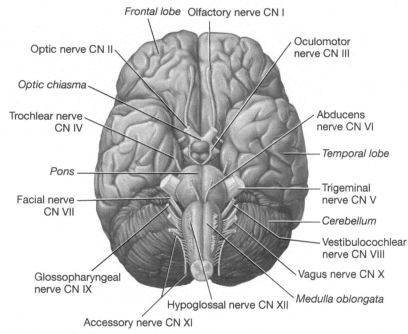

FIGURE **8.24** Inferior view of the brain with the cranial nerves.

HINTS & TIPS

Remembering the Order of the Cranial Nerves

Many cranial nerve mnemonics have been created over the years to help students remember their correct order. Following is one of my favorite mnemonics, but if this one doesn't stick for you, try making up your own or doing an Internet search for "cranial nerve mnemonics":

Oh (Olfactory)

Once (Optic)

One (Oculomotor)

Takes (Trochlear)

The (Trigeminal)

Anatomy (Abducens)

Final (Facial)

Very (Vestibulocochlear)

Good (Glossopharyngeal)

Vacations (Vagus)

Are (Accessory)

Happening (Hypoglossal)

You also can help yourself remember the olfactory and optic nerves by reminding yourself that you have one nose (CN I, the *olfactory* nerve) and two eyes (CN II, the *optic* nerve).

Procedure **1** Testing the Cranial Nerves

A component of every complete physical examination performed by healthcare professionals is the cranial nerve exam. In this procedure, you will put on your "doctor hat" and perform the same tests of the cranial nerves done during a physical examination. Pair up with another student, and take turns performing the following tests. For each test, first document your observations (in many cases, this will be "able to perform" or "unable to perform"). Then state which cranial nerve(s) you have checked with each test. Keep in mind that some tests check more than one cranial nerve, some cranial nerves are tested more than once, and each nerve is tested in this procedure at least once.

1 Have your partner perform the following actions one at a time (not all at once—your partner may have difficulty smiling and frowning at the same time): smile, frown, raise the eyebrows, and puff the cheeks.

Observations: _____

CN(s) tested: _____

2 Have your partner open and close his or her jaw and clench his or her teeth.

Observations: _____

CN(s) tested: _____

3 Have your partner elevate and depress his or her shoulders and turn his or her head to the right and the left.

Observations: _____

CN(s) tested: _____

4 Draw a large, imaginary "Z" in the air with your finger. Have your partner follow your finger with his or her eyes without moving the head. Repeat the procedure, this time drawing the letter "H" in the air with your finger.

Observations: _____

CN(s) tested: _____

5 Test pupillary response:

  a Dim the lights in the room about halfway.

  b Place your hand vertically against the bridge of your partner's nose as illustrated in Figure 8.25. This forms a light shield to separate the right and left visual fields.

  c Shine the penlight indirectly into the left eye from an angle, as illustrated in Figure 8.25. Watch what happens to the pupil in the left eye.

  d Move the penlight away, and watch what happens to the left pupil.

  e Shine the light into the left eye again, and watch what happens to the pupil in the right eye. Move the penlight away, and watch what happens to the right pupil.

  f Repeat this process with the right eye.

  g Record your results in Table 8.5 on the following page.

  CN(s) tested: _____

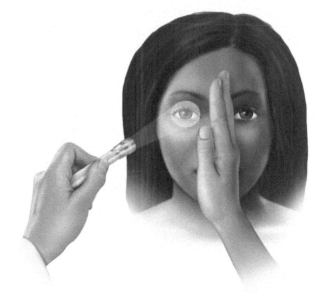

FIGURE **8.25** Method for testing the pupillary response.

TABLE **8.5** Pupillary Response Results

Action	Response of Left Pupil	Response of Right Pupil
Light shined into left eye		
Light removed from left eye		
Light shined into right eye		
Light removed from right eye		

6 Place your hand lightly on your partner's throat, and have him or her swallow and speak. Feel for symmetrical movement of the larynx (throat).

Observations: _____

CN(s) tested: _____

7 Have your partner protrude his or her tongue. Check for abnormal deviation or movement (e.g., does the tongue move straight forward or does it move to one side?).

Observations: _____

CN(s) tested: _____

8 To test vision, have your partner stand 20 feet from a Snellen chart and cover the right eye. Your partner should read the chart with the left eye, starting with the largest line and stopping with the smallest line he or she can read clearly. Record the ratio (e.g., 20/30) next to that smallest line as your observation. Now have your partner cover the left eye and repeat the process, reading the chart with the right eye.

Observations: _____

CN(s) tested: _____

9 Hold a tuning fork by its handle, and tap the tines lightly on the lab table to start ringing it. Touch the stem to the top of your partner's head along the midsagittal line in the manner shown in Figure 8.26. This is called the Weber test. Ask your partner whether the vibration is heard better in one ear, or whether the sound is heard equally well in both ears.

Observations: _____

CN(s) tested: _____

FIGURE **8.26** Weber test for hearing.

8

10 Have your partner stand with his or her eyes closed and arms at the sides for several seconds. Evaluate the ability to remain balanced.

Observations: _____

CN(s) tested: _____

11 Hold an unknown sample near your partner's nose, and fan the odor toward the nose by waving your hand over the container. Have your partner identify the substance by its scent.

Observations: _____

CN(s) tested: _____

12 Evaluate your partner's ability to taste by having your partner place a piece of PTC paper on the tongue (the ability to taste PTC is genetically determined; about half of the population can taste it). If your partner cannot taste the PTC, try the thiourea paper instead (a word of warning—thiourea tastes bad). If your partner cannot taste either of these papers, try the sodium benzoate paper.

Observations: _____

CN(s) tested: _____

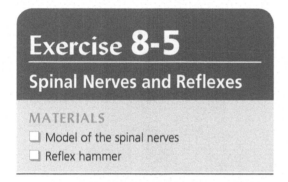

Exercise 8-5

Spinal Nerves and Reflexes

MATERIALS
- ☐ Model of the spinal nerves
- ☐ Reflex hammer

Axons enter and exit the spinal cord as a group of fibers called **nerve roots**. The **anterior roots** carry motor impulses from the anterior horn of the spinal cord to the PNS, and the **posterior roots** are axons that transmit sensory impulses from the PNS to the posterior horn of the spinal cord. Each of the 31 pairs of **spinal nerves** forms from the fusion of the anterior and posterior roots. Each spinal nerve carries both motor and sensory fibers, so all spinal nerves are mixed nerves.

Shortly after the anterior and posterior roots fuse to form the spinal nerve, it splits into three branches: a **posterior ramus**, an anterior **ramus**, and a small **meningeal branch**. The posterior rami serve the skin, joints, and musculature of the posterior trunk. The meningeal branches reenter the vertebral canal to innervate spinal structures. The larger anterior rami travel anteriorly to supply the muscles of the upper and lower limbs, the anterior thorax and abdomen, and part of the back. The general distribution of the anterior rami is illustrated in Figures 8.27 and 8.28.

The anterior rami of the thoracic spinal nerves travel between the ribs as 11 separate pairs of **intercostal nerves** that innervate the intercostal muscles, the abdominal muscles, and the skin of the chest and abdomen. The anterior rami of the cervical, lumbar, and sacral nerves combine to form four large **plexuses,** or networks, of nerves: the *cervical, brachial, lumbar,* and *sacral plexuses.* The major nerves of these plexuses are as follows:

1. **Cervical plexuses.** The **cervical plexuses** consist of the anterior rami of C1–C4 with small contributions from C5. The branches of each cervical plexus serve the skin of the head and the neck and certain neck muscles. The major branch is the **phrenic nerve** (FREN-ik; C3–C5), which serves the diaphragm.

2. **Brachial plexuses.** The complicated-looking **brachial plexuses** consist of the anterior rami of C5–T1 (Fig. 8.29). The first structures formed in each brachial plexus are its large **trunks.** Each trunk splits into an anterior division and a posterior division that become the plexuses' **cords.** Several nerves originate from the brachial plexuses' cords and trunks, including the following:

    a. **Axillary nerve.** The **axillary nerve** serves structures near the axilla, including the deltoid and teres minor muscles and the skin around this region.

    b. **Musculocutaneous nerve.** The **musculocutaneous nerve** (musk-yoo-loh-kyoo-TAY-nee-us) is located in the lateral arm. It serves the anterior arm muscles that flex the forearm (such as the biceps brachii muscle) and the skin of the lateral forearm.

c. **Radial nerve.** The **radial nerve** is located in the posterior upper limb. It serves the muscles that extend the forearm and hand as well as the skin in the lateral hand.

d. **Ulnar nerve.** The **ulnar nerve**, which you likely know as the "funny bone nerve," begins posteriorly but then crosses over to the anterior side of the forearm as it curves around the medial epicondyle of the humerus. At this point, the nerve is superficial and so is easily injured when you smack your elbow on something. The ulnar nerve supplies certain muscles in the forearm that flex the hand, most of the intrinsic muscles of the hand, and the skin over the medial hand.

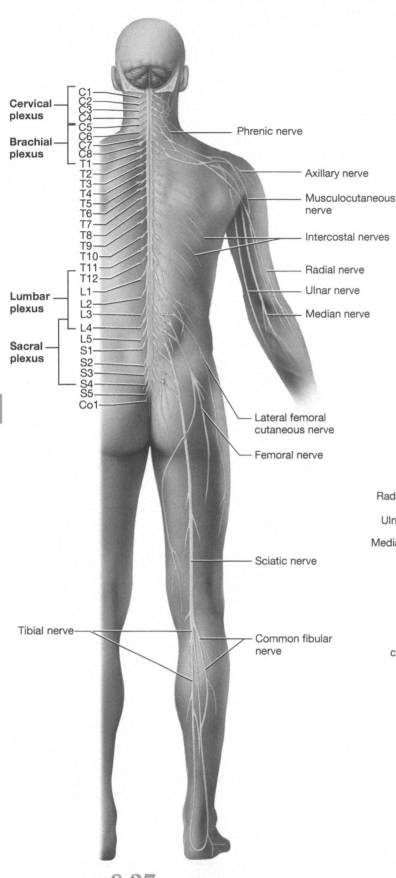

Cervical plexus — C1, C2, C3, C4, C5

Brachial plexus — C6, C7, C8, T1

T2, T3, T4, T5, T6, T7, T8, T9, T10, T11, T12

Lumbar plexus — L1, L2, L3, L4, L5

Sacral plexus — S1, S2, S3, S4, S5, Co1

Phrenic nerve
Axillary nerve
Musculocutaneous nerve
Intercostal nerves
Radial nerve
Ulnar nerve
Median nerve
Lateral femoral cutaneous nerve
Femoral nerve
Sciatic nerve
Tibial nerve
Common fibular nerve

FIGURE **8.27** Nerve plexuses and anterior rami of the spinal nerves.

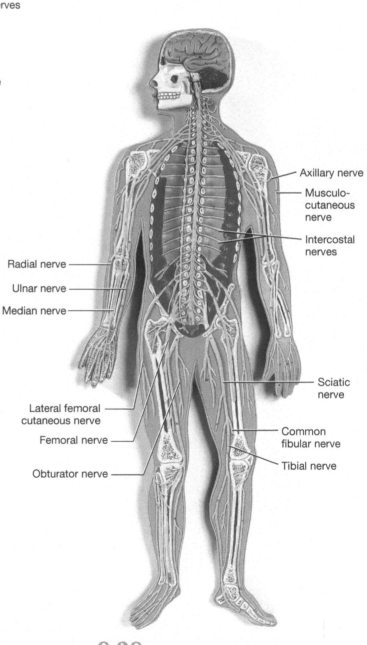

Axillary nerve
Musculo-cutaneous nerve
Intercostal nerves
Radial nerve
Ulnar nerve
Median nerve
Lateral femoral cutaneous nerve
Femoral nerve
Obturator nerve
Sciatic nerve
Common fibular nerve
Tibial nerve

FIGURE **8.28** Anatomical model of the anterior rami of the spinal nerves.

e. **Median nerve.** The **median nerve** is named as such because it travels approximately down the middle of the arm and forearm. It supplies most of the muscles of the forearm that flex the hand, certain intrinsic hand muscles, and the skin over the anterior and lateral hand. As the median nerve enters the wrist, it travels under a band of connective tissue called the *flexor retinaculum*. Occasionally, the median nerve becomes trapped and inflamed under the flexor retinaculum, which results in *carpal tunnel syndrome*.

3. **Lumbar plexuses.** Each **lumbar plexus** consists of the anterior rami of L1–L4 with a small contribution from T12 (Fig. 8.30). The branches of the lumbar plexus include the following:

   a. **Femoral nerve.** The largest nerve of this plexus is the **femoral nerve,** which provides motor innervation to most of the anterior thigh muscles that extend the knee and sensory innervation to the skin of the anterior and medial thigh.

   b. **Lateral femoral cutaneous nerve.** As implied by its name, the **lateral femoral cutaneous nerve** provides mainly sensory innervation to the skin of the anterolateral thigh.

   c. **Obturator nerve.** The **obturator nerve** (AHB-too-ray-tur) is located in the medial thigh, where it supplies the medial thigh muscles and the skin of the proximal, medial thigh.

   Other nerves of the lumbar plexus are labeled in Figure 8.30 for your reference.

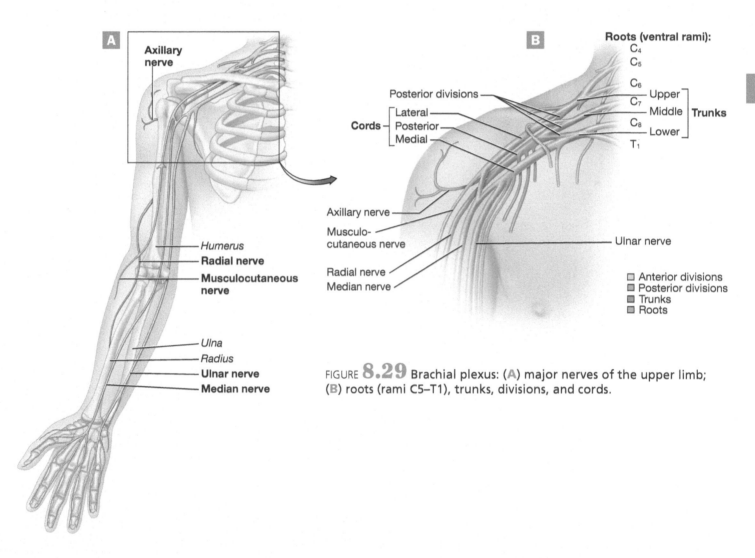

FIGURE **8.29** Brachial plexus: (**A**) major nerves of the upper limb; (**B**) roots (rami C5–T1), trunks, divisions, and cords.

4. **Sacral plexuses.** Each **sacral plexus** forms from anterior rami of L4–S4 (Fig. 8.31). The largest nerve of the sacral plexus, and indeed the largest nerve in the body, is the **sciatic nerve** (sy-AEH-tik). The sciatic nerve travels in the posterior thigh, where it splits into two branches: the tibial nerve and the common fibular nerve. The **tibial nerve** provides motor innervation to the posterior muscles of the thigh, posterior leg, and foot. The **common fibular nerve** provides motor innervation to the lateral leg, the anterior muscles of the leg that dorsiflex the foot, and the intrinsic muscles of the foot. For reference, the other major nerves of the sacral plexuses are labeled in Figure 8.31.

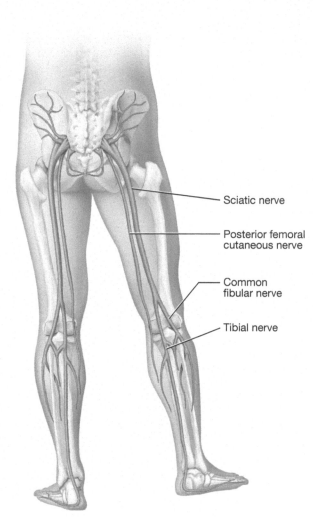

Sciatic nerve

Posterior femoral cutaneous nerve

Common fibular nerve

Tibial nerve

FIGURE **8.30** Nerves of the sacral plexus.

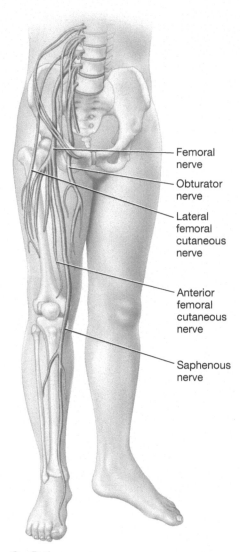

Femoral nerve

Obturator nerve

Lateral femoral cutaneous nerve

Anterior femoral cutaneous nerve

Saphenous nerve

FIGURE **8.31** Nerves of the lumbar plexus.

Procedure 1 Model Inventory for Nerves and Nerve Plexuses

Identify the following nerves and nerve plexuses on models and diagrams, using this unit and your textbook for reference. As you examine the anatomical models and diagrams, record the name of the model and the structures you were able to identify on the model inventory in Table 8.6.

1. Intercostal nerves
2. Cervical plexus
   a. Phrenic nerve
3. Brachial plexus
   a. Axillary nerve
   b. Musculocutaneous nerve
   c. Radial nerve
   d. Ulnar nerve
   e. Median nerve

4. Lumbar plexus
   a. Femoral nerve
   b. Lateral femoral cutaneous nerve
   c. Obturator nerve
5. Sacral plexus
   a. Sciatic nerve
      (1) Tibial nerve
      (2) Common fibular nerve

TABLE **8.6** Model Inventory for Spinal Nerve Anatomy

Model/Diagram	Structures Identified

Spinal Reflexes

A **reflex** is an involuntary, predictable motor response to a stimulus. The pathway through which information travels, shown below and in Figure 8.32, is called a **reflex arc**:

sensory receptor detects the stimulus ⟶ sensory neurons bring the stimulus to the CNS ⟶ the CNS processes and integrates the information ⟶ the CNS sends its output via motor neurons to an effector ⟶ the effector performs the triggered action

The human body has many different reflex arcs, one of the simplest being the **stretch reflex**. Stretch reflexes are important in maintaining posture and balance and are initiated when a muscle is stretched. The stretch is detected by **muscle spindles**, specialized stretch receptors in the muscles, and these stimuli are sent via sensory neurons to the CNS. The CNS then sends impulses down the motor neurons to the muscle that trigger a muscle contraction to counter the stretch.

You can demonstrate the stretch reflex easily: Sit down with your knees bent and relaxed, and palpate (feel) the musculature of your posterior thigh. How do the muscles feel (taut or soft)? Now stand up, and bend over to touch your toes. Palpate the muscles of your posterior thigh again. How do they feel? The reason they feel different in this position is that you stretched the hamstring muscles when you bent over to touch your toes. This triggered a stretch reflex that resulted in shortening (or tightening) of those muscles.

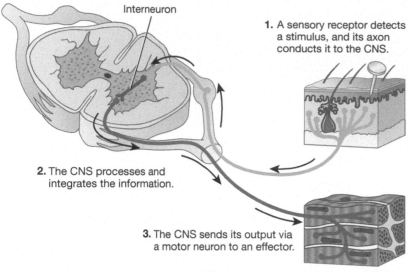

Interneuron

1. A sensory receptor detects a stimulus, and its axon conducts it to the CNS.

2. The CNS processes and integrates the information.

3. The CNS sends its output via a motor neuron to an effector.

FIGURE **8.32** Reflex arc.

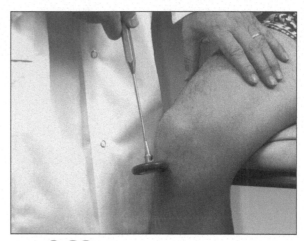

FIGURE **8.33** Patellar tendon reflex.

Procedure 2 Testing Spinal Reflexes

Technically, stretch reflexes can be carried out without the help of the cerebral cortex and can be mediated solely by the spinal cord. But does this mean that the cortex plays no role in these reflexes at all? Let's find out, using the simplest example of a stretch reflex—the patellar tendon (knee-jerk) reflex, shown in Figure 8.33.

1 Have your lab partner sit in a chair with legs dangling freely.

2 Palpate your partner's patellar tendon between the tibial tuberosity and the patella.

3 Tap this area with the flat end of a reflex hammer (sometimes a few taps are necessary to hit the right spot). What is the result?

4 Now give your partner a difficult math problem to work (long division with decimals usually does the trick). As your partner works the problem, tap the tendon again. Is this response different from the original response? If yes, how, and why?

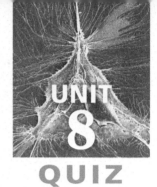

1 Label the following structures on **Figure 8.34.**

❑ Axon   ❑ Axon terminals   ❑ Dendrites

❑ Axon hillock   ❑ Cell body   ❑ Telodendria

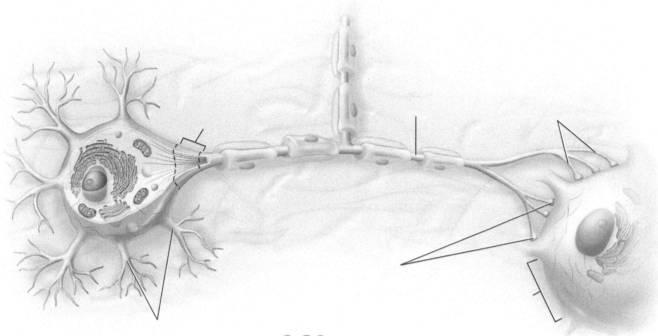

FIGURE **8.34** Structure of a neuron.

2 Which part of the neuron generates action potentials?

a. Cell body.

b. Axon.

c. Dendrites.

d. All of the above can generate action potentials.

3 *Matching:* Match the neuroglial cell with its correct function.

_____ Oligodendrocytes   A. Create the myelin sheath in the PNS

_____ Astrocytes   B. Ciliated cells in the CNS that form and circulate cerebrospinal fluid

_____ Microglial cells   C. Surround the cell bodies of neurons in the PNS

_____ Schwann cells   D. Anchor neurons and blood vessels; maintain extracellular environment around neurons; assist in the formation of the blood brain barrier

_____ Satellite cells   E. Phagocytic cells of the CNS

_____ Ependymal cells   F. Form the myelin sheath in the CNS

4 Multiple sclerosis is a *demyelinating* disease, in which the patient's immune system attacks and destroys the cells that form the myelin sheath in the central nervous system. What types of symptoms would you expect from such a disease? Why? Would Schwann cells or oligodendrocytes be affected? Explain.

_____

_____

_____

_____

5 Label the following parts of the brain in **Figures 8.35** and **8.36**.

- ❏ Cerebellum
- ❏ Corpus callosum
- ❏ Frontal lobe
- ❏ Hypothalamus
- ❏ Medulla oblongata
- ❏ Midbrain
- ❏ Occipital lobe
- ❏ Parietal lobe
- ❏ Pons
- ❏ Temporal lobe
- ❏ Thalamus

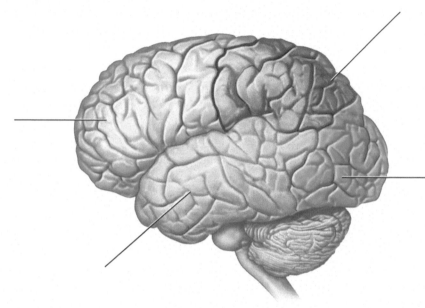

FIGURE **8.35** Brain, lateral view.

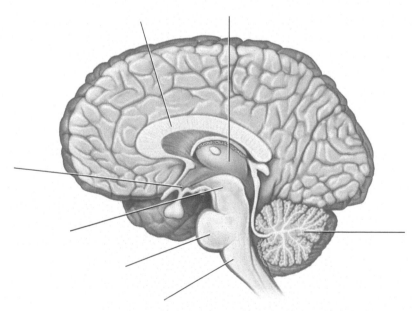

FIGURE **8.36** Brain, midsagittal section.

8

 **UNIT 8 QUIZ**
(continued)

6 Label the following parts of the spinal cord in **Figure 8.37**.

- ❏ Anterior horn
- ❏ Central canal
- ❏ Lateral horn
- ❏ Posterior horn
- ❏ Spinal arachnoid mater
- ❏ Spinal dura mater

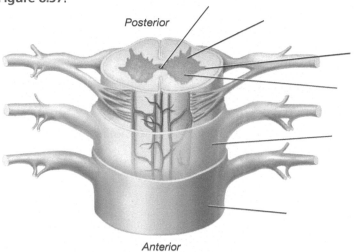

7 Cerebrospinal fluid is produced by the

a. ventricles of the brain.

b. arachnoid granulations.

c. dural sinuses.

d. choroid plexuses.

FIGURE **8.37** Spinal cord, transverse section.

8 *True/False:* Mark the following statements as true (T) or false (F). If the statement is false, correct it to make it a true statement.

_____ The diencephalon consists of the midbrain, pons, and medulla oblongata.

_____ The outer, tough meninx around the brain is the dura mater.

_____ The spinal cord terminates at the sacrum.

_____ The cerebral cortex is the outer 2 mm of white matter of the cerebrum.

_____ The corpus callosum is a large tract of white matter that connects the right and left cerebral hemispheres.

9 Predict the effects of injuries to the following areas:

a Cerebral cortex

_____

_____

b Brainstem

_____

_____

c Cerebellum

_____

_____

d Hypothalamus

_____

_____

10 A common way to deliver anesthesia for surgery and childbirth is to inject the anesthetic agent into the epidural space (*epidural anesthesia*). Why do you think the anesthetic is injected into the epidural space instead of the subarachnoid space?

_____

_____

_____

_____

11 An individual sustains injuries to only the ascending tracts of the spinal cord. Will this person experience deficits in movement, sensation, both, or neither? Explain.

_____

_____

_____

_____

12 *Matching:* Match the cranial nerve with its main functions.

_____ CN I   A. Sensory to the face; motor to the muscles of mastication

_____ CN II   B. Motor to the trapezius and sternocleidomastoid muscles

_____ CN III   C. Hearing and equilibrium

_____ CN IV   D. Olfaction (smell)

_____ CN V   E. Motor to the muscles of swallowing; taste to the posterior one-third of the tongue

_____ CN VI   F. Motor to one of six extrinsic eye muscles (superior oblique muscle)

_____ CN VII   G. Motor to the tongue

_____ CN VIII   H. Vision

_____ CN IX   I. Motor to the muscles of facial expression; taste to the anterior two-thirds of the tongue

_____ CN X   J. Motor to four of six extrinsic eye muscles; dilates the pupil; opens the eye; changes the shape of the lens

_____ CN XI   K. Motor to the muscles of swallowing and speaking; motor to the thoracic and abdominal viscera

_____ CN XII   L. Motor one of six extrinsic eye muscles (lateral rectus muscle)

13 The receptor that detects the stretch in a stretch reflex is called a(an)

a. mitotic spindle.

b. muscle spindle.

c. capsular receptor.

d. efferent neuron.

UNIT 8 QUIZ
(continued)

14 Label the following nerves and plexuses on Figure 8.38.

- ❏ Brachial plexus
- ❏ Cervical plexus
- ❏ Femoral nerve
- ❏ Lumbar plexus
- ❏ Median nerve
- ❏ Musculocutaneous nerve
- ❏ Radial nerve
- ❏ Sacral plexus
- ❏ Sciatic nerve
- ❏ Ulnar nerve

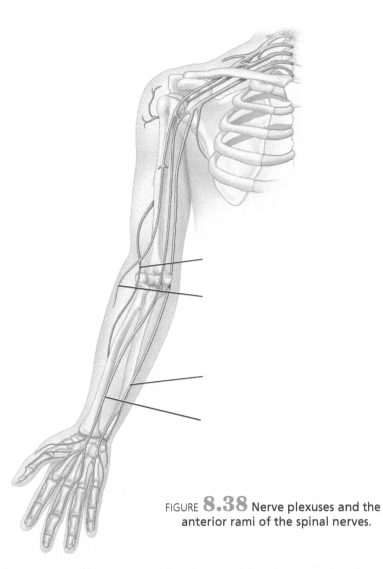

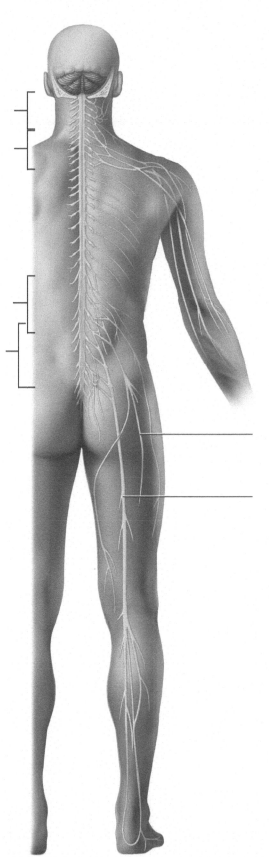

FIGURE **8.38** Nerve plexuses and the anterior rami of the spinal nerves.

8

15 Damage to which cranial nerve(s) might produce the following results?

a Inability to move the tongue

_____

b Inability to taste

_____

c Inability to move the eyes in any direction

_____

d Inability to shrug the shoulders

_____

e Inability to swallow

_____

General and Special Senses

When you have completed this unit, you should be able to:

1 Describe and identify structures of the eye.

2 Compare and contrast the functions of the rods and cones.

3 Describe and identify structures of the ear.

4 Perform tests of hearing and equilibrium.

5 Identify structures of the olfactory and taste senses.

6 Describe the relative concentration of cutaneous sensory receptors in different regions of the body.

PRE-LAB EXERCISES

Complete the following exercises prior to coming to lab, using your lab manual and textbook for reference.

Pre-Lab Exercise 9-1

✎ Key Terms

You should be familiar with the following terms before coming to lab.

Term	Definition

Structures of the Eye

Conjunctiva

Lacrimal gland

Sclera

Cornea

Iris

Pupil

Lens

Choroid

Retina

Structures of the Ear

Auricle

External auditory canal

Tympanic membrane

Auditory ossicles

Pharyngotympanic tube

Vestibule

Semicircular canals

Cochlea

Structures of Taste and Smell

Chemosenses

Olfactory epithelium

Tongue papillae

Pre-Lab Exercise 9-2

Anatomy of the Eye

Color the structures of the eye in Figure 9.1, and label them with the following terms from Exercise 9-1 (p. 243). Use Exercise 9-1 in this unit and your text for reference.

Fibrous Tunic
- ❑ Sclera
- ❑ Cornea

Vascular Tunic (Uvea)
- ❑ Choroid
- ❑ Ciliary body
- ❑ Iris
- ❑ Pupil
- ❑ Lens

Sensory Tunic
- ❑ Retina
- ❑ Optic nerve (cranial nerve II)

Other Structures
- ❑ Anterior cavity
- ❑ Posterior cavity

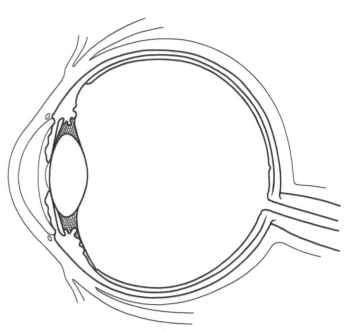

FIGURE **9.1** Eyeball, sagittal section.

Anatomy of the Ear

Color the structures of the ear in Figure 9.2, and label them with the following terms from Exercise 9-2 (p. 249). Use Exercise 9-2 in this unit and your text for reference.

❏ Outer ear
- ☐ Auricle (pinna)
- ☐ External auditory canal

❏ Middle ear
- ☐ Tympanic membrane
- ☐ Ossicles
  - ▪ Malleus
  - ▪ Incus
  - ▪ Stapes
- ☐ Pharyngotympanic tube

❏ Inner ear
- ☐ Vestibule
- ☐ Semicircular canals
- ☐ Cochlea
- ☐ Vestibulocochlear nerve (cranial nerve VIII)

9

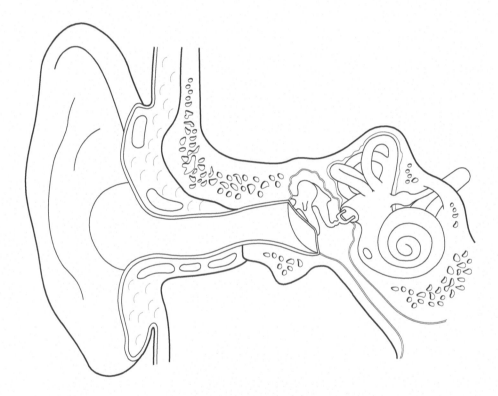

FIGURE **9.2** Anatomy of the ear.

Colorized SEM of rods and cones in the retina.

Sensation is broadly defined as the detection of changes in the internal and external environments. Sensation may be conscious or subconscious, depending on the destination of the sensory information. For example, certain blood vessels have receptors that detect blood pressure. These stimuli are transmitted to the brainstem, which makes changes as necessary to ensure that blood pressure adjusts to meet the body's needs. This information never makes it to the cerebral cortex, so you are not consciously aware of it. However, information eventually taken to the cerebral cortex for integration and interpretation (e.g., the taste of your food or the level of light in a room) is something of which you are consciously aware. This is called **perception**, and it is the focus of this unit.

The following exercises ask you to examine the anatomy and physiology of the **special senses**: vision, hearing and equilibrium, taste, and smell. You also will examine the **general senses** in this unit, which include touch, pain, and temperature.

Exercise 9-1

Anatomy of the Eye and Vision

MATERIALS
- ❏ Eye models
- ❏ Preserved eyeballs
- ❏ Dissection equipment
- ❏ Dissection trays
- ❏ Snellen vision chart
- ❏ Dark green or blue paper
- ❏ Ruler

The eye is a complex organ consisting of two main components: external structures such as the eyelids and the eyeball itself. Many of the external structures of the eye protect the delicate eyeball, and others move the eyeball.

Anteriorly, the eyeball is covered by the eyelids, or **palpebrae** (pal-PEE-bree; Fig. 9.3A). The palpebrae meet medially and laterally at the **medial** and **lateral canthi** (or *commissures*), respectively. There are several structures in and around the palpebrae that contain sebaceous or mucous glands to lubricate the palpebrae and anterior surface of the eyeball. One such structure is the **lacrimal caruncle** (LAK-rih-muhl kar-UN-kuhl), which is located at the medial canthus.

One of the most prominent external structures of the eye is the **lacrimal apparatus**, which produces and drains tears. The lacrimal apparatus consists of the **lacrimal gland**, located in the superolateral orbit, and the ducts that drain the tears it produces. As you can see in Figure 9.3A, tears drain into small **lacrimal canals** near the medial canthus, which then drain into the **lacrimal sac**, which is found in a depression in the lacrimal bone. From here, tears travel through the **nasolacrimal duct** and, finally, empty into the nasal cavity just inferior to the inferior nasal concha (which is why your nose runs when you cry).

9

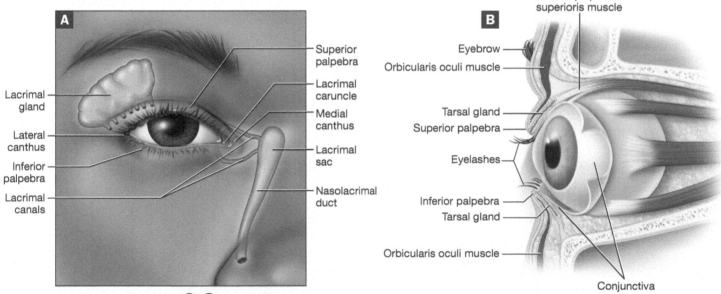

FIGURE **9.3** External structures of the eye: **(A)** anterior view; **(B)** lateral view.

Other accessory structures associated with the palpebrae are shown in Figure 9.3B. For example, the internal surfaces of the palpebrae and parts of the eyeball are lined with a thin mucous membrane called the **conjunctiva** (kon-junk-TY-vah). In addition, two muscles that open and close the eye insert into and around the palpebrae: the *levator palpebrae superioris muscle* and the *orbicularis oculi muscle*. Finally, we have the six **extrinsic eye muscles** that move the eyeball. These muscles are labeled in Figure 9.4.

The **eyeball** itself is a hollow organ that is divided into two main portions: the anterior cavity and the posterior cavity. The boundary for the cavities is a crystalline structure known as the **lens**, which is one of the structures in the eye that refracts (bends) light coming into the eye to focus it. The **anterior cavity** is anterior to the lens and is filled with a watery fluid called **aqueous humor**. Aqueous humor is produced relatively constantly and drained by a structure called the **scleral venous sinus**. The **posterior cavity** is posterior to the lens and contains a thicker fluid called **vitreous humor** (VIT-ree-us). Unlike aqueous humor, vitreous humor is present at birth and remains relatively unchanged throughout life. Both fluids help to refract light coming into the eye.

The eyeball has three distinct tissue layers, or tunics (Fig. 9.5):

1. **Fibrous tunic.** The outermost layer of the eyeball is the **fibrous tunic**, which consists mostly of dense irregular collagenous connective tissue. It is avascular (lacks a blood supply) and consists of two parts:

   a. **Sclera.** The **sclera** (SKLEHR-ah) is the white part of the eyeball, and makes up the posterior five-sixths of the fibrous tunic. It is white because of numerous collagen fibers that contribute to its thickness and toughness (in the same way a joint capsule or a ligament is tough and white).

   b. **Cornea.** The clear **cornea** (KOHR-nee-ah) makes up the anterior one-sixth of the fibrous tunic. It is the fourth refractory medium of the eyeball.

2. **Vascular tunic.** Also called the **uvea** (YOO-vee-uh), the **vascular tunic** carries most of the blood supply to the tissues of the eyeball. It is composed of three main parts:

   a. **Choroid.** The highly vascular **choroid** (KOHR-oyd) makes up the posterior part of the vascular tunic. The choroid is brown in color to prevent light scattering in the eye.

   b. **Ciliary body.** The **ciliary body** (SILL-ee-ehr-ee) is located at the anterior aspect of the vascular tunic. It is made chiefly of the **ciliary muscle**, smooth muscle fibers that control the shape of the lens. The muscle attaches to the lens via small **suspensory ligaments**. The ciliary body also produces aqueous humor.

   c. **Iris.** The pigmented **iris** is the most anterior portion of the vascular tunic. It consists of muscle fibers arranged around an opening called the **pupil**.

3. **Sensory tunic.** The **sensory tunic** consists of the retina and the optic nerve. The **retina** (RET-in-ah) is a thin, delicate structure that contains **photoreceptors** called rods and cones that detect light.

   a. **Rods.** The photoreceptors known as **rods** are scattered throughout the retina and are responsible for vision in dim light and for peripheral vision. They can produce vision in black and white only.

   b. **Cones.** Cones, the second type of photoreceptor, are concentrated at the posterior portion of the retina. They are found in highest numbers in an area called the **macula lutea** (MAK-yoo-lah LOO-tee-ah). At the center of the macula lutea is the **fovea centralis** (FOH-vee-uh sin-TRAL-iss), which contains only cones. Cones are responsible for color and high-acuity (sharp) vision in bright light.

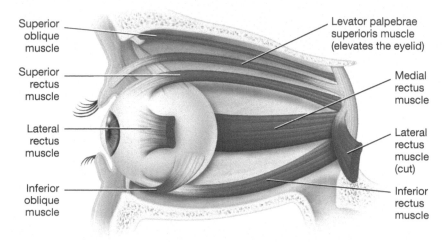

FIGURE **9.4** Extrinsic eye muscles.

Note that there are no rods or cones at the posteriormost aspect of the eyeball where the optic nerve leaves the eyeball. This location is called the **optic disc**. It is also known as the *blind spot* because its lack of photoreceptors means that this region can produce no images.

Notice in Figure 9.5 that light has to pass through four refractive media before it hits the retina: the cornea, the aqueous humor, the lens, and the vitreous humor. Of these four, the cornea and the lens have the greatest refractive power. The cornea, by itself, accounts for about two-thirds of the eye's refractive power, and, indeed, the cornea alone provides most of the necessary refraction when viewing distant objects.

When viewing nearer objects, however, additional "fine-tuning" refraction is needed by the lens. This is accomplished with the help of the ciliary muscle—when its smooth muscle fibers contract, the ciliary body moves closer to the lens and removes tension on the suspensory ligaments. This causes the lens to become rounder, an adjustment called *accommodation* which allows the lens to provide the additional refraction necessary to focus light on the retina. When the eye switches to a distant object again, the ciliary muscle relaxes, which moves the ciliary body farther away from the lens and puts tension on the suspensory ligaments. This flattens the lens and allows the cornea to again become the primary refractive medium.

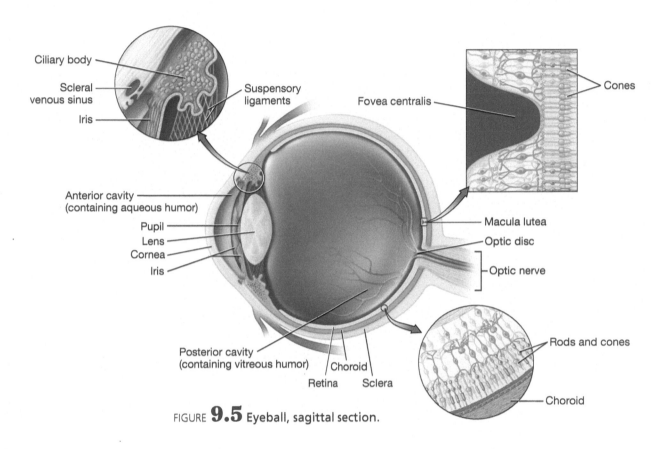

FIGURE **9.5** Eyeball, sagittal section.

Procedure 1 Model Inventory for the Eye

Identify the following structures of the eye on models and diagrams using this unit and your textbook for reference. As you examine the anatomical models and diagrams, record the name of the model and the structures you were able to identify on the model inventory in Table 9.1.

Accessory Structures
1. Palpebrae
2. Medial and lateral canthi
3. Lacrimal caruncle
4. Lacrimal apparatus
   a. Lacrimal gland
   b. Lacrimal canals
   c. Nasolacrimal duct
5. Conjunctiva
6. Extrinsic eye muscles

Eyeball
1. Lens
2. Anterior cavity
   a. Aqueous humor
3. Posterior cavity
   a. Vitreous humor

4. Fibrous tunic
   a. Sclera
   b. Cornea
5. Vascular tunic (uvea)
   a. Choroid
   b. Ciliary body
      (1) Ciliary muscle
      (2) Suspensory ligaments
   c. Iris
      (1) Pupil
6. Sensory tunic
   a. Retina
   b. Optic nerve (cranial nerve II)
   c. Macula lutea
   d. Fovea centralis
   e. Optic disc

TABLE **9.1** Model Inventory for the Eye

Model/Diagram	Structures Identified

Procedure 2 Eyeball Dissection

In this procedure, you will examine the structures of the eyeball on a fresh or preserved eyeball. Eyeball dissection isn't as gross as it sounds—I promise.

1 Examine the external anatomy of the eyeball (Fig. 9.6), and record the structures you can identify.

Safety Note

Goggles and gloves are required!

2 Use scissors to remove the adipose tissue surrounding the eyeball. Identify the optic nerve.

3 Hold the eyeball at its anterior and posterior poles, and use a sharp scalpel or scissors to make an incision in the frontal plane. Watch out, as aqueous humor and vitreous humor are likely to spill everywhere.

4 Complete the incision, and separate the anterior and posterior portions of the eyeball (Fig. 9.7). Take care to preserve the fragile retina—the thin, delicate yellow-tinted inner layer.

5 List the structures you can identify in the anterior half of the eyeball (Fig. 9.8):

6 List the structures you can identify in the posterior half of the eyeball:

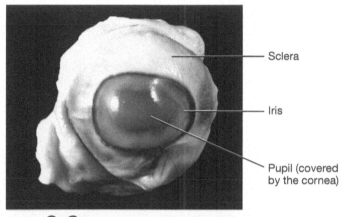

FIGURE **9.6** Anterior view of an eyeball.

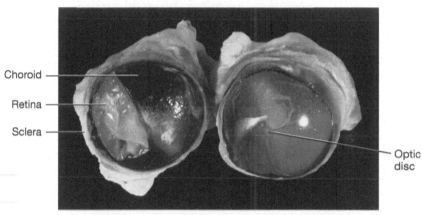

FIGURE **9.7** Frontal section of an eyeball showing the tunics.

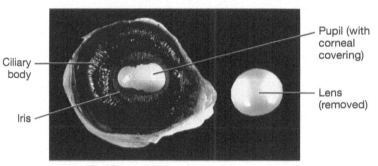

FIGURE **9.8** Posterior view of the anterior portion of an eyeball and lens.

Procedure 3 Comparing the Distribution of Rods and Cones

We discussed earlier the unequal distribution of the photoreceptors in the retina. In this procedure, you will see (no pun intended) firsthand the differences in vision produced by the rods and the vision produced by the cones.

1 On a small sheet of paper, write the phrase "Anatomy is fun" in your regular-size print.

2 Hold this piece of paper about 25 cm (10 inches) directly in front of your lab partner's eyes, and have your lab partner read the phrase.

   a Can your partner read the phrase clearly?

   b Which photoreceptors are producing the image?

3 Now write a second phrase on the paper, and don't tell your partner what the phrase says. Hold the paper about 10 inches from your partner's peripheral vision field. Have him or her continue to stare forward and attempt to read what you have written.

   a Can your partner read the phrase clearly?

   b Which photoreceptors are producing the image?

4 For the next test, dim the lights in the room. Have your partner stand 20 feet in front of a Snellen eye chart and read down the chart, stopping with the smallest line read accurately. (You should stand next to the chart to verify your partner has read the correct letters.) The numbers next to the line indicate your partner's vision relative to someone with perfect vision, a measurement known as **visual acuity**. For example, a person with 20/40 vision can see at 20 feet what someone with perfect vision could see at 40 feet. Record your partner's ratio (e.g., 20/40).

Visual acuity:

5 With the lights still dimmed and your partner standing in the same place, hold a piece of dark green or dark blue paper over the Snellen chart. Ask your partner to identify the color of the paper you are holding. Record your partner's response.

Paper color:

6 Repeat the above processes with the lights illuminated, and record your partner's responses.

Visual acuity:

Paper color:

7 In which scenario were visual acuity and color vision better? Explain your findings.

Procedure 4 Finding the Blind Spot

Recall that the location where the optic nerve exits the eyeball lacks photoreceptors and so leaves a "blind spot"— a place that can produce no images. You can find the blind spot with the simple diagram shown in Figure 9.9.

1 Close your left eye, and hold Figure 9.9 about 56 cm (18 inches) in front of your right eye so that the "+" is directly in line with your right eye.

2 Slowly move Figure 9.9 toward your right eye while staring at the " + " until the large dot disappears. When it disappears, it is in your blind spot.

3 Have your lab partner measure with a ruler the distance from the page to your eye with a ruler.

Distance in centimeters: _____

4 Repeat the process for your left eye.

Distance in centimeters: _____

FIGURE **9.9** Blind spot test.

9

Procedure 5 Testing for Astigmatism

The condition called **astigmatism** is characterized by irregularities in the surfaces in either the cornea or the lens, or both. These irregularities decrease visual acuity because these structures are unable to focus light precisely on the retina. You can test for astigmatism by using a chart like the one shown in **Figure 9.10**.

1 Hold **Figure 9.10** a comfortable reading distance from your eyes, about 35 cm (14 inches).

2 Cover one eye, and stare at the center of **Figure 9.10**. All lines should appear equally distinct and black. If any of the lines appear blurry or gray, astigmatism may be present.

3 Cover the other eye and repeat the process.

4 Was astigmatism present in either eye? If yes, which eye(s)? _____

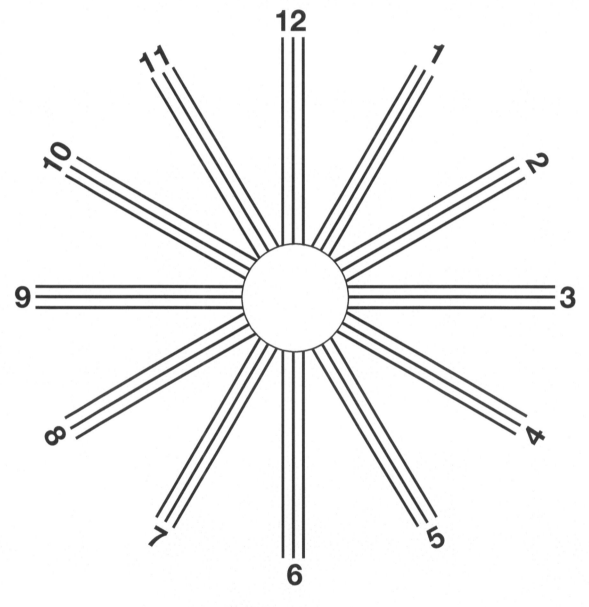

FIGURE **9.10** Astigmatism test chart.

Exercise 9-2

Anatomy of the Ear, Hearing, and Equilibrium

MATERIALS
- ❏ Ear models
- ❏ Tuning fork (500–1,000 Hz)
- ❏ Whiteboard marker

The ear contains structures both for hearing and equilibrium. It is divided into three regions: the outer, middle, and inner ear (Fig. 9.11).

1. **Outer ear.** The **outer ear** begins with the **auricle** (OHR-ih-kuhl), or *pinna*, a shell-shaped structure composed primarily of elastic cartilage that surrounds the opening to the **external auditory canal.** The external auditory canal extends about 2.5 cm into the temporal bone, where it ends in the **tympanic membrane,** a thin sheet of epithelium and connective tissue that separates the outer ear from the middle ear.

2. **Middle ear.** The **middle ear** is a small air-filled cavity within the temporal bone that houses tiny bones called the **auditory ossicles** (AW-sih-kullz): the **malleus** (MAL-ee-us; hammer), **incus** (ING-kus; anvil), and **stapes** (STAY-peez; stirrup). The ossicles transmit vibrations from the tympanic membrane to the inner ear through a structure called the **oval window,** to which the stapes is attached. An additional structure in the middle ear is the **pharyngotympanic tube** (fah-ring-oh-tim-PAN-ik; also called the *auditory tube*), which connects the middle ear to the pharynx (throat) and equalizes pressure in the middle ear.

3. **Inner ear.** The **inner ear** contains the sense organs for hearing and equilibrium. It consists of cavities collectively called the **bony labyrinth** filled with a fluid called **perilymph** (PEHR-ee-limf). Within the perilymph is a series of membranes, called the **membranous labyrinth,** that contain a thicker fluid called **endolymph** (EN-doh-limf; Fig. 9.12). The bony labyrinth has three regions:

   a. **Vestibule.** The **vestibule** is an egg-shaped bony cavity that houses structures responsible for *static equilibrium,* which refers to maintaining balance when the body is not moving.

   b. **Semicircular canals.** Situated at right angles to one another, the **semicircular canals** work together with the organs of the vestibule to maintain equilibrium. Their orientation allows them to be responsible for a type of equilibrium called *rotational equilibrium.* The structures of both the vestibule and the semicircular canals transmit their stimuli via the vestibular portion of the vestibulocochlear nerve (cranial nerve VIII).

   c. **Cochlea.** The **cochlea** (KOHK-lee-ah) is a spiral bony canal that contains a structure known as the **spiral organ** (or the **organ of Corti**) whose specialized **hair cells** transmit sound impulses. The cochlea has a hole in its lateral wall called the **round window,** which plays a role in allowing the perilymph in the cochlea to vibrate. The structures of the cochlea transmit sound stimuli via the cochlear portion of the vestibulocochlear nerve.

The structures of the membranous labyrinth are labeled in Figure 9.12 for your reference.

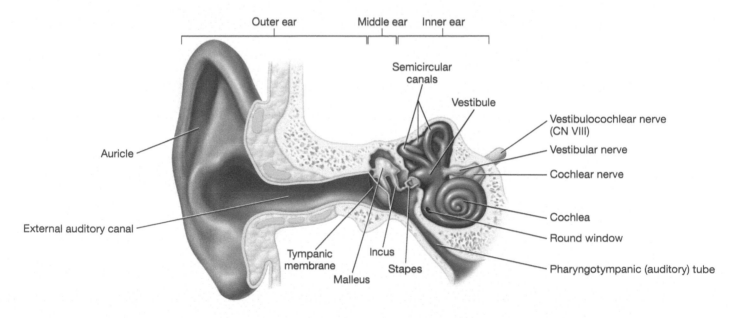

FIGURE **9.11** Anatomy of the ear.

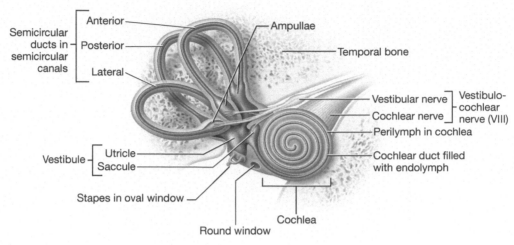

Semicircular ducts in semicircular canals
Anterior
Posterior
Lateral
Ampullae
Temporal bone
Vestibular nerve
Cochlear nerve
Vestibulo-cochlear nerve (VIII)
Perilymph in cochlea
Vestibule
Utricle
Saccule
Cochlear duct filled with endolymph
Stapes in oval window
Round window
Cochlea

FIGURE **9.12** Membranous labyrinth of the inner ear.

Procedure **1** Model Inventory for the Ear

Identify the following structures of the ear on models and diagrams, using this unit and your textbook for reference. As you examine the anatomical models and diagrams, record the name of the model and the structures you were able to identify on the model inventory in Table 9.2.

1. Outer ear
   a. Auricle (pinna)
   b. External auditory canal
2. Middle ear
   a. Tympanic membrane
   b. Ossicles
      (1) Malleus
      (2) Incus
      (3) Stapes

   c. Oval window
   d. Pharyngotympanic tube
3. Inner ear
   a. Vestibule
   b. Semicircular canals
   c. Cochlea
   d. Vestibulocochlear nerve (cranial nerve VIII)

TABLE **9.2** Model Inventory for the Ear

Model/Diagram	Structures Identified

Hearing Acuity

There are two possible types of hearing loss:

1. **Conductive hearing loss** results from interference of sound conduction through the outer and/or middle ear.

2. **Sensorineural hearing loss** results from damage to the inner ear or the vestibulocochlear nerve.

Two clinical tests can help a healthcare professional determine whether hearing loss is conductive or sensorineural: the Weber test and the Rinne (rinn-ay) test. Both tests use tuning forks that vibrate at specific frequencies when struck. The tuning forks are placed either directly on the bones of the skull to evaluate bone conduction—the ability to hear the vibrations transmitted through the bone—or in front of the ear, to evaluate air conduction—the ability to hear the vibrations transmitted through the air.

Procedure 2 Weber Test

1 Obtain a tuning fork with a frequency of 500–1,000 Hz (cycles per second).

2 Hold the tuning fork by the base, and tap it lightly on the edge of the table. The fork should begin ringing softly. If it is ringing too loudly, grasp the tines to stop it from ringing, and try again.

3 Place the base of the vibrating tuning fork on the midline of your partner's head, as shown in Figure 9.13.

4 Ask your partner whether the sound is heard better in one ear or whether the sound is heard equally in both ears. If the sound is heard better in one ear, this is called lateralization.

   a Was the sound lateralized? If yes, to which ear?

5 To illustrate what it would sound like if the sound were lateralized, have your partner place a finger in one ear. Repeat the test.

   a In which ear was the sound heard better?

   b If a patient has conduction deafness, in which ear do you think the sound will be heard most clearly (the deaf ear or the good ear)?

   Why? (If you are confused, think about your results when one ear was plugged.)

FIGURE **9.13** Weber test.

   c If a patient has sensorineural deafness, in which ear do you think the sound will be heard most clearly? Why?

Procedure **3** Rinne Test

1 Strike the tuning fork lightly to start it ringing.

2 Place the base of the tuning fork on your partner's mastoid process, as shown in Figure 9.14.

3 Time the interval during which your partner can hear the sound by having your partner tell you when the ringing stops.

Time interval in seconds: _____

4 After your partner can no longer hear the ringing, quickly move the still-vibrating tuning fork 1 to 2 cm lateral to the external auditory canal (the fork should not be touching your partner at this point).

5 Time the interval from the point when you moved the tuning fork in front of the external auditory canal to when your partner can no longer hear the sound.

Time interval in seconds: _____

Which situation tested bone conduction? _____

Which situation tested air conduction? _____

6 Typically, the air-conducted sound is heard twice as long as the bone-conducted sound. For example, if the bone-conducted sound was heard for 15 seconds, the air-conducted sound should be heard for 30 seconds.

Were your results normal? _____

What type of deafness is present if the bone-conducted sound is heard longer than the air-conducted sound? Explain.

_____

9

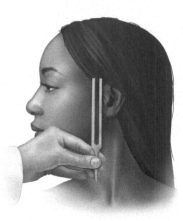

FIGURE **9.14** Rinne test.

Equilibrium

A common and simple test of equilibrium is the **Romberg test**, in which the person is asked to stand still, first with the eyes open, and then with the eyes closed. Under normal conditions, the vestibular apparatus should be able to maintain equilibrium in the absence of visual input. If the vestibular apparatus is impaired, however, the brain relies on visual cues to maintain balance.

 Procedure 4 Romberg Test

1 Have your partner stand erect with the feet together and the arms at the sides in front of a whiteboard.

2 Use a whiteboard marker to draw lines on the board on either side of your partner's torso. These lines are for your reference in the next part.

3 Have your partner stand in front of the whiteboard for 1 minute, staring forward with eyes open. Use the lines on either side of the torso to note how much your partner sways while standing. Below, record the amount of side-to-side swaying (i.e., minimal or significant):

4 Now have your partner stand in the same position for 1 minute with eyes closed. Again note the amount of side-to-side swaying, using the marker lines for reference.

Was the amount of swaying more or less with the eyes closed? _____

Why do you think this is so?

What do you predict would be the result for a person with an impaired vestibular apparatus? Explain.

9

Exercise 9-3

Olfactory and Taste Senses

MATERIALS
☐ Head and neck models
☐ Tongue model

Both olfaction and taste are sometimes referred to as the **chemosenses,** because they both rely on chemoreceptors to relay information about the environment to the brain. The chemoreceptors of the olfactory sense are located in a small patch in the roof of the nasal cavity called the **olfactory epithelium** (Fig. 9.15). The olfactory epithelium contains bipolar neurons called **olfactory receptor cells.** Their axons, which are collectively called **cranial nerve I** or the **olfactory nerve,** penetrate the holes in the cribriform plate to synapse on the **olfactory bulb,** which then sends the impulses down the axons of the **olfactory tract** to the olfactory cortex.

Taste receptors are located on **taste buds** housed on projections from the tongue called **papillae** (Fig. 9.16). Of the four types of papillae—**filiform, fungiform, foliate,** and **circumvallate**—all but filiform papillae house taste buds. **Fungiform papillae** are scattered over the surface of the tongue, whereas the large **circumvallate papillae** are located at the posterior aspect of the tongue, arranged in a "V" shape. **Foliate papillae** contain taste buds primarily during childhood; they are located on the lateral aspects of the tongue.

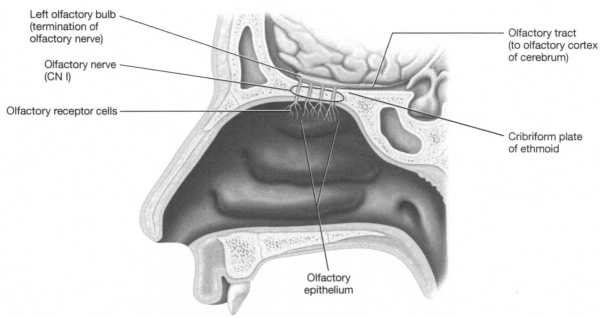

FIGURE **9.15** Nasal cavity and olfactory epithelium.

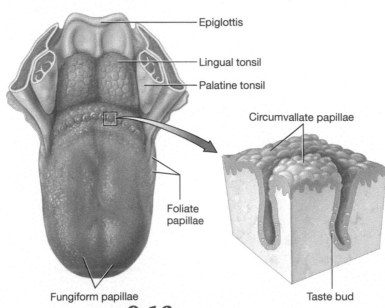

FIGURE **9.16** Surface of the tongue.

Procedure 1 Model Inventory for Olfaction and Taste

Identify the following structures of the olfactory and taste senses on anatomical models and charts. As you examine the anatomical models and diagrams, record the name of the model and the structures you were able to identify on the model inventory in Table 9.3.

Olfaction
1. Nasal cavity
   a. Cribriform plate
2. Olfactory epithelium
   a. Olfactory receptor cells
3. Olfactory nerve (cranial nerve I)
4. Olfactory bulbs
5. Olfactory tract

Taste
1. Taste buds
2. Papillae
   a. Fungiform papillae
   b. Circumvallate papillae
   c. Foliate papillae

TABLE **9.3** Model Inventory for Olfaction and Taste

Model/Diagram	Structures Identified

Exercise 9-4

The General Senses: Cutaneous Sensation

MATERIALS

- ❑ Water-soluble marking pens (two colors)
- ❑ Ruler
- ❑ 2 wooden applicator sticks (or toothpicks)
- ❑ Alcohol swabs

Sensory receptors in the skin respond to different stimuli, including temperature, touch, and pain. These receptors are not distributed throughout the skin equally, but instead are concentrated in certain regions of the body. The following experiments will allow you to determine the relative distribution of the receptors for touch in the skin by performing two tests: the error of localization and two-point discrimination.

Error of Localization

Every region of the skin corresponds to an area of the primary somatosensory cortex of the cerebrum. Some regions are better represented than others and, therefore, are capable of localizing stimuli with greater precision than are less well-represented areas. The **error of localization** (also called tactile localization) tests the ability to determine the location of the skin touched and demonstrates how well represented each region of the skin is in the cerebral cortex.

Procedure 1 Testing Error of Localization

1 Have your partner sit with his or her eyes closed.

2 Use a water-soluble marking pen to place a mark on your partner's anterior forearm.

3 Using a different color of marker, have your partner, still with his or her eyes closed, place a mark as close as possible to where he or she believes the original spot is located.

4 Use a ruler to measure the distance between the two points in millimeters. This is your error of localization.

5 Repeat this procedure for each of the following locations:
  a Anterior thigh
  b Face
  c Palm of hand
  d Fingertip

6 Record your data in Table 9.4.

TABLE **9.4** Error of Localization

Location	Error of Localization (mm)
Anterior forearm	
Anterior thigh	
Face	
Palm of hand	
Fingertip	

Two-Point Discrimination

The **two-point discrimination test** assesses the ability to perceive the number of stimuli ("points") placed on the skin. Areas that have a higher density of touch receptors are better able to distinguish between multiple stimuli than those with fewer touch receptors.

Procedure 2 Testing Two-Point Discrimination

1 Have your partner close his or her eyes.

2 Place the ends of two wooden applicator sticks close together (they should be nearly touching) on your partner's skin on the anterior forearm. Ask how many points your partner can discriminate—one or two.

3 If your partner can sense only one point, move the sticks farther apart. Repeat this procedure until your partner can distinguish two separate points touching his or her skin.

4 Use a ruler to measure the distance between the two sticks in millimeters. This is your two-point discrimination.

5 Repeat this procedure for each of the following locations:

   a Anterior thigh

   b Face (around the lips and/or eyes)

   c Vertebral region

   d Fingertip

6 Record your data in Table 9.5.

TABLE **9.5** Two-Point Discrimination

Location	Two-Point Discrimination (mm)
Anterior forearm	
Anterior thigh	
Face	
Vertebral region	
Fingertip	

7 What results did you expect for the error of localization and two-point discrimination tests? Explain.

8 Did your observations agree with your expectations? Interpret your results.

_____

_____

_____

9

1 Label the following parts of
   the eyeball on **Figure 9.17.**
   - ❏ Choroid
   - ❏ Conjunctiva
   - ❏ Cornea
   - ❏ Iris
   - ❏ Lens
   - ❏ Optic nerve
   - ❏ Posterior cavity
   - ❏ Pupil
   - ❏ Retina
   - ❏ Sclera

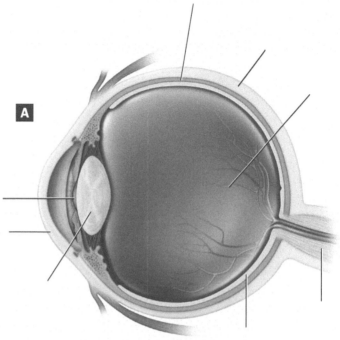

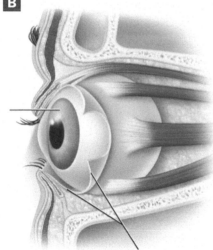

FIGURE **9.17** Eyeball: (**A**) sagittal section; (**B**) external structures.

2 The lacrimal gland is located in the _____ and produces _____ .
   a. superolateral orbit; mucus
   b. inferomedial orbit; tears
   c. superolateral orbit; tears
   d. inferomedial orbit; mucus

3 The rods are responsible for _____, whereas the cones are responsible for _____.

   a. peripheral and dim light vision; high-acuity color vision

   b. high-acuity color vision; peripheral and dim light vision

   c. peripheral and color vision; high-acuity and dim light vision

   d. high-acuity and dim light vision; peripheral and color vision

4 *Fill in the blanks:* The fluid in the anterior cavity is known as _____, whereas the fluid in

the posterior cavity is known as _____.

5 Label the following parts of the ear on **Figure 9.18**.

  ❏ Auricle   ❏ Malleus   ❏ Tympanic membrane

  ❏ Cochlea   ❏ Pharyngotympanic tube   ❏ Vestibule

  ❏ External auditory canal   ❏ Semicircular canals   ❏ Vestibulocochlear nerve

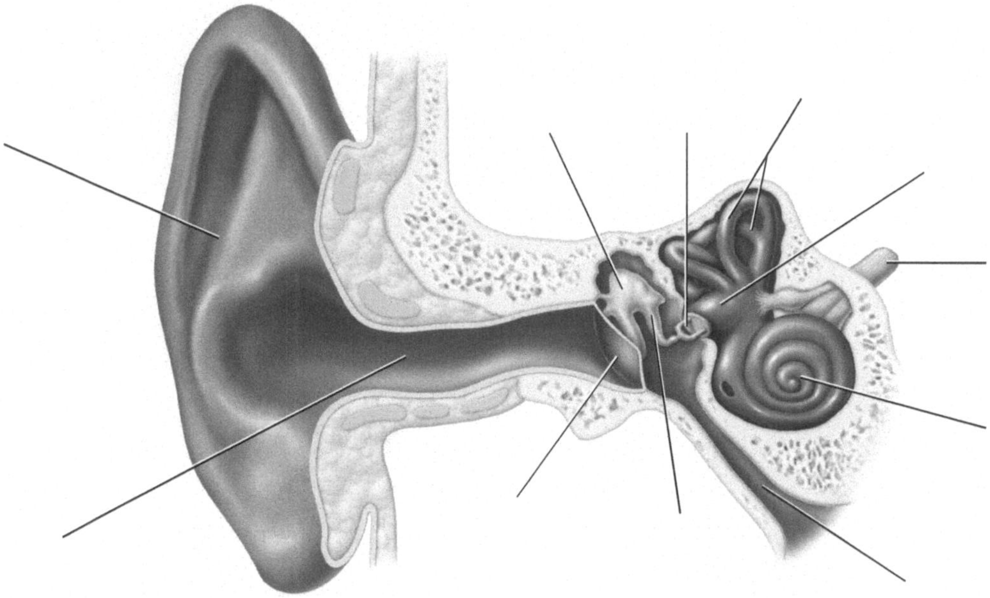

  ❏ Incus   ❏ Stapes

FIGURE **9.18** Anatomy of the ear.

6 The pharyngotympanic tube connects the _____ to the _____.

   a. middle ear; inner ear

   b. external auditory canal; middle ear

   c. inner ear; pharynx

   d. middle ear; pharynx

7 The structures of the cochlea are responsible for _____, whereas the structures of the vestibule and semicircular canals are responsible for _____.

   a. equilibrium; balance

   b. equilibrium; hearing

   c. hearing; equilibrium

   d. hearing; audition

8 The receptors for smell are located in the
   a. gustatory mucosa.
   b. olfactory epithelium.
   c. olfactory fossa.
   d. squamous epithelium.

9 Which of the following types of papillae of the tongue does not house taste buds?
   a. Fungiform papillae.
   b. Filiform papillae.
   c. Circumvallate papillae.
   d. Foliate papillae.

10 You would expect the error of localization and the two-point discrimination threshold to be lowest on the
   a. back.
   b. forearm.
   c. fingertip.
   d. thigh.

9

11 The disease *macular degeneration* is characterized by a gradual loss of vision as a result of degeneration of the macula lutea. Considering the type of cells located in the macula lutea, which type of vision do you think a sufferer of macular degeneration would lose? Why?

_____

_____

_____

_____

12 How would the signs and symptoms differ from those in question 11 in a condition that caused degeneration of rods?

_____

_____

_____

_____

13 LASIK, or *laser-assisted in situ keratomileusis*, is a surgical procedure on the eye that millions of people have undergone to improve visual acuity. It involves the use of a laser to reshape the cornea in individuals suffering from astigmatism, near-sightedness (myopia), or far-sightedness (hyperopia). How could changing the shape of the cornea affect one's visual acuity?

_____

_____

_____

_____

14 Explain why infectious *otitis media* (inflammation of the middle ear) may result in a simultaneous *pharyngitis* (inflammation of the throat).

_____

_____

_____

_____

15 The most common cause of hearing loss is exposure to loud noise such as music, which can permanently damage the hair cells of the cochlea. Would you expect this to result in conductive or in sensorineural hearing loss? What results would you expect from the Rinne and Weber tests in an individual with noise-induced hearing loss?

_____

_____

_____

_____

9

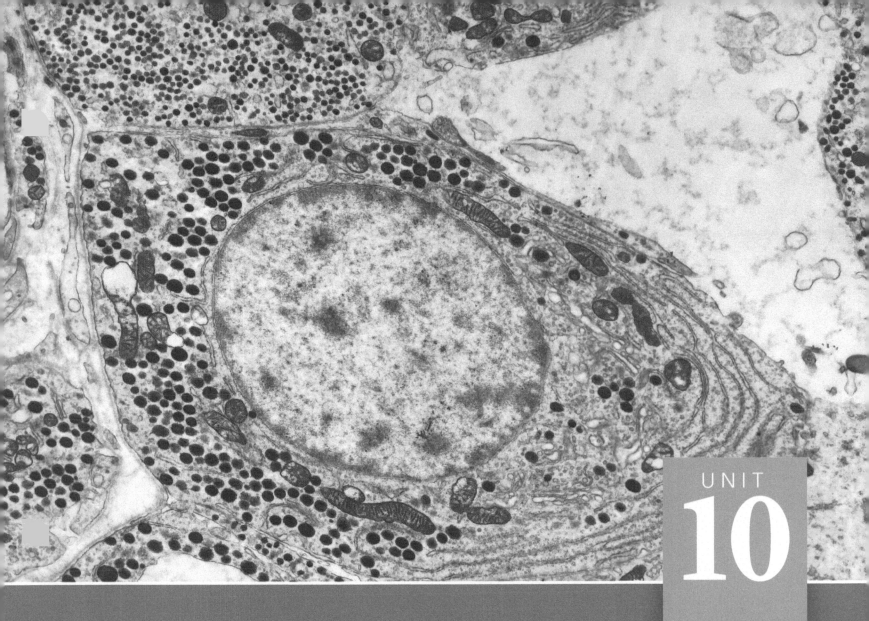

Endocrine System

When you have completed this unit, you should be able to:

1 Describe and identify endocrine organs and structures.

2 Trace the functions, stimulus for secretion, and target tissues of various hormones.

3 Describe the negative feedback mechanisms that control hormone secretion.

4 Apply principles of the endocrine system to clinical cases.

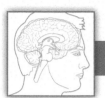

Name _____ Section _____ Date _____

PRE-LAB EXERCISES

Complete the following exercises prior to coming to lab, using your lab manual and textbook for reference.

Pre-Lab Exercise **10-1**

 Key Terms

You should be familiar with the following terms before coming to lab. Please note that key hormones are covered in Pre-Lab Exercise 10-3 (p. 268).

Term	Definition
General Terms	
Endocrine organ (gland)	
Hormone	
Target tissue	
Negative feedback	
Endocrine Organs	
Hypothalamus	
Anterior pituitary	
Posterior pituitary	
Thyroid gland	
Parathyroid glands	
Pineal gland	
Pancreas	

10

Adrenal glands

Ovaries

Testes

10

Pre-Lab Exercise **10-2**

Endocrine System Anatomy

Color the structures of the endocrine system in Figure 10.1, and label them with the following terms from Exercise 10-1 (p. 270). Use Exercise 10-1 in this unit and your text for reference.

❑ Hypothalamus
❑ Pituitary gland
  ☐ Anterior pituitary
  ☐ Posterior pituitary
❑ Pineal gland
❑ Thyroid gland
  ☐ Isthmus
  ☐ Right and left lobes

❑ Parathyroid glands
❑ Thymus gland
❑ Adrenal glands
  ☐ Adrenal cortex
  ☐ Adrenal medulla

❑ Pancreas
  ☐ Pancreatic islets
❑ Ovaries
❑ Testes

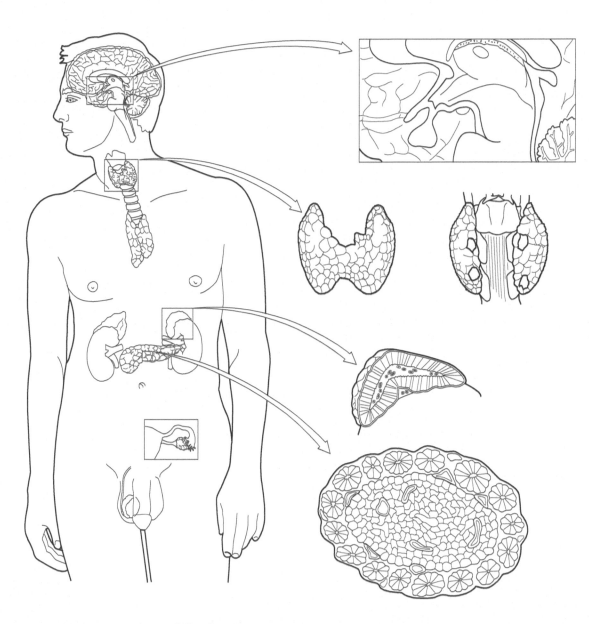

FIGURE **10.1** Organs and tissues of the endocrine system.

10

Pre-Lab Exercise 10-3

✎ Hormones: Target Tissues and Effects

Complete Table 10.1 with the organ that secretes each hormone, the hormone's target tissue(s), and its main effects.

TABLE **10.1** Properties of Hormones

Hormone	Organ that Secretes the Hormone	Target Tissue(s)	Main Effects
Antidiuretic hormone			
Oxytocin			
Thyroid-stimulating hormone			
Adrenocorticotropic hormone			
Growth hormone			
Thyroxine and triiodothyronine (T4 and T3)			
Parathyroid hormone			
Cortisol			
Aldosterone			
Insulin			
Glucagon			

EXERCISES

Color-enhanced TEM of a hormone-secreting pituitary gland cell.

The **endocrine system** is a diverse group of ductless glands that plays a major role in maintaining the homeostasis of multiple physiological variables. The glands work closely with the other system that maintains the homeostasis of multiple physiological variables—the *nervous system*. Although these two systems both work toward the same goal, you will notice that the methods by which they do so differ. The nervous system functions via action potentials (nerve impulses), and releases **neurotransmitters** that directly affect target cells. The effects are nearly immediate, but they are very short in duration.

In contrast, the endocrine system brings about its effects via the secretion of **hormones**—chemicals released into the bloodstream that typically act on distant targets. The effects of hormones are not immediate, but they are longer-lasting than those of the nervous system.

In general, hormones regulate the processes of other cells, including inducing the production of enzymes or other hormones, changing the metabolic rate of a cell, and altering permeability of the plasma membrane. You might think of hormones as the "middle managers" of the body, because they communicate the messages from their "bosses" (the endocrine glands) and tell the "workers" (other cells) what to do. Some endocrine glands (e.g., the thyroid and anterior pituitary glands) secrete hormones as their primary function. Others, however, secrete hormones as a secondary function, examples of which are the heart (atrial natriuretic peptide), adipose tissue (leptin), the kidneys (erythropoietin), and the stomach (gastrin).

This unit introduces you to the anatomy, histology, and physiology of the endocrine organs and hormones. To close out this unit, you will play "endocrine detective" and solve two "endocrine mysteries."

Exercise **10-1**

Endocrine System Anatomy

MATERIALS
- ❏ Endocrine system models
- ❏ Human torso models
- ❏ Head and neck models
- ❏ Fetal models

The 10 organs in the body that have hormone secretion as a primary function are the hypothalamus, the pituitary gland, the pineal gland, the thyroid gland, the parathyroid glands, the thymus, the adrenal gland, the pancreas, and the ovaries or testes (Fig. 10.2). Let's take a closer look at each of these organs (note that we discuss the ovaries and testes only briefly here; they are discussed further in Unit 16).

1. **Hypothalamus and pituitary gland.** The **hypothalamus**, the inferior part of the diencephalon, is known as a **neuroendocrine organ** (Fig. 10.3). It can be likened to the endocrine system's chief executive officer (CEO). It has a close working relationship with the **pituitary gland** (pih-TOO-ih-tehr-ee), to which it is attached by a stalk called the **infundibulum** (in-fun-DIB-yoo-lum). Notice in Figure 10.3 that the pituitary gland is actually two separate structures:

The **anterior pituitary gland**, or **adenohypophysis** (ad-in-oh-hy-PAWF-ih-sis), is composed of glandular epithelium and secretes a variety of hormones that affect other tissues in the body. The **posterior pituitary,** or **neurohypophysis** (noor-oh-hy-PAWF-ih-sis), is actually composed of nervous tissue rather than glandular tissue.

The hypothalamus produces two types of hormones. The first, called **inhibiting** and **releasing hormones**, are those that inhibit and stimulate secretion from the anterior pituitary gland, respectively. In response to hypothalamic-releasing hormones, the anterior pituitary gland secretes hormones that stimulate other endocrine and exocrine glands in the body. In Figure 10.3A, you can see how the hypothalamus communicates with the anterior pituitary gland via a specialized set of blood vessels collectively called the **hypothalamic-hypophyseal portal system** (hy-PAW-fih-see-uhl). The releasing and inhibiting hormones are synthesized by hypothalamic neurons and enter capillaries in the hypothalamus, after which they travel through small veins in the infundibulum. They then enter a second capillary bed in the anterior pituitary, where they exit the blood and interact with anterior pituitary cells to influence their functions.

In addition to inhibiting and releasing hormones, the hypothalamus makes the hormone **oxytocin** (awks-ee-TOH-sin), which triggers milk ejection from the mammary gland and uterine contractions. It also produces **antidiuretic hormone** (an-tee-dy-yoo-RET-ik; **ADH**), which causes water retention from the kidneys. These hormones are produced by hypothalamic neurons that extend the length of the infundibulum down into the posterior pituitary, where they are stored (Fig. 10.3B).

10

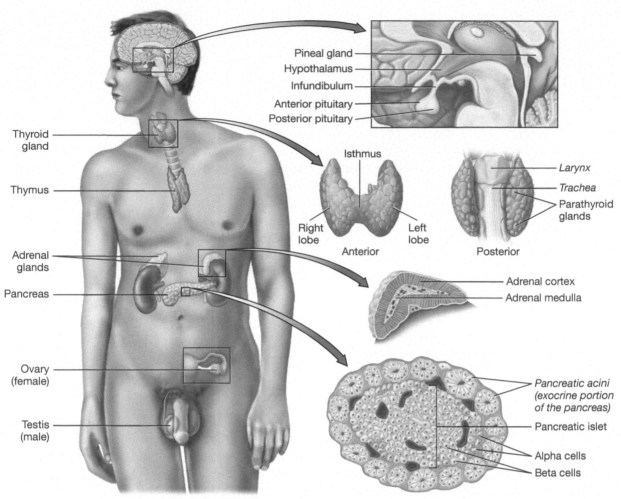

FIGURE **10.2** Organs and tissues of the endocrine system.

   a. The **anterior pituitary gland** mostly produces hormones known as *tropic hormones,* or those that influence the functions of other glands. Examples include **thyroid-stimulating hormone** (**TSH**), which stimulates growth of and secretion from the thyroid, and **adrenocorticotropic hormone** (ah-dree-noh-kohr-tih-koh-TROH-pik; **ACTH**), which stimulates secretion from the adrenal cortex. An exception is **growth hormone** (**GH**), which increases the rate of cell division and protein synthesis in all tissues and has both tropic and nontropic effects.

   b. The posterior pituitary doesn't produce any hormones at all and functions merely as a place to store the oxytocin and ADH produced by the hypothalamus.

2. **Pineal gland.** Recall from Unit 8 (p. 211) that the tiny **pineal gland** (pin-EE-uhl) is located in the posterior and superior diencephalon. This neuroendocrine organ secretes the hormone **melatonin** (mel-uh-TOH-nin) in response to decreased light levels and acts on the reticular formation of the brainstem to trigger sleep.

3. **Thyroid gland.** The **thyroid gland** is located in the anterior and inferior neck superficial to the larynx. It consists of right and left lobes connected by a thin band of tissue called the **isthmus.** Microscopically, it is composed of hollow spheres called **thyroid follicles** (Fig. 10.4). The cells that line the thyroid follicles are simple cuboidal cells called **follicle cells,** and they surround a gelatinous, iodine-rich substance known as **colloid** (KAWL-oyd). The follicle cells respond to TSH from the anterior pituitary by secreting a chemical into the colloid that reacts with iodine to produce two different hormones: **thyroxine** or **T4,** which has four iodine atoms, and **triiodothyronine** (try-aye-oh-doh-THY-roh-neen) or **T3,** which has three iodine atoms. T3 is the most active of the two hormones and acts on essentially all cells in the body to increase the metabolic rate, increase protein synthesis, and regulate the heart rate and blood pressure, among other things.

   Between the follicles we find another cell type called the **parafollicular cells.** These cells produce the hormone **calcitonin** (kal-sih-TOH-nin), which plays a role in calcium ion homeostasis. Calcitonin is secreted when calcium ion levels in the blood rise, and it triggers osteoblast activity and bone deposition.

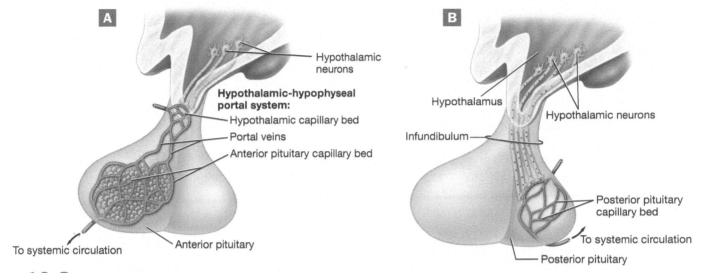

FIGURE **10.3** Hypothalamus and pituitary gland: (**A**) hypothalamus and anterior pituitary; (**B**) hypothalamus and posterior pituitary.

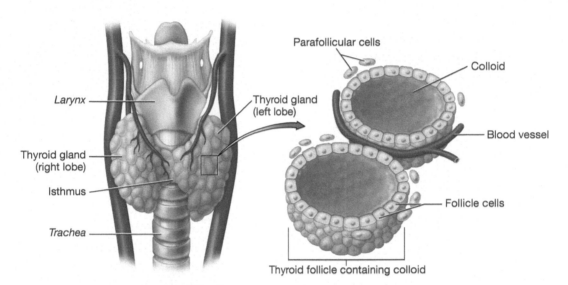

FIGURE **10.4** Thyroid gland and thyroid follicles.

4. **Parathyroid glands.** Refer back to Figure 10.2 where you can see the small **parathyroid glands** on the posterior thyroid gland. They secrete the hormone **parathyroid hormone (PTH)**, which is secreted in response to a decreased level of calcium ions in the blood. It triggers osteoclast activity and resorption of bone tissue, increased calcium ion absorption from the intestines, and increased calcium ion reabsorption from the kidneys. Hormones, that have opposite actions, such as PTH and calcitonin, are called **antagonists**.

5. **Thymus.** The **thymus** sits in the superior mediastinum. It is large and active in infancy and early childhood, during which time it secretes hormones that stimulate the development of T lymphocytes within the thymus. In adults, most of the thymic tissue is gradually replaced by fat and other connective tissue.

6. **Adrenal gland.** As the name implies, the **adrenal glands** (uh-DREE-nuhl) sit atop the superior pole of each kidney (*ad-* = next to; *renal* = kidney). Like the pituitary gland, the adrenal gland is actually composed of two separate structures, the adrenal cortex and the adrenal medulla, which are surrounded by a connective tissue capsule (Fig. 10.5).

   a. **Adrenal cortex.** The superficial region of the adrenal gland consists of glandular tissue called the **adrenal cortex,** which secretes **steroid hormones** in response to stimulation by ACTH and other factors. The outermost zone of the adrenal cortex secretes steroids, such as **aldosterone** (al-DAHS-tur-ohn), that maintain fluid, electrolyte, and acid-base homeostasis. The middle zone secretes steroids, such as **cortisol**, that regulate the stress response, blood glucose, fluid homeostasis, and inflammation. The innermost zone secretes cortisol and steroids called *androgens* that affect the gonads and other tissues.

b. **Adrenal medulla.** The deep region of the adrenal gland, called the **adrenal medulla** (meh-DOOL-uh), consists of modified postsynaptic sympathetic neurons. When stimulated by the sympathetic nervous system, the cells of the adrenal medulla secrete the *catecholamines* **epinephrine** and **norepinephrine** into the blood. These adrenal catecholamines have the same effects on target cells as the norepinephrine and epinephrine released by sympathetic neurons. Secretion of adrenal catecholamines prolongs the sympathetic response, and these hormones are able act on target cells that are not innervated by sympathetic neurons.

7. **Pancreas.** The **pancreas** (PAYN-kree-us) has both endocrine and exocrine functions. Its exocrine functions are carried out by groups of cells called *pancreatic acini*. Embedded within the pancreatic acini are small, round "islands" called **pancreatic islets** (AYE-lets; **Fig. 10.6**) that carry out the pancreas' endocrine functions. The cells within the pancreatic islets secrete the hormones insulin and glucagon (GLOO-kah-gawn), which play a major role in regulating blood glucose levels. **Insulin,** which is produced by cells called **beta (β) cells,** triggers the uptake of glucose by cells, which decreases the concentration of glucose in the blood. The hormone **glucagon,** which is produced by **alpha (α) cells,** is insulin's antagonist—it triggers the release of stored glucose from the liver and the production of new glucose, which increases the concentration of glucose in the blood.

8. **Testes.** The **testes** are the male reproductive organs that produce sperm cells, the male gametes. Cells within the testes called **interstitial cells** produce the steroid hormone **testosterone.** This hormone promotes the production of sperm cells and the development of male secondary sex characteristics, such as a deeper voice, greater bone and muscle mass, and facial hair.

9. **Ovaries.** The **ovaries** are the female reproductive organs that produce oocytes, the female gametes. The ovaries produce the steroid hormones **estrogens** and **progesterone.** Estrogens play a role in the development of oocytes, female secondary sex characteristics such as breasts and subcutaneous fat stores, and a variety of other processes. Progesterone has many effects that prepare the body for pregnancy.

Let's now examine these structures on models and charts. Note that human torso models are typically a good place to start when studying the endocrine system because most of the organs are easy to find. The one exception is the thymus; many torsos and models choose not to show this structure because it is fairly inactive in adults. Fetal models, however, typically show well-developed thymus glands.

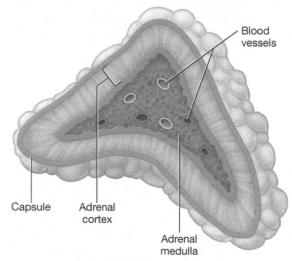

FIGURE **10.5** Adrenal gland.

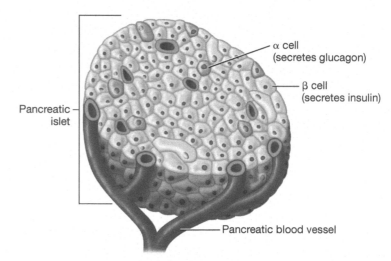

FIGURE **10.6** Pancreatic islet.

Procedure 1 Model Inventory of the Endocrine System

Identify the following structures of the endocrine system on models and diagrams, using this unit and your textbook for reference. As you examine the anatomical models and diagrams, record the name of the model and the structures you were able to identify on the model inventory in Table 10.2.

Endocrine Glands

1. Hypothalamus
   a. Infundibulum
2. Pituitary gland
   a. Anterior pituitary
   b. Posterior pituitary
3. Pineal gland
4. Thyroid gland
   a. Right and left lobes
   b. Isthmus

5. Parathyroid glands
6. Thymus gland
7. Adrenal glands
   a. Adrenal cortex
   b. Adrenal medulla
8. Pancreas
   a. Pancreatic islets
9. Testes
10. Ovaries

TABLE **10.2** Model Inventory for the Endocrine System

Model/Diagram	Structures Identified

10

Exercise 10-2

Time to Trace: Negative Feedback Loops

Earlier in this unit, we pointed out that each hormone has its own stimulus for secretion. The stimulus for secretion is generally a disturbance in the homeostasis of a particular physiological variable, such as blood pressure, the concentration of glucose in the blood, or body temperature. The hormone's response is to act on distant target cells to cause changes that restore the variable to its normal range. When homeostasis is restored, the activity of the glands and concentration of the hormone declines. This type of response is called a **negative feedback loop**.

In this exercise, you will be tracing a hormone's negative feedback loop from the initial homeostatic disturbance, through the hormone's effects on its target cells to restore homeostasis, and, finally, to the decline of the concentration of the hormone in the blood. Following is an example:

Start: the concentration of glucose in the blood falls → the pancreas releases glucagon → the level of the hormone glucagon in the blood rises → the hormone glucagon causes glycogenolysis by the liver → the concentration of glucose in the blood rises → the level of the hormone glucagon in the blood decreases → **End**

Procedure 1 Tracing Negative Feedback Loops

Now it's your turn! Complete the negative feedback loops that follow, using the example as a guide. Refer to Pre-Lab Exercise 10-3 and your textbook for help with the hormones and their actions.

1 **Start:** the concentration of the blood increases (i.e., there is *inadequate water* in the blood) → the hypothalamus

releases the hormone _____ and stores it in the _____ →

the stored hormone _____ is released and so its level in the blood rises →

the hormone _____ causes _____ →

the concentration of the blood _____ → the level of the hormone

_____ in the blood _____ → **End**

2 **Start:** the concentration of glucose in the blood increases → the pancreas releases the hormone

_____ → its concentration in the blood rises → the hormone

_____ causes _____ → blood glucose levels

_____ → the level of the hormone _____

in the blood _____ → **End**

3 **Start:** the concentration of calcium ions in the blood decreases → the parathyroid glands release the hormone

_____ → its level in the blood rises → the hormone _____

causes _____ → the concentration of calcium ions

in the blood _____ → the level of the hormone

_____ in the blood _____ → **End**

10

Exercise 10-3

Endocrine "Mystery Cases"

In this exercise, you will be playing the role of "endocrine detective" to solve endocrine disease mysteries. In each case, you will have a victim who has suddenly fallen ill with a mysterious malady. You will be presented a set of "witnesses," each of whom will give you a clue as to the nature of the illness. Other clues will come from samples you send off to the lab for analysis. You will solve the mystery by providing the victim a diagnosis. (*Note:* You may wish to use your textbook for assistance with these cases.)

Case 1: The Cold Colonel

You are called upon to visit the ailing Col. Lemon. Before you see him, you speak with three witnesses who were with him when he fell ill.

Witness Statements

- *Ms. Magenta:* "Col. Lemon has been hot-blooded for as long as I've known him. But I noticed that he couldn't seem to keep warm. He kept complaining about being cold . . ."

- *Mr. Olive:* "Just between you and me, I've noticed that the old chap has put on quite a bit of weight lately."

- *Professor Purple:* "The colonel and I used to go on major expeditions together. Now he just doesn't seem to have the energy to do much of anything."

What are your initial thoughts about the witnesses' statements? Does one hormone come to mind that may be the cause?

You see the Colonel and collect some blood to send off to the lab. The analysis comes back as follows:

T3 (triiodothyronine): 0.03 ng/dl (normal: 0.2–0.5 ng/dl)

T4 (thyroxine): 1.1 μg/dl (normal: 4–7 μg/dl)

TSH (thyroid-stimulating hormone): 86 mU/l (normal: 0.3–4.0 mU/l)

Analyze the Results

Why do you think the T3 and T4 are low and the TSH is elevated? (*Hint:* Think about negative feedback loops.)

Based upon the witness statements and the laboratory analysis, what is your final diagnosis?

10

Case 2: The Parched Professor

Your next call is to the aid of Professor Purple. Three witnesses are present from whom to take statements.

Witness Statements

- *Mr. Olive*: "I swear I saw him drink a full glass of water every half an hour today. He kept saying how thirsty he was!"

- *Mrs. Blanc*: "He must be going to the . . . well, you know, the little boys' room, two or three times every hour!"

- *Ms. Feather*: "He's been saying lately that his mouth is dry and that he feels weak. Personally, I think he's just not following a healthy diet! He should be drinking some of my herbal teas!"

Based upon the witnesses' statements, what are your initial thoughts? Does one hormone come to mind that could produce these effects?

You interview Professor Purple and collect blood and urine specimens to be sent off to the lab for analysis. The lab reports that the urine osmolality is 150 mOsm/kg, which means the urine is overly dilute (too *much* water in the urine). The blood osmolality is 300 mOsm/kg, meaning the blood is overly concentrated (too *little* water in the blood). The lab also reports that his blood glucose is completely *normal*. What is the significance of these clues?

Analyze the Results

Based upon the witness statements and the laboratory analysis, what is your final diagnosis? (**Hint:** Think of the hormone that is supposed to trigger water retention from the kidneys. Is there a disease where this hormone is deficient?)

10

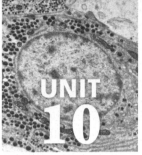

1 Label **Figure 10.7** with the terms below.

❑ Adrenal gland   ❑ Ovary   ❑ Posterior pituitary

❑ Anterior pituitary   ❑ Pancreas   ❑ Testis

❑ Hypothalamus   ❑ Parathyroid glands   ❑ Thyroid gland

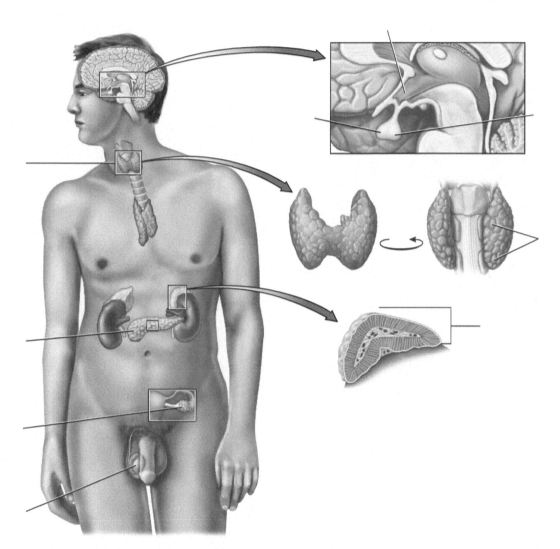

FIGURE **10.7** Endocrine system.

2 A hormone is best described as a

a. chemical released into the extracellular fluid that affects neighboring cells.

b. chemical released by a neuron directly onto a target cell.

c. chemical released into the bloodstream that typically acts on distant targets.

d. All of the above describe hormones.

3 Which of the following is not a function of the hypothalamus?

   a. Produces antidiuretic hormone and oxytocin.

   b. Stimulates production and release of hormones from the anterior pituitary.

   c. Stimulates production and release of hormones from the posterior pituitary.

   d. Inhibits the production and release of hormones from the anterior pituitary.

4 Which of the following sets of hormones are antagonists?

   a. T3 and T4.

   b. Glucagon and insulin.

   c. Epinephrine and cortisol.

   d. Growth hormone and thyroxine.

5 *Matching:* Match the following endocrine organs with the hormone(s) each secretes.

_____ Adrenal medulla	A. Parathyroid hormone
_____ Thyroid gland	B. Insulin and glucagon
_____ Pancreas	C. Steroid hormones
_____ Pineal gland	D. Epinephrine and norepinephrine
_____ Parathyroid glands	E. Thyroxine and triiodothyronine
_____ Adrenal cortex	F. Thyroid-stimulating hormone and growth hormone
_____ Anterior pituitary	G. Melatonin

6 Which of the following is not true regarding endocrine organ histology?

   a. The thyroid gland consists of rings of simple cuboidal follicle cells surrounding colloid.

   b. The pancreas has an exocrine portion consisting of pancreatic islets and an endocrine portion consisting of acinar cells.

   c. The adrenal cortex has three zones of cells that secrete three different types of hormones.

   d. The adrenal medulla is modified nervous tissue of the sympathetic nervous system.

7 What is a negative feedback loop? Cite an example of a negative feedback loop in the endocrine system.

_____

_____

_____

_____

8 *True/False:* Mark the following statements as true (T) or false (F). If the statement is false, correct it to make it a true statement.

   _____ a. Insulin triggers actions that raise the concentration of glucose in the blood.

   _____ b. Interstitial cells within the testes produce testosterone.

   _____ c. Aldosterone is the adrenal hormone responsible for the stress response.

   _____ d. Parathyroid hormone triggers actions that lower the concentration of calcium ions in the blood.

   _____ e. The posterior pituitary produces no hormones of its own.

10

9 In a negative feedback loop, you would expect secretion from an endocrine gland to _____ after homeostasis has been restored.

   a. decrease

   b. increase

   c. remain the same

10 Explain why an increase in thyroid hormone generally leads to a decrease in thyroid-stimulating hormone (TSH).

11 Tumors of the parathyroid gland often result in secretion of excess parathyroid hormone. Considering the function of this hormone, predict the effects of such a tumor.

12 The disease *diabetes mellitus* is due to either destruction of the cells that produce insulin or a decrease in sensitivity of target cells to circulating insulin. How would this affect the level of glucose in the blood? Why?

13 A type of tumor in the pancreas called a *glucagonoma* produces and releases excess glucagon into the blood. How would this affect the level of glucose in the blood? Would the overall effect on blood glucose differ from that of type I diabetes mellitus? Why or why not?

10

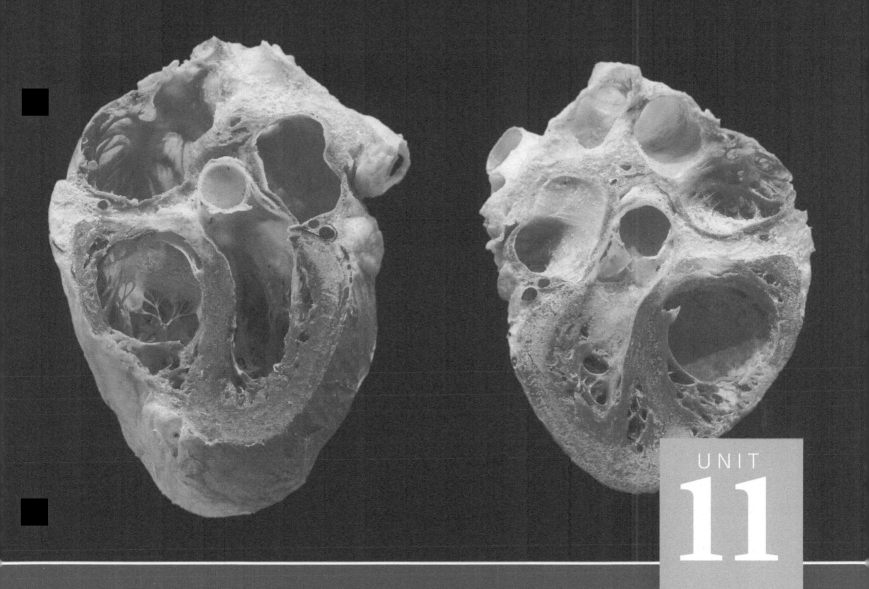

Cardiovascular System

When you have completed this unit, you should be able to:

1 Describe and identify structures of the heart.

2 Trace the pathway of blood flow through the heart.

3 Describe and identify selected arteries and veins.

4 Describe and demonstrate common physical examination tests of the heart and blood vessels.

5 Measure blood pressure using a stethoscope and a sphygmomanometer.

6 Describe the waves and intervals on an ECG, and perform basic ECG interpretation.

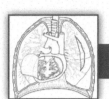

PRE-LAB EXERCISES

Complete the following exercises prior to coming to lab, using your lab manual and textbook for reference.

Pre-Lab Exercise **11-1**

✎ Key Terms

You should be familiar with the following terms before coming to lab.

Term	Definition

Structures of the Heart

Pericardium

Atria (right and left)

Ventricles (right and left)

Tricuspid valve

Mitral (bicuspid) valve

Pulmonary valve

Aortic valve

Great Vessels

Superior vena cava

Inferior vena cava

Pulmonary trunk

Pulmonary veins

Aorta

Cardiovascular Physiology

S1 _____

S2 _____

Pulse point _____

Systolic pressure _____

Diastolic pressure _____

Electrocardiogram _____

Pre-Lab Exercise **11-2**

Pathway of Blood Flow through the Heart

Answer the following questions about the pathway of blood flow through the heart. Use Exercise 11-1 (p. 287) in this unit and your textbook for reference.

1. Regarding veins:

   a. Where do veins carry blood? _____

   b. Is this blood generally oxygenated or deoxygenated? _____

   c. Does this rule have any exceptions? If yes, where? _____

2. Regarding arteries:

   a. Where do arteries carry blood? _____

   b. Is this blood generally oxygenated or deoxygenated? _____

   c. Does this rule have any exceptions? If yes, where? _____

3. Where does each atrium pump blood when it contracts?

   a. Right atrium: _____

   b. Left atrium: _____

4. Where does each ventricle pump blood when it contracts?

   a. Right ventricle: _____

   b. Left ventricle: _____

Pre-Lab Exercise **11-3**

Anatomy of the Heart

Color the two views of the heart in Figure 11.1, and label them with the following terms from Exercise 11-1 (pp. 289–290). Use Exercise 11-1 in this unit and your text for reference.

Structures of the Heart's Chambers
❑ Right atrium
❑ Left atrium
❑ Right ventricle
❑ Left ventricle
❑ Interventricular septum
❑ Chordae tendineae
❑ Papillary muscles

Atrioventricular Valves
❑ Tricuspid valve
❑ Mitral valve

Semilunar Valves
❑ Pulmonary valve
❑ Aortic valve

Great Vessels
❑ Superior vena cava
❑ Inferior vena cava
❑ Pulmonary trunk
❑ Pulmonary veins
❑ Aorta

Coronary Arteries
❑ Right coronary artery
❑ Anterior interventricular artery
❑ Circumflex artery

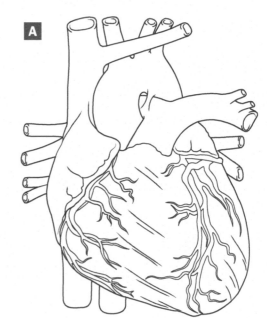

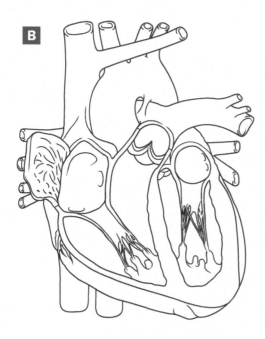

FIGURE **11.1** Heart: **(A)** anterior view; **(B)** frontal dissection of the heart.

11

Pre-Lab Exercise **11-4**

Arterial Anatomy

Label the arterial diagrams in Figure 11.2 with the following terms from Exercise 11-2 (pp. 298 and 300). Use Exercise 11-2 in this unit and your text for reference. Note that these diagrams are presented in color to facilitate identification of the vessels.

Arteries of the Trunk

❑ Aorta
   ☐ Ascending aorta
   ☐ Aortic arch
   ☐ Abdominal aorta
❑ Brachiocephalic artery
❑ Celiac trunk
   ☐ Splenic artery
   ☐ Left gastric artery
   ☐ Common hepatic artery
❑ Superior mesenteric artery
❑ Renal artery
❑ Inferior mesenteric artery
❑ Common iliac artery

Arteries of the Head and Neck

❑ Common carotid artery
❑ Vertebral artery

Arteries of the Upper Limbs

❑ Subclavian artery
❑ Axillary artery
❑ Brachial artery
❑ Radial artery
❑ Ulnar artery

Arteries of the Lower Limbs

❑ External iliac artery
❑ Femoral artery
❑ Popliteal artery
   ☐ Anterior tibial artery
      ▪ Dorsalis pedis artery
   ☐ Posterior tibial artery

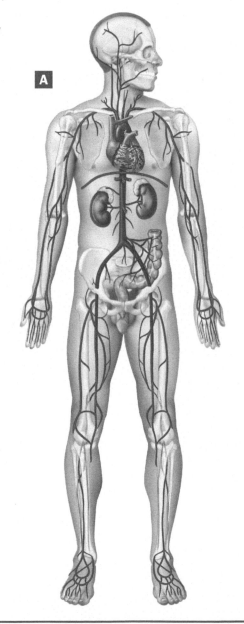

A

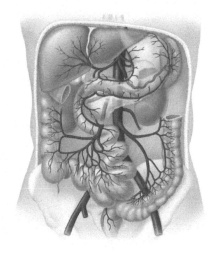

B

FIGURE **11.2** Major arteries:
(A) of the body; **(B)** of the abdomen.

11

Pre-Lab Exercise **11-5**

Venous Anatomy

Label the venous diagrams in Figure 11.3 with the following terms from Exercise 11-3 (pp. 303 and 305). Use Exercise 11-3 in this unit and your text for reference. Note that these diagrams are presented in color to facilitate identification of the vessels.

Veins of the Trunk
- ❏ Superior vena cava
- ❏ Inferior vena cava
- ❏ Brachiocephalic vein
- ❏ Hepatic portal vein
- ❏ Superior mesenteric vein
- ❏ Inferior mesenteric vein
- ❏ Renal vein
- ❏ Common iliac vein

Veins of the Head and Neck
- ❏ Internal jugular vein
- ❏ Vertebral vein

Veins of the Upper Limbs
- ❏ Ulnar vein
- ❏ Radial vein
- ❏ Median cubital vein
- ❏ Brachial vein
- ❏ Basilic vein
- ❏ Cephalic vein
- ❏ Axillary vein
- ❏ Subclavian vein

Veins of the Lower Limbs
- ❏ Great saphenous vein
- ❏ Femoral vein
- ❏ External iliac vein

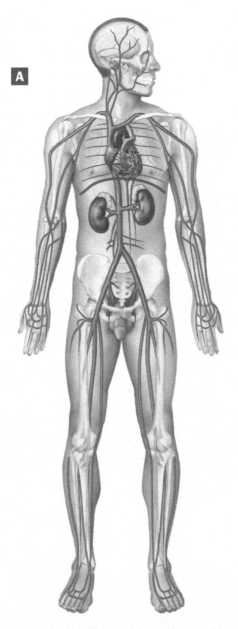

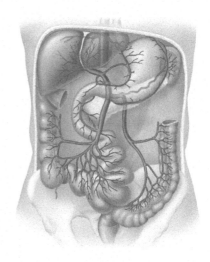

FIGURE **11.3** Major veins: (**A**) of the body; (**B**) of the abdomen and the hepatic portal system.

11

Plastinated and sectioned human heart.

The cardiovascular system transports oxygen, nutrients, wastes, other solutes, and cells throughout the body in the blood. The heart is the remarkable organ that drives this transport, tirelessly beating more than 100,000 times per day to pump more than 8,000 liters of blood around the body.

Blood is delivered to and from the heart by a series of organs known as **blood vessels**. The blood vessels are a closed system of tubes that carries blood around the body. The heart pumps blood away from the heart through a series of **arteries**. Arteries branch as they pass through organs and tissues to form progressively smaller vessels until they branch into tiny **capillary beds**, where gas, nutrient, and waste exchange take place. The blood is drained from the capillaries via a series of **veins** that returns the blood to the heart.

The three major circuits of blood flow in the body are the following:

1. **Systemic.** The **systemic circuit** delivers oxygenated blood to most organs and tissues in the body.

2. **Coronary.** The **coronary circuit** delivers oxygenated blood to the heart.

3. **Pulmonary.** The **pulmonary circuit** delivers deoxygenated blood to the lungs.

In this unit, we examine the anatomy and physiology of this remarkable system. In the upcoming exercises, you will identify the parts of the heart and the body's major blood vessels and trace various pathways of blood flow through the body. In the final exercises, you will investigate cardiovascular physiology by performing common clinical tests and basic interpretation of the *electrocardiogram*.

Exercise 11-1

Anatomy of the Heart

MATERIALS
- ❑ Heart models
- ❑ Preserved heart
- ❑ Dissection equipment
- ❑ Dissection tray
- ❑ Water-soluble marking pens (red and blue)
- ❑ Laminated outline of the heart and lungs
- ❑ Colored pencils

The heart is located in the *mediastinum* and is, on average, about the size of a fist (Fig. 11.4A). Its **apex** is its pointy inferior tip, and its **base** is its flattened posterior side (Fig. 11.4B). The heart is composed of four chambers—the small, superior *right* and *left atria* and the larger, inferior *right* and *left ventricles*. The chambers are separated visually by grooves on the heart's surface: The **atrioventricular sulcus** (ay-tree-oh-ven-TRIK-yoo-lur) is between the atria and ventricles, and the **interventricular sulcus** is between the two ventricles.

As you can see in Figure 11.5, the heart is surrounded by a double-layered membrane called the **pericardium** (pehr-ee-KAR-dee-um). The outermost layer of the pericardium, called the **fibrous pericardium**, anchors the heart to surrounding structures. It is made of dense irregular collagenous connective tissue that is not very distensible, which helps to prevent the heart from overfilling. The inner layer, called the **serous pericardium**, is itself composed of two layers. The outer portion, called the **parietal pericardium**, is functionally fused to the fibrous pericardium. Notice that at the edges of the heart, the parietal pericardium folds over on itself to attach to the heart muscle and form the inner layer of the serous membrane called the **visceral pericardium**, also known as the *epicardium*. Between the parietal and visceral layers, we find a thin layer of serous fluid that occupies a narrow potential space called the **pericardial cavity**. The fluid within the pericardial cavity helps the heart to beat with minimal friction.

The heart itself is an organ that consists of three tissue layers:

1. **Epicardium.** The **epicardium** (ep-ih-KAR-dee-um), or visceral pericardium, is considered the outermost layer of the heart wall. It consists of a layer of simple squamous epithelial tissue and loose connective tissue.

2. **Myocardium.** The middle **myocardium** (my-oh-KAR-dee-um) is the actual muscle of the heart. It consists of cardiac muscle tissue and its fibrous skeleton.

3. **Endocardium.** The innermost **endocardium** is a type of simple squamous epithelium called *endothelium*. It is continuous with the endothelium lining all blood vessels in the body.

11

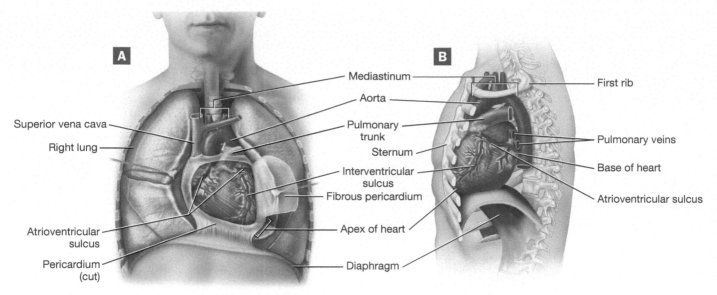

FIGURE **11.4** Thoracic cavity: (**A**) anterior view; (**B**) left lateral view.

Let's now look at the external anatomy of the heart, shown in Figure 11.6. As you can see in the figure, the atria receive blood from *veins*, which are vessels bringing blood to the heart. The ventricles eject blood into *arteries*, which carry blood away from the heart. The vessels entering and exiting the heart are the largest in the body and so are called **great vessels**. The four sets of great vessels include the following:

1. **Superior and inferior venae cavae.** The **superior vena cava** (VEE-nah KAY-vah) is a large vein that drains deoxygenated blood from structures located, in general, above the diaphragm. The **inferior vena cava,** however, drains structures located, in general, below the diaphragm. Both empty into the right atrium and drain the systemic circuit.

2. **Pulmonary trunk.** The **pulmonary trunk** is a large artery that exits from the right ventricle. Shortly after it forms, it splits into **right** and **left pulmonary arteries,** which deliver deoxygenated blood to the lungs through the pulmonary circuit.

3. **Pulmonary veins.** The **pulmonary veins** are the portion of the pulmonary circuit that brings oxygenated blood back to the heart. There are generally four pulmonary veins, and they empty into the left atrium.

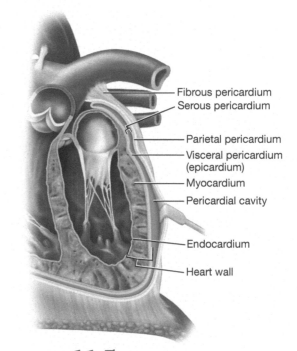

FIGURE **11.5** Layers of the heart wall.

4. **Aorta.** The large **aorta** (ay-OHR-tah) is the first and largest artery of the systemic circuit. It stems from the left ventricle, after which it branches repeatedly to deliver oxygenated blood to the body's cells.

The other set of blood vessels visible on the external surface of the heart consists of the vessels collectively called the **coronary circulation** (KOHR-oh-nehr-ee). The **coronary arteries** branch off the base of the aorta and bring oxygenated blood to the cells of the myocardium. The myocardium is drained by a set of **cardiac veins.** The first coronary artery, the **right coronary artery,** travels along the right side of the atrioventricular sulcus. It terminates as the **posterior interventricular artery,** which serves the posterior heart.

The other coronary artery is the **left coronary artery,** which splits into two branches shortly after it forms. Its first branch is the **anterior interventricular artery** (also known as the *left anterior descending artery*), which travels along the interventricular sulcus to supply the anterior heart. Its second branch is the **circumflex artery** (SIR-kum-flex), which travels in the left side of the atrioventricular sulcus to supply the left atrium and posterior left ventricle. When a coronary artery is blocked, the reduced blood flow to the myocardium causes a situation known as *myocardial ischemia*. Severe blockage may result in hypoxic injury and death to the tissue, a condition termed **myocardial infarction** (commonly called a heart attack).

11

The anatomy of the cardiac veins often varies from person to person, but the following three main veins generally are present: (1) the *small cardiac vein*, which drains the right inferolateral heart; (2) the *middle cardiac vein*, which drains the posterior heart; and (3) the *great cardiac vein*, which drains most of the left side of the heart. All three veins drain into the large **coronary sinus** located on the posterior right atrium. The coronary sinus drains into the right atrium.

On a dissection of the heart, as shown in Figure 11.7, we can see that the atria and ventricles are divided by muscular walls called *septa*. In between the atria is a thin wall called the **interatrial septum** (in-ter-AY-tree-uhl). This wall has a small dent in it called the **fossa ovalis** (FAWS-ah oh-VAL-is), which is a remnant of a hole called the *foramen ovale* that was present during fetal life. The much thicker **interventricular septum** separates the two ventricles.

HINTS & TIPS

Red or Blue?

On anatomical models, vessels that carry oxygenated blood are red, whereas those that carry deoxygenated blood are blue. Systemic arteries carry oxygenated blood to the body's cells and so are red on anatomical models. Systemic veins, on the other hand, carry deoxygenated blood back to the right atrium and so are blue. But be sure to remember that the reverse is true in the pulmonary circuit: The pulmonary arteries carry deoxygenated blood to the lungs, and the pulmonary veins carry oxygenated blood to the heart. So, in the pulmonary circuit, the arteries are blue and the veins are red. Do be aware, though, that sometimes on anatomical models the pulmonary trunk and arteries are painted more of a purple color to differentiate them from systemic veins.

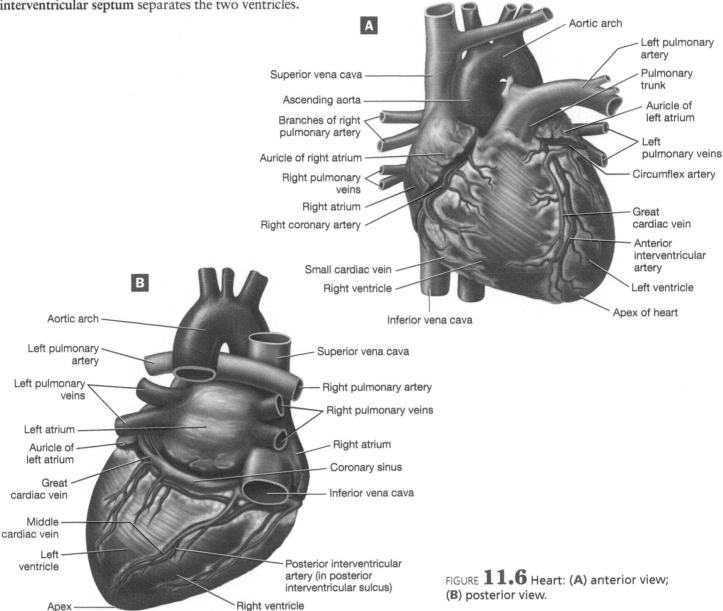

FIGURE **11.6** Heart: (**A**) anterior view; (**B**) posterior view.

11

Let's now look more closely at the heart's four chambers.

1. **Right atrium.** The **right atrium** (AY-tree-um) is the superior right chamber. It receives deoxygenated blood from the superior vena cava, the inferior vena cava, and the coronary sinus—the openings for which we find on the right atrium's posterior side. Externally, it has a large pouch called the **right auricle** (OHR-ih-kuhl) that allows the right atrium to expand and fill with more blood.

2. **Right ventricle.** The **right ventricle** is a wide, crescent-shaped, thin-walled chamber inferior to the right atrium, from which it receives deoxygenated blood. It ejects blood into the pulmonary trunk.

3. **Left atrium.** The superior left chamber is the **left atrium.** It receives oxygenated blood returning from the pulmonary circuit via the pulmonary veins. Externally, it has a **left auricle,** although it is much smaller than the right auricle.

4. **Left ventricle.** The **left ventricle** is a thick, long, circular chamber that receives oxygenated blood from the left atrium and pumps it into the aorta. Notice in the figure that the left ventricle is considerably thicker than the right ventricle. This reflects the fact that the pressure is much higher in the systemic circuit than it is in the pulmonary circuit. The higher pressure requires the left ventricle to pump harder, and so it has greater muscle mass and is thicker.

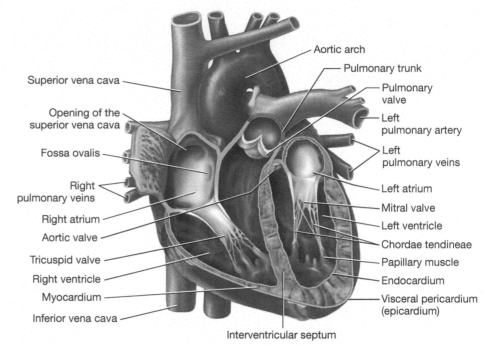

FIGURE **11.7** Frontal dissection of the heart.

Structures called **valves** ensure that blood in the heart flows only in a single direction (Figs. 11.7 and 11.8). There are two types of valves in the heart. First are the valves between the atria and ventricles, which are called **atrioventricular valves.** The three-cusped **tricuspid valve** is between the right atrium and right ventricle, and the two-cusped **mitral** (MY-trul) or **bicuspid valve** is between the left atrium and left ventricle. Each cusp of the atrioventricular valves is attached to collagenous "strings" called **chordae tendineae** (KOHR-dee tin-din-EE-ee), which are themselves attached to muscles within the ventricular wall called **papillary muscles.** When the ventricles contract, the papillary muscles pull the chordae tendineae taut, which puts tension on the cusps and prevents them from everting into the atria, a condition called *prolapse.*

Second are the valves between the ventricles and their arteries, which are called **semilunar valves.** The **pulmonary valve** sits between the right ventricle and the pulmonary trunk, and the **aortic valve** sits between the left ventricle and the aorta. Note that there are no chordae tendineae or papillary muscles attached to the semilunar valves.

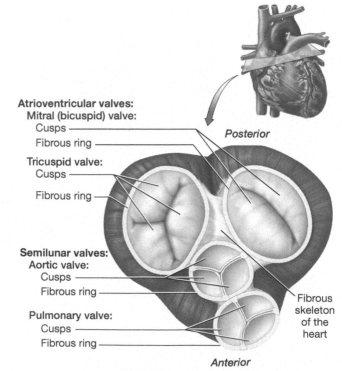

FIGURE **11.8** Transverse section of the heart showing all four valves.

Procedure 1 Model Inventory for the Heart

Identify the following structures of the heart on models and diagrams, using this unit and your textbook for reference. As you examine the anatomical models and diagrams, record the name of the model and the structures you were able to identify on the model inventory in Table 11.1.

1. General structures
   a. Mediastinum
   b. Apex of the heart
   c. Base of the heart
   d. Atrioventricular sulcus
   e. Interventricular sulcus
   f. Pericardium
      (1) Fibrous pericardium
      (2) Serous pericardium
         (a) Parietal pericardium
         (b) Visceral pericardium (epicardium)
         (c) Pericardial cavity
   g. Myocardium
   h. Endocardium
2. Great vessels
   a. Superior vena cava
   b. Inferior vena cava
   c. Pulmonary trunk
   d. Right and left pulmonary arteries
   e. Pulmonary veins
   f. Aorta
3. Coronary arteries
   a. Right coronary artery
   b. Posterior interventricular artery
   c. Left coronary artery
   d. Anterior interventricular artery
   e. Circumflex artery

4. Cardiac veins
   a. Coronary sinus
5. Interatrial septum
   a. Fossa ovalis
6. Interventricular septum
7. Right atrium
   a. Opening of the superior vena cava
   b. Opening of the inferior vena cava
   c. Opening of the coronary sinus
   d. Right auricle
8. Right ventricle
9. Left atrium
   a. Left auricle
   b. Openings of the pulmonary veins
10. Left ventricle
11. Atrioventricular valves
   a. Tricuspid valve
   b. Mitral valve
   c. Chordae tendineae
   d. Papillary muscles
12. Semilunar valves
   a. Pulmonary valve
   b. Aortic valve

11

TABLE **11.1** Model Inventory for the Heart

Model/Diagram	Structures Identified

Procedure 2 Time to Draw

In the space provided, draw, color, and label one of the heart models that you examined. Draw both the anterior view and the frontal dissection. In addition, write the function or definition of each structure that you label.

Procedure 3 Heart Dissection

You will now examine a preserved heart or a fresh heart, likely from a sheep or a cow. Follow the procedure detailed here to find the structures you just studied on models.

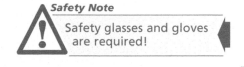

Safety Note

Safety glasses and gloves are required!

1 Orient yourself by first determining the superior aspect and the inferior aspect of the heart. The superior aspect of the heart is the broad end, and the inferior aspect (apex) is the pointy tip. Now orient yourself to the anterior and posterior sides. The easiest way to do this is to locate the pulmonary trunk—the vessel directly in the middle of the anterior side. Find the side from which the pulmonary trunk originates, and you will be on the anterior side, which you can see in Figure 11.9. Structures to locate at this time are the following:

a Parietal pericardium (may not be attached).

b Visceral pericardium (shiny layer over the surface of the heart).

c Aorta.

d Pulmonary trunk.

e Superior vena cava.

f Inferior vena cava.

g Pulmonary veins.

h Ventricles.

i Atria.

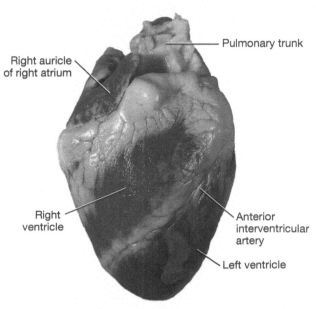

FIGURE **11.9** Anterior view of a sheep heart.

Finding the coronary vessels tends to be difficult because the superficial surface of the heart is covered with adipose tissue. To see the coronary vessels, carefully dissect the adipose tissue.

2 Locate the superior vena cava. Insert scissors or a scalpel into the superior vena cava, and cut down into the right atrium. Before moving on to step 3, note the structure of the tricuspid valve, and draw it in the space provided. How many flaps do you see? What is the function of this valve?

3 After the right atrium is exposed, continue the cut down into the right ventricle, which is shown in Figure 11.10. Structures to locate at this time include:

a Tricuspid valve.

b Chordae tendineae.

c Papillary muscles.

d Myocardium.

e Endocardium (shiny layer on the inside of the heart).

4 Insert the scissors into the pulmonary trunk. Note the structure of the pulmonary valve, and draw it in the space provided. How does it differ structurally from the tricuspid valve? What is the function of this valve?

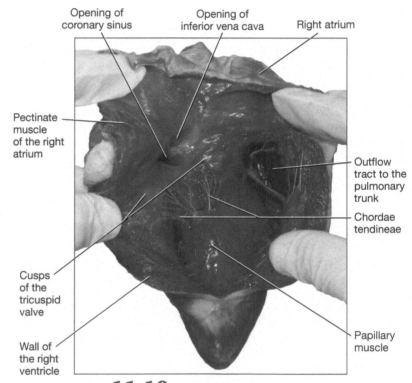

FIGURE **11.10** Right ventricle of a sheep heart.

5 Insert the scissors into a pulmonary vein. Cut down into the left atrium. Note the structure of the mitral valve, and draw it in the space provided. What is the function of this valve? How does its structure differ from that of the pulmonary and tricuspid valves?

6 Continue the cut into the left ventricle. Note the thickness of the left ventricle, as shown in Figure 11.11. How does it compare with the thickness of the right ventricle? Why is there a difference?

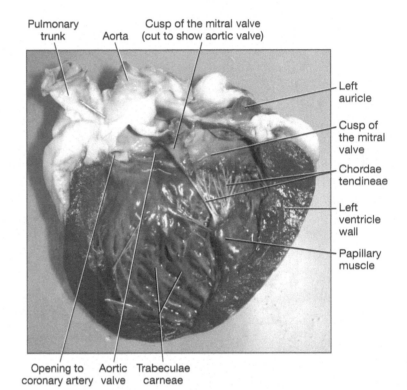

FIGURE **11.11** Left ventricle of a sheep heart.

7 Insert the scissors into the aorta. Extend the cut until you can see the aortic valve. Draw the aortic valve in the space provided. Is it structurally more similar to the pulmonary valve or the mitral valve? What is the function of this valve?

8 Your instructor may wish you to identify other structures on the heart. List any additional structures.

Procedure 4 Tracing Blood through the Heart

Now that you've learned the parts of the heart, put the pieces together and trace the pattern of blood as it flows through the heart and lungs. First, write out the entire pathway, beginning with the main systemic veins that drain into the heart and ending with the main systemic artery into which the heart pumps. Don't forget to include the valves in your tracing. Then, use water-soluble markers and a laminated outline of the heart to draw the pathway of blood as it flows through the heart and pulmonary circulation. Use a blue marker to indicate areas that contain deoxygenated blood and a red marker to indicate areas that contain oxygenated blood. If no laminated outline is available, use colored pencils and Figure 11.12.

Start: Systemic venous circuit →

→ **End:** Systemic arterial circuit

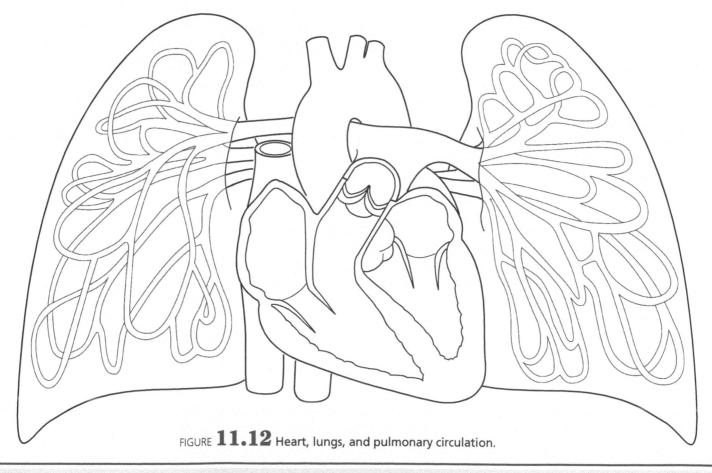

FIGURE **11.12** Heart, lungs, and pulmonary circulation.

Exercise 11-2

Major Arteries of the Body

MATERIALS

❏ Blood vessel models:
☐ Human torsos
☐ Brain
☐ Head and neck
☐ Abdomen
☐ Upper limb
☐ Lower limb

The systemic arterial circuit begins with the largest artery in the body, the **aorta**. An overview of the systemic arterial circuit is shown in Figure 11.13. The aorta originates from the left ventricle as the **ascending aorta,** which ascends until it curves around to form the **aortic arch**. There are three major branches off the aortic arch: the *brachiocephalic trunk*, the *left common carotid artery*, and the *left subclavian artery*. The **brachiocephalic trunk** (bray-kee-oh-seh-FAL-ik) is a small trunk that veers to the right. Shortly after passing deep to the clavicle, it splits into the right common carotid and right subclavian arteries.

The arterial supply of the head and neck comes primarily from the **right** and **left common carotid arteries** (kuh-RAWT-id; Fig. 11.14). Notice that the right and left common carotid arteries have different origins—the right side branches from the brachiocephalic trunk, and the left side from the aortic arch. In the neck, these arteries branch into the internal and external carotid arteries. The **external carotid artery** gives off many branches that supply the structures of the head, neck, and face. One of its terminal branches is the large **superficial temporal artery**, which crosses the temporal bone to supply the scalp.

The **internal carotid arteries** pass through the carotid canals to enter the cranial cavity, where they give off branches that supply the brain. Notice in Figure 11.15 that two of these vessels, the *anterior cerebral arteries*, are connected by a small *anterior communicating artery*. Also supplying the brain are the **vertebral arteries**, which branch from the subclavian arteries. The vertebral arteries then pass through the vertebral foramina of the cervical vertebrae and enter the cranial cavity through the foramen magnum. At the brainstem, the two vertebral arteries fuse to become the single **basilar artery** (BAY-zih-lur). Shortly after it forms, the basilar artery splits again into branches that give off small *posterior communicating arteries*, which connect the circulation of the basilar artery with that of the internal carotid arteries. As you can see in Figure 11.15, the anterior cerebral arteries, anterior communicating artery, and posterior communicating arteries form a continuous structure known as the **cerebral arterial circle**. This vascular "roundabout" provides alternate routes of circulation to the brain if one of the arteries supplying the brain becomes blocked.

The arterial supply to the upper limb begins with the right and left **subclavian arteries** (sub-KLAY-vee-in; Fig. 11.16). The subclavian artery becomes the **axillary artery** near the axilla. In the arm, the axillary artery becomes the **brachial artery** (BRAY-kee-uhl), which splits into the **radial artery** and the **ulnar artery** just distal to the antecubital fossa. As you would expect, these arteries travel alongside the bones for which they are named—the radial artery is lateral and the ulnar artery is medial.

Let's now turn to the thorax and abdomen. Near the left superior sternal border, the aortic arch curves inferiorly to become the **thoracic aorta**. The thoracic aorta descends through the thoracic cavity posterior to the heart, giving off *posterior intercostal arteries*, after which it passes through the diaphragm to become the **abdominal aorta**. The major branches of the abdominal aorta, shown in Figure 11.17, include the following:

1. **Celiac trunk.** The short, stubby **celiac trunk** (SEE-lee-ak) is the first branch off the abdominal aorta. It splits almost immediately into the **common hepatic artery**, which supplies the liver, stomach, pancreas, and duodenum (part of the small intestine); the **splenic artery** (SPLEN-ik), which supplies the spleen, stomach, and pancreas; and the **left gastric artery**, which supplies the stomach.

2. **Renal arteries.** Inferior to the celiac trunk, we find the two **renal arteries** (REE-nuhl), which serve the kidneys. Note that the kidneys are posterior to the other abdominal organs in Figure 11.17 and so the renal arteries are slightly obscured by other organs and vessels. They are seen more clearly in Figure 11.13.

3. **Superior mesenteric artery.** In the same vicinity of the renal arteries is another branch called the **superior mesenteric artery** (mez-en-TEHR-ik). As its name implies, it travels through the membranes of the intestines (called the *mesentery*) and supplies the small and much of the large intestines.

4. **Inferior mesenteric artery.** The last large branch off the abdominal aorta is the **inferior mesenteric artery**, which supplies the remainder of the large intestine.

The abdominal aorta terminates by bifurcating into two **common iliac arteries** (ILL-ee-ak), which themselves bifurcate into an internal iliac artery and an external iliac artery (Figs. 11.17 and 11.18). The **internal iliac artery** supplies structures of the pelvis, and the **external iliac artery** passes deep to the inguinal ligament to enter the anterior thigh, where it becomes the **femoral artery**. The femoral artery remains anterior until about midway down the thigh when it passes posteriorly to become

11

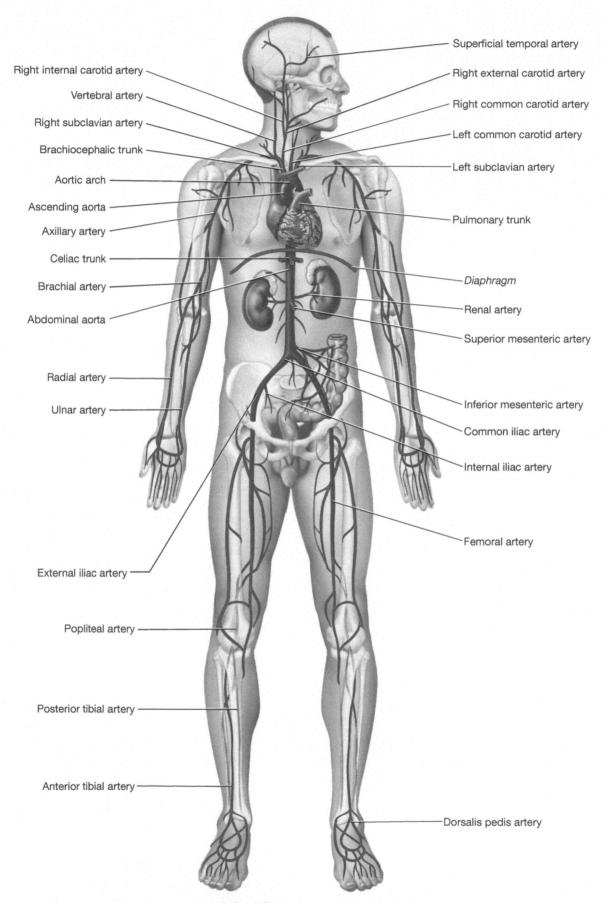

Right internal carotid artery

Vertebral artery

Right subclavian artery

Brachiocephalic trunk

Aortic arch

Ascending aorta

Axillary artery

Celiac trunk

Brachial artery

Abdominal aorta

Radial artery

Ulnar artery

External iliac artery

Popliteal artery

Posterior tibial artery

Anterior tibial artery

Superficial temporal artery

Right external carotid artery

Right common carotid artery

Left common carotid artery

Left subclavian artery

Pulmonary trunk

Diaphragm

Renal artery

Superior mesenteric artery

Inferior mesenteric artery

Common iliac artery

Internal iliac artery

Femoral artery

Dorsalis pedis artery

FIGURE **11.13** Major arteries of the body.

the **popliteal artery** (pahp-lih-TEE-uhl; named for the popliteal fossa, or the posterior knee). Just distal to the popliteal fossa, the popliteal artery divides into its two main branches: the **anterior tibial artery**, which continues in the anterior foot as the **dorsalis pedis artery** (dohr-SAL-iss PEE-diss), and the **posterior tibial artery**, which curls underneath the medial malleolus and continues to the plantar surface of the foot.

Many of the models in your lab likely show both arteries and veins. For this reason, we have included a few photos of anatomical models with arteries and veins to help guide you as you do your model inventories. These photos are in Figures 11.26 through 11.28 on pp. 308–309 (after we discuss veins).

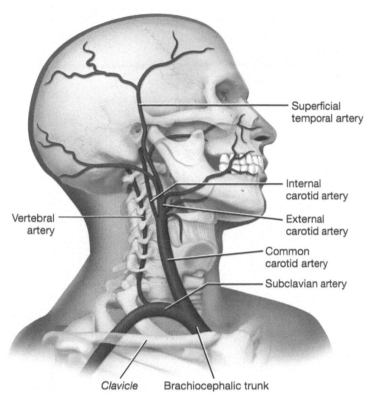

FIGURE **11.14** Arteries of the head and neck.

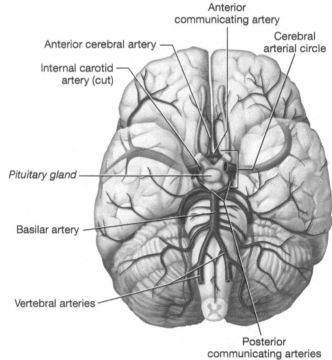

FIGURE **11.15** Arteries of the brain, inferior view.

11

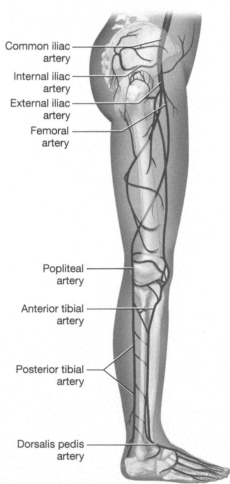

Arteries of the right upper limb and thorax:
- Vertebral artery
- Common carotid arteries
- Subclavian artery
- Left subclavian artery
- Axillary artery
- Brachiocephalic trunk
- Posterior intercostal arteries
- Thoracic aorta
- Brachial artery
- Abdominal aorta
- Radial artery
- Ulnar artery

FIGURE **11.16** Arteries of the right upper limb and thorax.

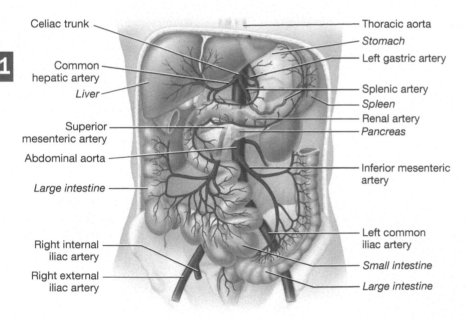

Arteries of the abdomen:
- Celiac trunk
- Common hepatic artery
- Liver
- Superior mesenteric artery
- Abdominal aorta
- Large intestine
- Right internal iliac artery
- Right external iliac artery
- Thoracic aorta
- Stomach
- Left gastric artery
- Splenic artery
- Spleen
- Renal artery
- Pancreas
- Inferior mesenteric artery
- Left common iliac artery
- Small intestine
- Large intestine

FIGURE **11.17** Arteries of the abdomen.

Arteries of the right lower limb, lateral view:
- Common iliac artery
- Internal iliac artery
- External iliac artery
- Femoral artery
- Popliteal artery
- Anterior tibial artery
- Posterior tibial artery
- Dorsalis pedis artery

FIGURE **11.18** Arteries of the right lower limb, lateral view.

11

Procedure 1 Model Inventory for Arteries

Identify the following arteries on models and diagrams, using this unit and your textbook for reference. As you examine the anatomical models and diagrams, record the name of the model and the structures you were able to identify on the model inventory in Table 11.2.

Arteries of the Trunk

1. Aorta
   a. Ascending aorta
   b. Aortic arch
   c. Thoracic aorta
   d. Abdominal aorta
2. Brachiocephalic trunk
3. Subclavian artery
4. Celiac trunk
   a. Common hepatic artery
   b. Splenic artery
   c. Left gastric artery
5. Renal arteries
6. Superior mesenteric artery
7. Inferior mesenteric artery
8. Common iliac artery
   a. Internal iliac artery
   b. External iliac artery

Arteries of the Head and Neck

1. Common carotid arteries
   a. Right common carotid artery
   b. Left common carotid artery
2. External carotid artery
   a. Superficial temporal artery
3. Internal carotid artery
4. Vertebral artery
5. Basilar artery
6. Cerebral arterial circle

Arteries of the Upper Limbs

1. Axillary artery
2. Brachial artery
3. Radial artery
4. Ulnar artery

Arteries of the Lower Limbs

1. Femoral artery
2. Popliteal artery
   a. Anterior tibial artery
      (1) Dorsalis pedis artery
   b. Posterior tibial artery

TABLE **11.2** Model Inventory for Arteries

Model/Diagram	Structures Identified

Exercise 11-3

Major Veins of the Body

MATERIALS
- ☐ Blood vessel models:
  - ☐ Human torsos
  - ☐ Brain
  - ☐ Head and neck
  - ☐ Abdomen
  - ☐ Upper limb
  - ☐ Lower limb
  - ☐ Dural sinuses

Arteries of the systemic circuit deliver oxygenated, nutrient-rich blood to capillary beds. Here, gases, nutrients, and wastes are exchanged. The deoxygenated, carbon dioxide-rich blood is then drained from the capillary beds by a series of veins. The two largest veins in the body are the **superior vena cava**, which drains most structures superior to the diaphragm, and the **inferior vena cava**, which drains most structures inferior to the diaphragm. An overview of the major veins of the body is shown in Figure 11.19.

The head and the neck are drained primarily by the internal and external jugular veins, with a small contribution from the **vertebral vein** in the posterior neck (Fig. 11.20). The much smaller and more lateral **external jugular vein** (JUG-yoo-lur) drains the face and the scalp, and the larger **internal jugular vein,** which travels in a sheath with the common carotid artery, drains the brain. Note, however, that venous blood from the brain does not simply drain into one vein and exit the head. Instead, blood from brain capillaries drains into cerebral veins, and then into the **dural sinuses,** which are spaces between the two layers of the dura mater. From here, blood drains into the internal jugular vein. (The individual dural sinuses are labeled in Figure 11.21 for your reference.)

Blood from the deep structures of the upper limb is drained by the **radial** and **ulnar veins,** both of which parallel the bones for which they are named (Fig. 11.22). These two veins merge in the arm to form the **brachial vein** near the antecubital fossa. The superficial structures of the upper limb are drained by three veins: the lateral **cephalic vein,** the middle **median antebrachial vein,** and the medial **basilic vein.** Notice in the figure that the cephalic vein and the basilic vein are united in the antecubital fossa by the **median cubital vein** (KYOO-bit-uhl). This is a frequent site for drawing blood with a syringe. Around the axilla, the basilic vein joins the brachial vein to form the **axillary vein.** Near the clavicle, the axillary vein becomes the **subclavian vein,** which drains into the **brachiocephalic vein,** and finally into the superior vena cava.

Also shown in Figure 11.22 is the venous drainage system of the posterior thoracic and abdominal walls, which drains into a set of veins called the **azygos system** (ay-ZY-gus). This system consists of three veins: the *azygos vein*, the *hemiazygos vein*, and the *accessory hemiazygos vein.*

Veins draining blood from the organs of the abdomen are named largely in parallel with the arteries that serve the organs: The **renal veins** drain the kidneys (Fig. 11.23), the **splenic vein** drains the spleen, the **gastric veins** drain the stomach, the **superior mesenteric vein** drains the small intestine and much of the large intestine, and the **inferior mesenteric vein** drains the remainder of the large intestine. Although the renal veins empty into the inferior vena cava, the blood from the latter four veins does not drain into the inferior vena cava directly. Instead, note in Figure 11.24 that each vein drains into a common vein called the **hepatic portal vein.** Here, nutrient-rich blood passes through the sinusoids of the liver, where it is processed and detoxified. In this way, everything we ingest (except lipids, which we discuss in Unit 14) must travel through the liver before entering the systemic circulation. After the blood has filtered through the hepatic portal system, it exits via **hepatic veins** and drains into the inferior vena cava.

The deep structures of the leg are drained by the **femoral vein** (Figs. 11.19 and 11.25). The largest superficial vein of the lower limb is the **great saphenous vein,** which drains the medial leg and thigh and empties into the femoral vein in the proximal thigh. The femoral vein becomes the **external iliac vein** after it passes deep to the inguinal ligament. In the pelvis, the external iliac vein merges with the **internal iliac vein,** which drains pelvic structures, forming the **common iliac vein.** The two common iliac veins unite to form the inferior vena cava near the superior part of the pelvis.

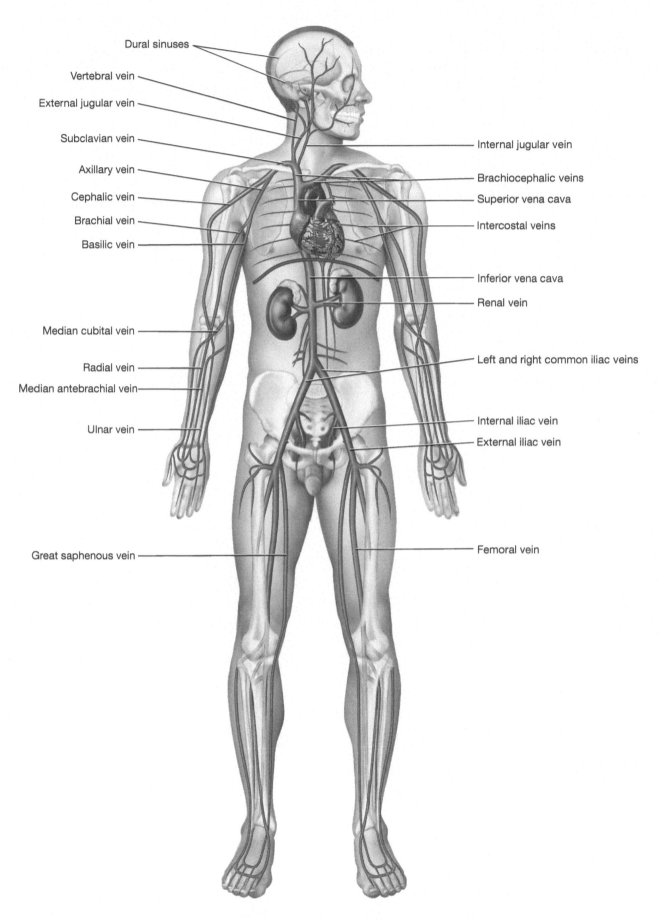

Dural sinuses

Vertebral vein

External jugular vein

Subclavian vein

Axillary vein

Cephalic vein

Brachial vein

Basilic vein

Median cubital vein

Radial vein

Median antebrachial vein

Ulnar vein

Great saphenous vein

Internal jugular vein

Brachiocephalic veins

Superior vena cava

Intercostal veins

Inferior vena cava

Renal vein

Left and right common iliac veins

Internal iliac vein

External iliac vein

Femoral vein

FIGURE **11.19** Major veins of the body.

11

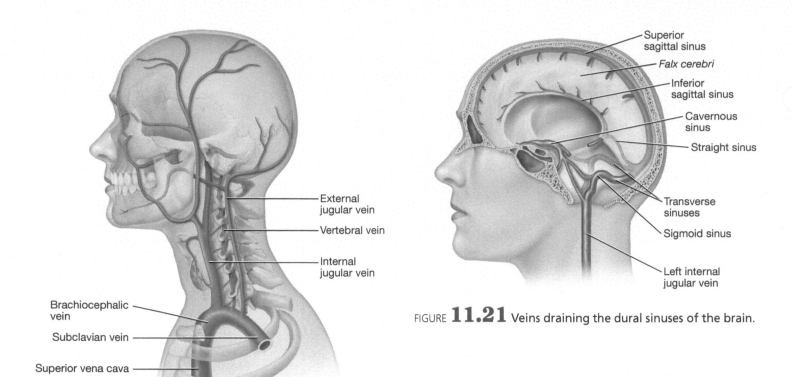

FIGURE **11.20** Superficial veins of the head and neck.

FIGURE **11.21** Veins draining the dural sinuses of the brain.

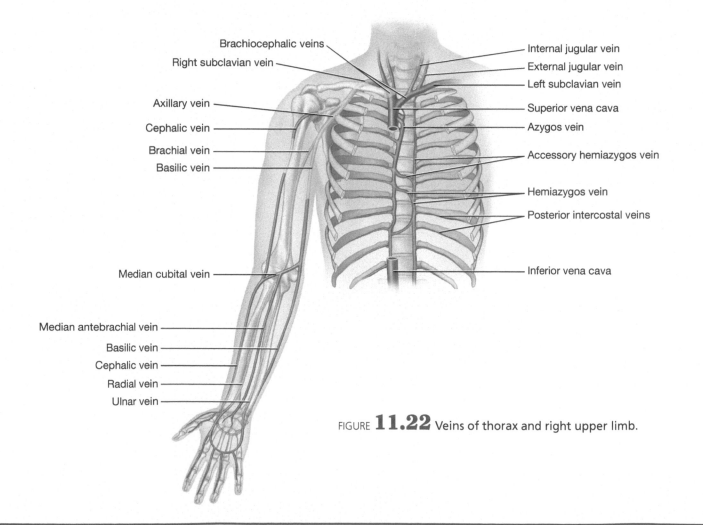

FIGURE **11.22** Veins of thorax and right upper limb.

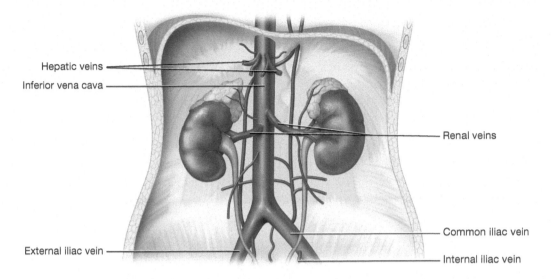

FIGURE **11.23** Veins draining the abdominal area (excluding the hepatic portal system).

Hepatic veins
Inferior vena cava
Renal veins
External iliac vein
Common iliac vein
Internal iliac vein

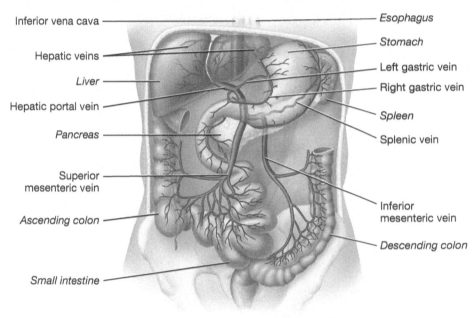

Inferior vena cava
Hepatic veins
Liver
Hepatic portal vein
Pancreas
Superior mesenteric vein
Ascending colon
Small intestine

Esophagus
Stomach
Left gastric vein
Right gastric vein
Spleen
Splenic vein
Inferior mesenteric vein
Descending colon

FIGURE **11.24** Veins of the abdomen and the hepatic portal system.

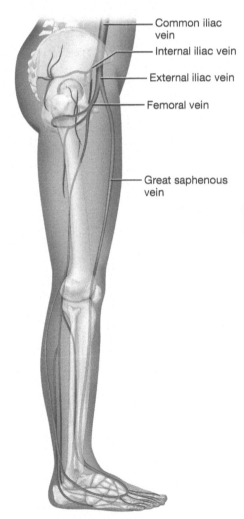

Common iliac vein
Internal iliac vein
External iliac vein
Femoral vein
Great saphenous vein

FIGURE **11.25** Veins of the lower limb, lateral view.

Procedure 1 Model Inventory for Veins

Identify the following veins on models and diagrams, using this unit and your textbook for reference. As you examine the anatomical models and diagrams, record the name of the model and the structures you were able to identify on the model inventory in Table 11.3. Figures 11.26 through 11.28 are photos of anatomical models that show both arteries and veins; you may use these to help guide your work as you do your model inventory.

Veins of the Trunk

1. Superior vena cava
2. Inferior vena cava
3. Azygos system
4. Renal vein
5. Splenic vein
6. Gastric veins
7. Superior mesenteric vein
8. Inferior mesenteric vein
9. Hepatic portal vein
10. Hepatic veins

Veins of the Head and Neck

1. Vertebral vein
2. External jugular vein
3. Internal jugular vein
4. Dural sinuses

Veins of the Upper Limbs

1. Brachial vein
2. Cephalic vein
3. Median antebrachial vein
4. Basilic vein
5. Median cubital vein
6. Axillary vein
7. Subclavian vein
8. Brachiocephalic vein

Veins of the Lower Limbs and Pelvis

1. Great saphenous vein
2. Femoral vein
3. External iliac vein
4. Internal iliac vein
5. Common iliac vein

11

TABLE **11.3** Model Inventory for Veins

Model/Diagram	Structures Identified

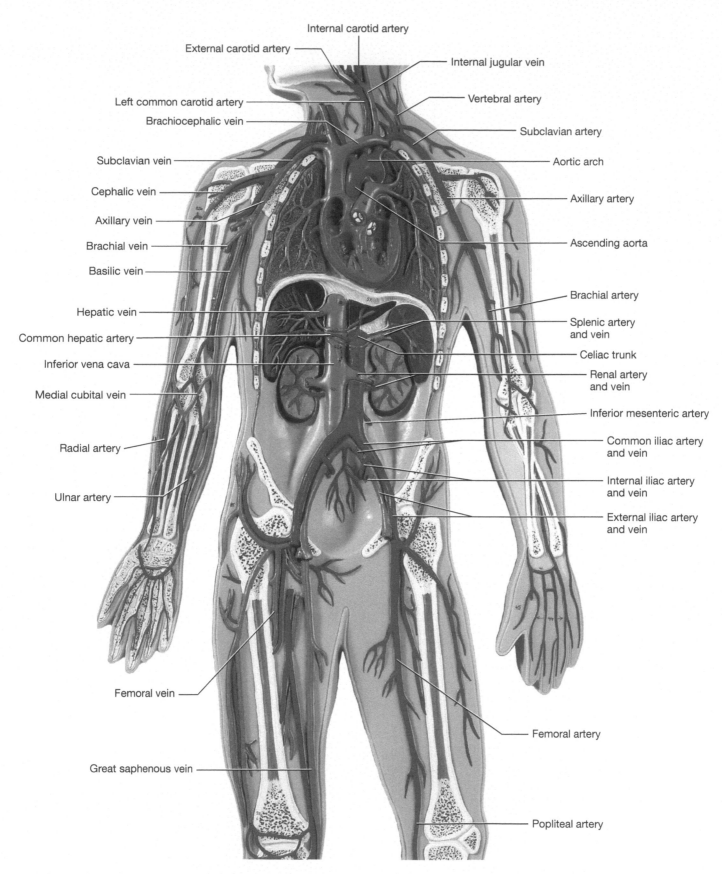

Internal carotid artery

External carotid artery

Internal jugular vein

Left common carotid artery

Vertebral artery

Brachiocephalic vein

Subclavian artery

Subclavian vein

Aortic arch

Cephalic vein

Axillary artery

Axillary vein

Ascending aorta

Brachial vein

Basilic vein

Brachial artery

Hepatic vein

Splenic artery and vein

Common hepatic artery

Celiac trunk

Inferior vena cava

Renal artery and vein

Medial cubital vein

Inferior mesenteric artery

Common iliac artery and vein

Radial artery

Internal iliac artery and vein

Ulnar artery

External iliac artery and vein

Femoral vein

Femoral artery

Great saphenous vein

Popliteal artery

FIGURE **11.26** Arteries and veins of the body, model photo.

11

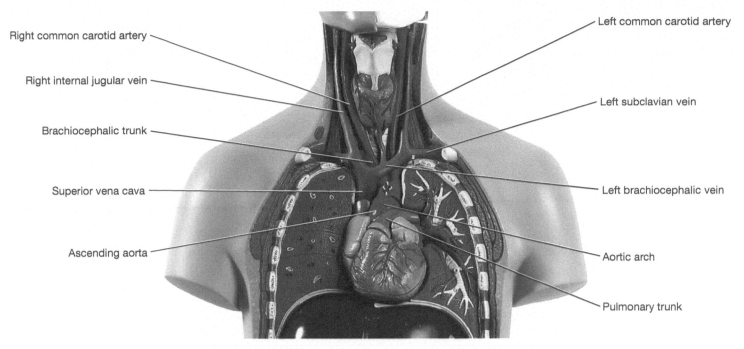

Right common carotid artery

Right internal jugular vein

Brachiocephalic trunk

Superior vena cava

Ascending aorta

Left common carotid artery

Left subclavian vein

Left brachiocephalic vein

Aortic arch

Pulmonary trunk

FIGURE **11.27** Arteries and veins of the neck and thorax, model photo.

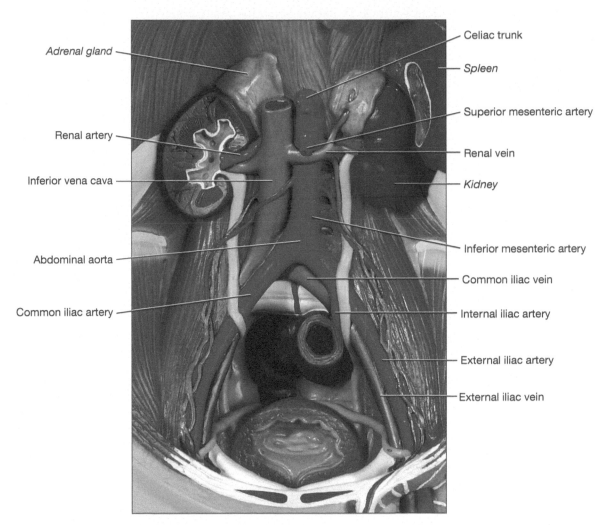

Adrenal gland

Renal artery

Inferior vena cava

Abdominal aorta

Common iliac artery

Celiac trunk

Spleen

Superior mesenteric artery

Renal vein

Kidney

Inferior mesenteric artery

Common iliac vein

Internal iliac artery

External iliac artery

External iliac vein

FIGURE **11.28** Arteries and veins of the posterior abdominal wall, model photo.

11

Exercise 11-4

Time to Trace!

MATERIALS

❏ Laminated outline of the human body
❏ Water-soluble marking pen

In this exercise, you will trace the blood flow through various circuits in the body. As you trace, keep the following hints in mind:

- Don't forget about the hepatic portal system. Remember that most venous blood coming from the abdominal organs has to go through the hepatic portal vein and the hepatic portal system before it can enter the general circulation.

- Don't forget that the venous blood in the brain drains first into the dural sinuses, then into the internal jugular vein.

- If you start in a vein, you have to go through the venous system, through the right heart, through the pulmonary circuit, and then through the left heart before you can get back to the arterial system.

- If you start in an artery, you have to go through the arterial system and then through a *capillary bed* before you can get to the venous system. You can't go backward through the arterial system—that's cheating.

- If you start in an artery and end in an artery, you likely will have to go through the arterial circuit, through a capillary bed, through the venous circuit, back to the heart and lungs, and *then* reenter the arterial circuit.

Following is an example in which we have started in the right popliteal vein and ended in the left internal carotid artery:

Start: right popliteal vein ➔ right femoral vein ➔ right external iliac vein ➔ right common iliac vein ➔ inferior vena cava ➔ right atrium ➔ tricuspid valve ➔ right ventricle ➔ pulmonary valve ➔ pulmonary trunk ➔ right or left pulmonary artery ➔ lungs ➔ pulmonary veins ➔ left atrium ➔ mitral valve ➔ left ventricle ➔ aortic valve ➔ ascending aorta ➔ aortic arch ➔ left common carotid artery ➔ left internal carotid artery ➔ **End**

Wasn't that easy?

Trace the path of blood flow through the following circuits, using the example on page 310 for reference. It is helpful to draw the pathway on a laminated outline of the human body and label each vessel as you trace. If a laminated outline is not available, use Figures 11.29 through 11.31 instead.

1 **Start:** Right renal vein

   End: Right radial artery

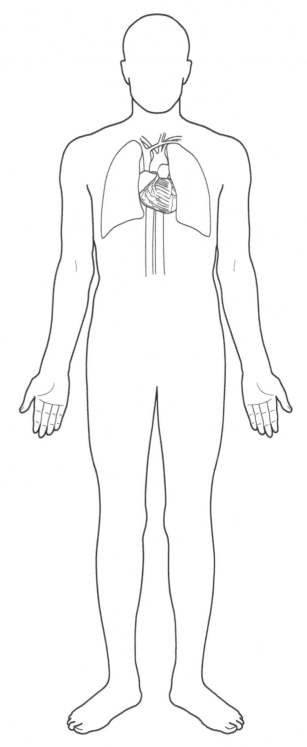

FIGURE **11.29** Outline of the human body.

11

2 **Start:** Superior mesenteric vein

End: Right dorsalis pedis artery

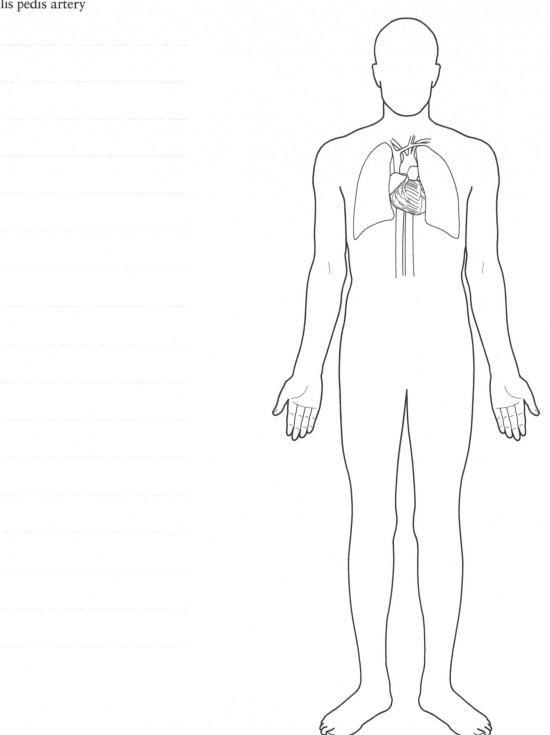

FIGURE **11.30** Outline of the human body.

3 **Start:** Internal carotid artery

End: Left ulnar artery

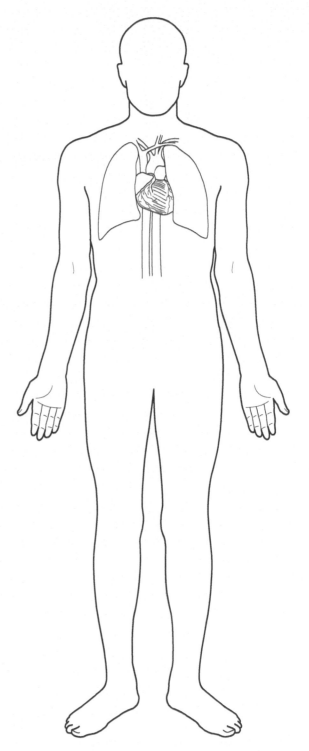

FIGURE **11.31** Outline of the human body.

11

Exercise 11-5

Heart and Vascular Examination

MATERIALS
❑ Stethoscope
❑ Alcohol and cotton ball

Examination of the heart and blood vessels is an important part of every physical examination. The process of listening to heart sounds is known as **auscultation** (aws-kuhl-TAY-shun). The two main sounds you hear when you listen to the heart are produced by the closing of valves at certain points during the cardiac cycle. The first heart sound is known as **S1**, and it is caused by closure of the mitral and tricuspid valves. The second heart sound, **S2**, is caused by closure of the aortic and pulmonary valves.

Heart sounds are typically auscultated in four areas, each of which is named for the valve best heard at that specific location. The position of each area is described relative to the sternum and the spaces between the ribs, known as *intercostal spaces*. The first intercostal space is located between the first and second rib, roughly below the clavicle. From the clavicle, you can count down to consecutive spaces to auscultate in the appropriate areas. The four areas are as follows (Fig. 11.32):

1. **Aortic area.** The **aortic area** is the location where the sounds of the aortic valve are best heard. It is located in the second intercostal space (between ribs two and three) at the right sternal border (to the right of the sternum).

2. **Pulmonic area.** The pulmonary valve is best heard over the **pulmonic area**, which is located at the second intercostal space at the left sternal border.

3. **Tricuspid area.** The sounds produced by the tricuspid valve are best heard over the **tricuspid area**, which is found in the fourth intercostal space at the left sternal border.

4. **Mitral area.** The **mitral area** is located in the fifth intercostal space at the left midclavicular line (draw an imaginary line down the middle of the clavicle).

The following variables are checked during heart auscultation:

▮ **Heart rate.** The **heart rate** refers to the number of heartbeats per minute. If the rate is more than 100, it is termed **tachycardia** (tak-ih-KAR-dee-uh). If the rate is below 60 beats per minute, it is termed **bradycardia** (bray-dih-KAR-dee-uh).

▮ **Heart rhythm.** The heart's **rhythm** refers to the pattern and regularity with which it beats. Some rhythms are *regularly irregular*, in which the rhythm is irregular but still follows a defined pattern. Others are *irregularly irregular*, in which the rhythm follows no set pattern.

▮ **Additional heart sounds.** Sometimes sounds in addition to S1 and S2 are heard, which could be a sign of pathology. These sounds are called **S3**, which occurs just after S2, and **S4**, which occurs immediately prior to S1.

▮ **Heart murmur.** A heart **murmur** is a clicking or "swooshing" noise heard between the heart sounds. Murmurs are caused by a valve leaking, called **regurgitation**, or by a valve that has lost its pliability, called **stenosis** (sten-OH-sis).

Heart sounds are auscultated with a **stethoscope** (STETH-oh-skohp), which you will use in this unit. Most stethoscopes contain the following parts (Fig. 11.33):

▮ **Earpieces** are gently inserted into the external auditory canal and allow you to auscultate the heart sounds.

▮ The **diaphragm** is the broad, flat side of the end of the stethoscope. It is used to auscultate higher-pitched sounds and is the side used most often in auscultation of heart sounds.

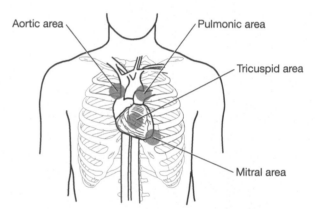

Aortic area · Pulmonic area · Tricuspid area · Mitral area

FIGURE **11.32** Areas of auscultation.

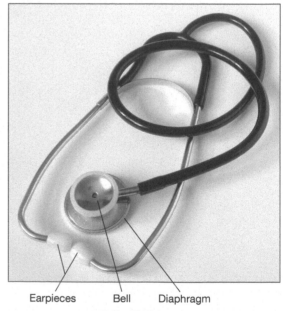

Earpieces · Bell · Diaphragm

FIGURE **11.33** Stethoscope.

11

■ The **bell** is the concave, smaller side of the end of the stethoscope. It is used to auscultate lower-pitched sounds.

Note that sounds are not audible through both the bell and the diaphragm at the same time. Typically, the end can be flipped from one side to the other by simply turning it clockwise or counterclockwise. Before auscultating with either side, lightly tap the end to ensure that you can hear sound through it. If the sounds are faint or muted, turn the end to the other side and try again. If the end of your stethoscope has only one side (the diaphragm), it works slightly differently. In these stethoscopes, placing light pressure on the end as you are auscultating yields sounds associated with the diaphragm, while placing heavier pressure yields sounds associated with the bell. If you are trying to auscultate with the diaphragm and the sounds are faint, try decreasing the amount of pressure you are placing on the end.

Procedure 1 Heart Auscultation

1 Obtain a stethoscope, and clean the earpieces and diaphragm with alcohol and cotton balls.

2 Place the earpieces in your ears, and gently tap the diaphragm to ensure that it is on the proper side. If it is not, flip it to the other side.

3 Lightly place the diaphragm on your partner's chest in the aortic area. (**Note:** You may wish to have your lab partner place the stethoscope on his or her chest under the shirt, because the sounds are heard best on bare skin.)

4 Auscultate several cardiac cycles, and determine which sound is S1 and which sound is S2.

5 Measure the heart rate by counting the number of beats for 15 seconds and multiplying by four. Also note the rhythm, and whether it is regular, regularly irregular, or irregularly irregular. Finally, note the presence of any extra heart sounds or murmurs.

6 Record your results in Table 11.4.

7 Move on to the next area, and repeat.

TABLE **11.4** Heart Auscultation Results

Area	Rate	Rhythm	Extra Heart Sounds? (Yes/No)	Murmurs Present? (Yes/No)
Aortic area				
Pulmonic area				
Tricuspid area				
Mitral area				

Pulse Palpation

Pulse palpation is the process of using the fingertips to feel **pulse points**—locations where the artery is superficial enough that the artery's pulsations with each systole can be felt. It is performed to assess the rate, rhythm, and regularity of the heartbeat, and to assess the arterial circulation to different parts of the body. The pulse points commonly assessed are those found at the radial, ulnar, brachial, carotid, temporal, femoral, popliteal, posterior tibial, and dorsalis pedis arteries, shown in Figure 11.34.

When pulses are palpated, they are **graded** according to a standard scale. This allows healthcare professionals to communicate about a patient unambiguously and to assess the progress or deterioration of a patient's condition. The scale utilizes the following four grades:

- Grade 0/4: The pulse is absent.
- Grade 1/4: The pulse is barely or only lightly palpable.
- Grade 2/4: The pulse is normal.
- Grade 3/4: The pulse is abnormally strong.
- Grade 4/4: The pulse is bounding and visible through the skin.

Notice that this scale has no negative numbers or decimal numbers (e.g., you would not use -1/4 or 2.5/4). In a healthy person, most pulses are grade 2/4 (read as, "two out of four"), although occasionally a pulse is weak or absent. This is simply normal anatomical variation and does not signify pathology. Students often mistakenly grade any strong pulse as 4/4. If a pulse were truly 4/4, however, this would be a sign of extremely high blood pressure in that artery and would possibly be a medical emergency. Most strong, healthy pulses are graded as 2/4.

Please note before you begin that you should never assess both of your lab partner's carotid pulses at the same time. Doing so might initiate the **baroreceptor reflex** (BEHR-oh-reh-sep-ter), in which the parasympathetic nervous system triggers a reflexive and often dramatic drop in blood pressure and heart rate. This could cause your lab partner to momentarily lose consciousness.

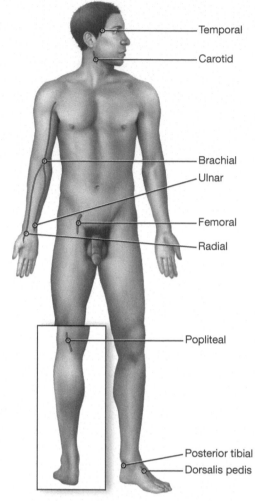

FIGURE **11.34** Common pulse points.

Procedure 2 Pulse Palpation

1 Wash your hands prior to palpating your lab partner's pulses.

2 On a model or diagram, locate the artery you are palpating.

3 Lightly place your index finger and middle finger over the artery. You may increase the pressure slightly, but be careful not to press too hard, because you could cut off blood flow through the artery and also could mistake the pulse in your fingertips for your partner's pulse. If you are unsure whether the pulse is yours or your partner's, feel the lab table. If the lab table "has a pulse," you are feeling the pulse in your own fingertips.

4 Palpate *only* one side (right or left) at a time, especially in the carotid artery.

5 Grade your partner's pulses according to the 0/4 to 4/4 scale, and record the results in Table 11.5.

TABLE **11.5** Pulse Points Grades

Artery	Right-Side Grade	Left-Side Grade
Carotid		
Temporal		
Brachial		
Radial		
Ulnar		
Dorsalis pedis		
Posterior tibial		

Exercise 11-6

Blood Pressure

MATERIALS
- ☐ Sphygmomanometer
- ☐ Stethoscope
- ☐ Alcohol and cotton ball
- ☐ Bucket of ice water

Blood pressure is defined as the pressure exerted by the blood on the walls of the blood vessels. It is determined by the following three factors:

1. **Cardiac output. Cardiac output** is the amount of blood each ventricle pumps in 1 minute. It is a product of heart rate and **stroke volume**, or the amount of blood pumped with each beat.

2. **Peripheral resistance.** Resistance is defined as any impedance to blood flow encountered in the blood vessels. It is determined largely by the degree of **vasoconstriction** or **vasodilation** in the systemic circulation. Vasoconstriction increases resistance, and vasodilation has the opposite effect. Resistance is highest away from the heart in the body's periphery, which is why we often refer to it more specifically as **peripheral resistance**.

3. **Blood volume.** The amount of blood found in the blood vessels at any given time is known as the **blood volume**. It is greatly influenced by overall fluid volume and is largely controlled by the kidneys and hormones of the endocrine system.

Note that cardiac output and peripheral resistance are factors that can be altered quickly to change blood pressure. Alterations to blood volume, however, occur relatively slowly and generally require several hours to days to have a noticeable effect.

Arterial blood pressure is measured clinically and experimentally using an instrument called a **sphygmomanometer** (sfig-moh-muh-NAH-muh-ter) and a stethoscope. This procedure yields two pressure readings:

1. **Systolic pressure.** The pressure in the arteries during the period of ventricular contraction, called **ventricular systole** (SIS-toh-lee), is known as the **systolic pressure** (sis-TAHL-ik). This is the larger of the two readings, averaging between 100 and 120 mmHg.

2. **Diastolic pressure.** The pressure in the arteries during ventricular relaxation, or **ventricular diastole** (dy-AEH-stoh-lee), is the **diastolic pressure** (dy-uh-STAHL-ik). This is the smaller of the two readings, averaging between 60 and 80 mmHg.

Arterial blood pressure is measured by placing the cuff of the sphygmomanometer around the upper arm. When the cuff is inflated, it compresses the brachial artery and cuts off blood flow. When the pressure is released to the level of the systolic arterial pressure, blood flow through the brachial artery resumes but is turbulent. This results in noises known as **Korotkoff sounds** (koh-RAWT-koff), which may be auscultated with a stethoscope.

Procedure 1 Measuring Blood Pressure

Practice is necessary to develop the skills to accurately measure arterial blood pressure. We have provided steps you may follow to practice using the sphygmomanometer and stethoscope together, which is shown in Figure 11.35. All readings should be taken with your lab partner seated and relaxed.

1 Obtain a stethoscope and sphygmomanometer of the appropriate size (about 80 percent of the circumference of the arm).

2 Clean the earpieces and diaphragm as in Exercise 11-5.

3 Wrap the cuff around your partner's arm. It should not be noticeably tight, but it should stay in place when you are not holding it. It should be about 1½ inches proximal to the antecubital fossa.

4 Place the stethoscope in your ears and tap the end gently to ensure that it is on the diaphragm side. Palpate for your partner's brachial pulse in the antecubital fossa. When you find it, set the diaphragm of your stethoscope over the pulse. You will *not* hear anything through the stethoscope at this point.

5 Support your partner's arm by cradling it in your arm, or have your partner rest the arm on the lab table.

6 Locate the screw of the sphygmomanometer near the bulb, and close it by turning it clockwise. Inflate the cuff by squeezing the bulb several times. Pay attention to the level of pressure you are applying by watching the pressure gauge. Do not inflate it beyond about 30 mmHg above your partner's normal systolic pressure (for most people, this is no higher than 180 mmHg). Your lab partner will not likely be happy with you if you inflate it above about 200 mmHg, as this can be uncomfortable.

7 Slowly open the screw by turning it counterclockwise. Watch the pressure gauge, and listen to the brachial artery with your stethoscope.

8 Eventually you will see the needle on the pressure gauge begin to bounce; at about the same time, you will begin to hear the pulse in the brachial artery. The pressure you read at that moment should be recorded as the systolic pressure.

9 Continue to listen and watch the gauge. Eventually, the sound of the pulse will become softer and then finally stop (at this point the needle usually stops bouncing, too). Record the pressure reading from when the sound ends as the diastolic pressure. The numbers should be recorded as a fraction (e.g., 110/70, where 110 is the systolic pressure and 70 is the diastolic pressure).

10 Take your partner's blood pressure twice, and record both readings here:

Blood pressure 1: _____

Blood pressure 2: _____

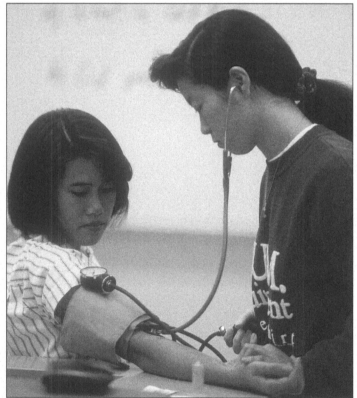

FIGURE **11.35** Measuring blood pressure.

The Autonomic Nervous System and Blood Pressure

The autonomic nervous system (ANS) exerts a great deal of control over blood pressure through its influence on cardiac output, peripheral resistance, and blood volume. Recall that the two divisions of the ANS are the sympathetic nervous system (SNS) and the parasympathetic nervous system (PSNS).

The SNS is the "fight or flight" branch that helps the body to maintain homeostasis during situations such as exercise, emotion, and emergency. Its neurons release norepinephrine onto cardiac muscle cells to increase the rate and force of their contraction, which increases cardiac output. They also release norepinephrine onto the smooth muscle cells lining the walls of blood vessels, which triggers most blood vessels to constrict (note that some blood vessels, such as those serving skeletal muscles, dilate). This vasoconstriction increases peripheral resistance. The combined effects of increased cardiac output and peripheral resistance leads to an increase in blood pressure.

The PSNS is the "rest and digest" branch of the ANS that helps the body to maintain homeostasis in between those bursts of sympathetic activity. As you might expect, the PSNS has the opposite effects on blood pressure as from the SNS. Parasympathetic neurons release acetylcholine onto cardiac muscle cells and slow the heart rate, which decreases cardiac output. The vast majority of blood vessels aren't innervated by PSNS neurons; however, blood vessels do dilate when sympathetic stimulation stops. So indirectly, vasodilation occurs when PSNS activity increases, which decreases peripheral resistance. Together, the decreased cardiac output and peripheral resistance cause a decrease in blood pressure.

Procedure 2 Measuring the Effects of the Autonomic Nervous System on Blood Pressure and Heart Rate

Here we will witness the effects of the ANS on blood pressure, although we are limited to measuring the effects on cardiac output and peripheral resistance.

1 Have your lab partner remain seated and relaxed for 3 minutes. After 3 minutes, measure your partner's pulse rate by palpating the radial artery and counting the number of beats. Then, take your partner's blood pressure. Record these data in Table 11.6.

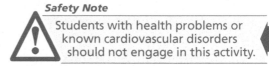

Safety Note
Students with health problems or known cardiovascular disorders should not engage in this activity.

2 Have your partner place one hand into a bucket of ice water.

3 Repeat the blood pressure and pulse measurements with your partner's hand still in the ice water. (*Note:* Be kind to your partner and do this quickly!) Record these data in Table 11.6.

4 Have your partner remove his or her hand from the ice water.

5 Wait 5 minutes, and then repeat the blood pressure and pulse measurements. Record these data in Table 11.6.

6 Interpret your results:

a In which situation(s) was the sympathetic nervous system dominant? How can you tell?

b In which situation(s) was the parasympathetic nervous system dominant? How can you tell?

TABLE **11.6** Blood Pressure and Pulse Readings: Ice-Water Experiment

Test Situation	Blood Pressure	Pulse Rate
At rest		
After immersing in ice water		
5 minutes after removing from ice water		

The electrical activity of the heart can be examined by taking an **electro-cardiogram**, or **ECG**, which is a recording of the changes that occur in the electrical activity of cardiac muscle cells over a period of time. Electrical activity is recorded by placing electrodes on the surface of the skin that record the changes in electrical activity. The changes in electrical activity are visible on the ECG as **waves**. Note that if there is no *net* change in electrical activity, the line on the ECG is flat. However, even when the ECG is flat between waves, the cells of the heart are in some phase of an action potential.

A standard ECG recording is shown in Figure 11.36. Notice that it consists of five waves, each of which represents the depolarization or repolarization of different parts of the heart.

- **P wave.** The initial **P wave** shows the depolarization of the cells of the right and left atria. It is fairly small because of the small number of cells in the atria.

- **QRS complex.** The **QRS complex** is actually a set of three waves that represent the depolarization of the right and left ventricles. The first wave is the **Q wave**, a downward deflection, the next is the **R wave**, a large upward deflection, and the last is the **S wave**, the final downward deflection. The large size of the QRS complex compared with the P wave results from the large number of ventricular cells.

- **T wave.** The small **T wave** is usually the final wave, and it represents the repolarization of the right and left ventricles.

The waves aren't the only important landmarks of the ECG. In addition, we consider the periods between the waves, which show the spread of electrical activity through the heart and phases of the contractile cells' action potentials. We look at two types of periods: **intervals**, which include one or more waves in the measurement, and **segments**, which do not include waves in the measurement.

One important interval is the **R-R interval**, or the period of time between two R waves. This interval represents the duration of the generation and spread of an action potential through the heart. It can be used to determine the heart rate, which we discuss shortly. Another interval we examine is the **P-R interval**, defined as the period from the beginning of the P wave to the beginning of the QRS complex. During the P-R interval, the depolarization spreads through the atria to the ventricles. A final key interval is the **Q-T interval**, the time from the beginning of the Q wave to the end of the T wave. During the Q-T interval, the ventricular cells are depolarizing and repolarizing.

An important segment that we examine is the **S-T segment**, which is between the end of the S wave and the beginning of the T wave. This segment is recorded during the ventricles' plateau phase. There is no net change in electrical activity during this phase, and for this reason, the S-T segment is usually flat.

The normal pattern seen on an ECG is known as a **normal sinus rhythm**, which means that a population of cells known as the **sinoatrial node** is setting the pace of the heart, generally at a rate between 60 and 100 beats per minute. Any deviation from the normal sinus rhythm is known as a **dysrhythmia** (dis-RITH-mee-uh) or **arrhythmia.**

An electrocardiograph records the tracing at a standard speed of 25 mm/second. This allows us to determine precisely the heart rate and the duration of the intervals we discussed. As you can see in Figure 11.36, each small box on the ECG tracing measures 0.04s, and each large box measures 0.20s. Five large boxes together measure 1 second. Determining the duration of most intervals is simple—just count the small or large boxes, and add the seconds together. Calculating the heart rate is equally simple: Count the number of large boxes, and divide 300 by this number. For example, if you count 4.2 boxes: 300/4.2 = 71 beats per minute. The normal values for the periods we discussed are given in Table 11.7.

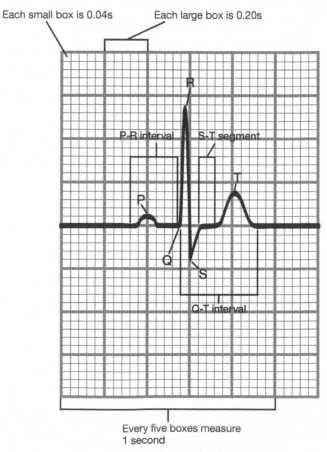

Each small box is 0.04s   Each large box is 0.20s

Every five boxes measure 1 second

FIGURE **11.36** Standard ECG recording.

TABLE **11.7** Normal Values for ECG Periods

Period	Normal Value
Heart rate	60–100 beats per minute
R-R interval	0.60–1.0s
P-R interval	0.12–0.20s
Q-T interval	0.42–0.44s
QRS complex duration	Less than or equal to 0.12s

Procedure **1** Interpreting an ECG

You will now perform some basic ECG interpretation. Following are two tracings for which you will calculate the heart rate and determine the duration of key periods of the ECG.

1 Identify and label the P wave, QRS complex, T wave, P-R interval, R-R interval, and Q-T interval on Tracings 1 and 2 in Figure 11.37.

2 Calculate the heart rate for each tracing. Are the values normal or abnormal?

Heart Rate Tracing 1: _____

Heart Rate Tracing 2: _____

3 Determine the R-R interval, QRS duration, P-R interval, and Q-T interval for each tracing, and record the values in Table 11.8.

Tracing 1

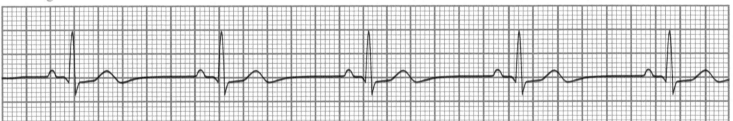

Tracing 2

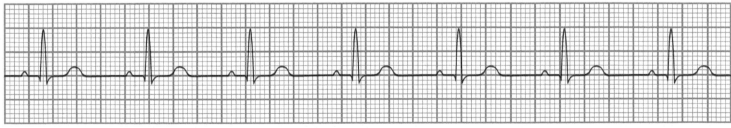

FIGURE **11.37** ECG tracings.

TABLE **11.8** Values for ECG Periods

Value	Tracing 1	Tracing 2
R-R interval		
QRS duration		
P-R interval		
Q-T interval		

11

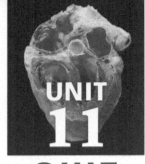

UNIT
11
QUIZ

1 Label the following parts of the heart on Figure 11.38.

- ❏ Anterior interventricular artery
- ❏ Aorta
- ❏ Circumflex artery
- ❏ Inferior vena cava
- ❏ Pulmonary trunk
- ❏ Pulmonary veins
- ❏ Right coronary artery
- ❏ Superior vena cava

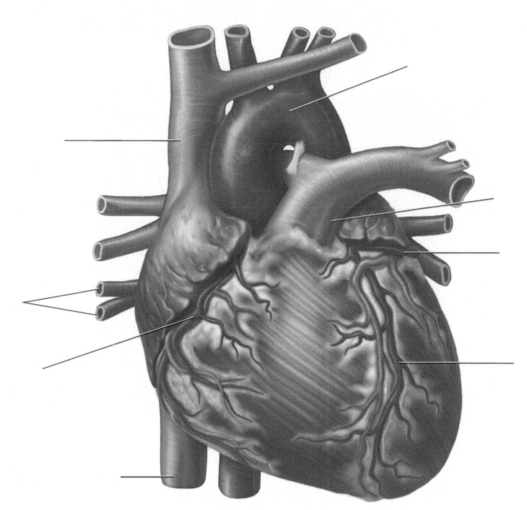

FIGURE **11.38** Heart, anterior view.

11

2 Label the following parts of the heart on **Figure 11.39.**

- ❏ Aortic valve
- ❏ Left atrium
- ❏ Left ventricle
- ❏ Mitral valve
- ❏ Papillary muscle
- ❏ Pulmonary valve
- ❏ Right atrium
- ❏ Right ventricle
- ❏ Tricuspid valve

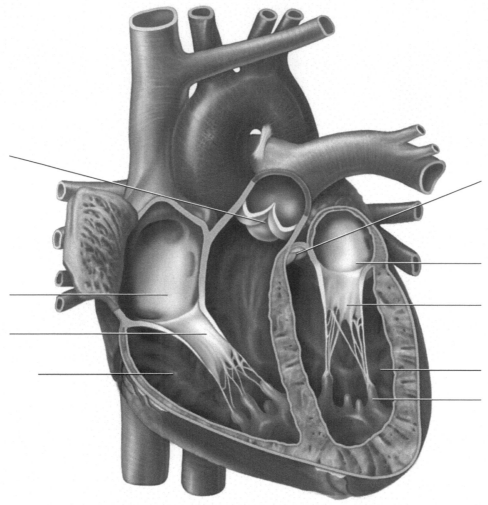

FIGURE **11.39** Frontal dissection of the heart.

3 The tricuspid and mitral valves are known as the

a. chordae tendineae.

b. semilunar valves.

c. coronary valves.

d. atrioventricular (AV) valves.

4 The arteries of the systemic circuit carry _____ blood, and the arteries of the pulmonary circuit carry _____ blood.

a. oxygenated; deoxygenated

b. oxygenated; oxygenated

c. deoxygenated; deoxygenated

d. deoxygenated; oxygenated

5 High pressure in the systemic and pulmonary circuits often results in *ventricular hypertrophy*, in which the ventricle enlarges to pump against greater force. Which side(s) of the heart would be affected by high pressure in the pulmonary circuit? Which side(s) of the heart would be affected by high pressure in the systemic circuit? Explain.

_____

_____

_____

_____

_____

6 Ms. F. visited her physician for a routine physical. During the exam, she explained that over the last two weeks, she had been feeling much more tired than normal and occasionally felt short of breath. Knowing that women usually present with atypical symptoms of a heart attack, her physician ran some diagnostic tests and found that Ms. F. was indeed having a heart attack.

a Imaging studies showed that Ms. F. had blockages in her anterior interventricular artery. What parts of the heart would be affected by these blockages?

_____

_____

_____

b Ms. F.'s heart attack damaged one of her papillary muscles. What is the normal function of a papillary muscle? Predict the consequences of a malfunctioning papillary muscle.

_____

_____

_____

7 *True/False:* Mark the following statements as true (T) or false (F). If the statement is false, correct it to make it a true statement.

_____ a. The three major circuits of blood flow in the body are the systemic, cerebral, and pulmonary circuits.

_____ b. The external carotid arteries and vertebral arteries supply blood to the brain.

_____ c. The cerebral arterial circle provides alternate routes of blood flow through the brain.

_____ d. Venous blood from the spleen, digestive tract, and pancreas drains into the inferior vena cava.

8 The venous blood of the brain drains into a set of _____ before draining into a vein.
a. coronary arteries
b. cerebral veins
c. dural sinuses
d. paranasal sinuses

11

9 Label the following arteries in Figures 11.40 and 11.41.

Figure 11.40

- ❏ Brachial artery
- ❏ Cerebral arterial circle
- ❏ Femoral artery
- ❏ Left common carotid artery

- ❏ Posterior tibial artery
- ❏ Radial artery
- ❏ Renal artery
- ❏ Right subclavian artery

- ❏ Ulnar artery
- ❏ Vertebral artery

Figure 11.41

- ❏ Celiac trunk
- ❏ Common hepatic artery

- ❏ Inferior mesenteric artery
- ❏ Splenic artery

- ❏ Superior mesenteric artery

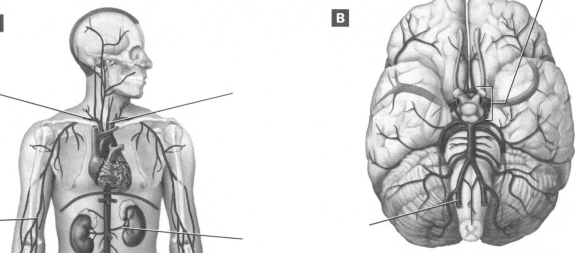

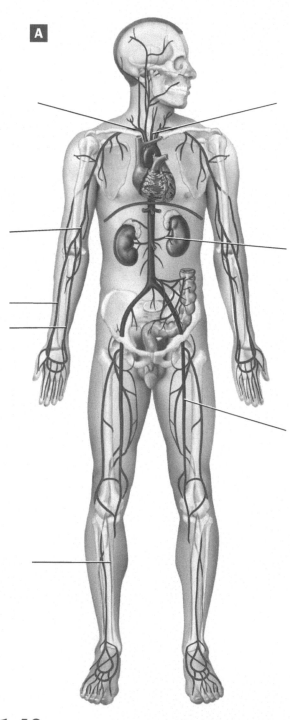

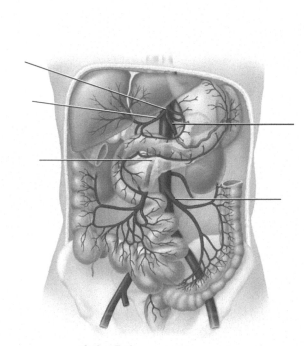

FIGURE **11.40** Major arteries of the (**A**) body; (**B**) brain.

FIGURE **11.41** Arteries of the abdomen.

10 Label the following veins on **Figures 11.42** and **11.43**.

Figure 11.42

❑ Brachiocephalic vein
❑ Cephalic vein
❑ Great saphenous vein

❑ Internal jugular vein
❑ Renal vein
❑ Subclavian vein

❑ Vertebral vein

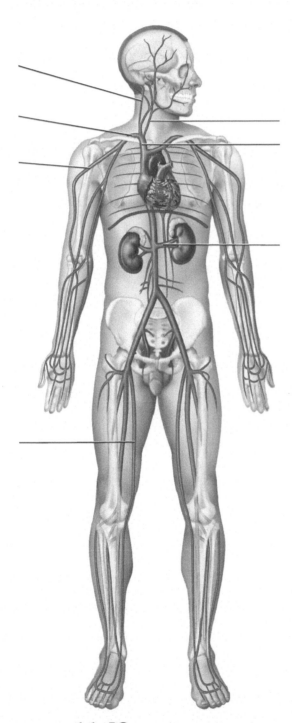

Figure 11.43

❑ Hepatic portal vein
❑ Inferior mesenteric vein
❑ Splenic vein
❑ Superior mesenteric vein

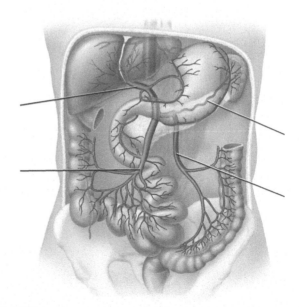

FIGURE **11.42** Major veins of the body.

FIGURE **11.43** Veins of the abdomen and the hepatic portal system.

11

11 A blood clot that forms along the wall of a blood vessel is called a *thrombus*. An *embolus* is a piece of a thrombus that breaks off and flows through the blood until it gets stuck in a small blood vessel downstream from the original thrombus.

  a Trace the pathway an embolus could take if it broke off from a thrombus in the external iliac artery (assume it gets stuck in arterioles, before it reaches capillary beds):

  _____

  _____

  _____

  _____

  b Trace the pathway an embolus could take if it broke off from a thrombus in the femoral vein (assume it gets stuck in arterioles, before it reaches capillary beds):

  _____

  _____

  _____

  _____

12 David has an ultrasound of his common carotid arteries, which shows that his right common carotid artery is nearly 90 percent blocked. He wonders how the right side of his brain is getting enough blood if this artery is so occluded. What do you tell him?

  _____

  _____

  _____

  _____

13 *Fill in the blanks:* The _____ is the pressure in the arteries during ventricular

  systole and averages about _____. The _____ is the

  pressure in the arteries during ventricular diastole and averages about _____.

14 The T wave on an ECG represents
  a. depolarization of the atria.
  b. repolarization of the atria.
  c. depolarization of the ventricles.
  d. repolarization of the ventricles.

15 Your patient has been admitted to the emergency room with an occupational injury from an industrial saw. He has lost a significant volume of blood. What effect has this blood loss likely had on his blood pressure, and why?

  _____

  _____

  _____

  _____

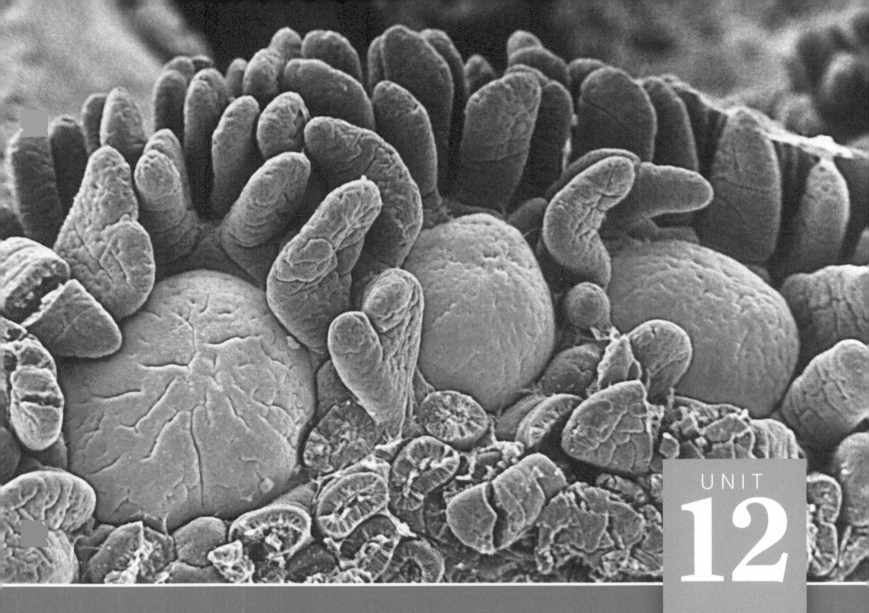

Blood and Lymphatics

When you have completed this unit, you should be able to:

1 Identify the formed elements of blood.

2 Perform blood typing of the ABO and Rh blood groups using simulated blood.

3 Explain the basis for blood typing and matching for blood donation.

4 Identify structures of the lymphatic system.

5 Trace the pathway of lymph as it is returned to the cardiovascular system.

PRE-LAB EXERCISES

Complete the following exercises prior to coming to lab, using your lab manual and textbook for reference.

Pre-Lab Exercise **12-1**

✎ Key Terms

You should be familiar with the following terms before coming to lab.

Term	Definition
Formed Elements	
Erythrocyte	
Neutrophil	
Eosinophil	
Basophil	
Lymphocyte	
Monocyte	
Platelets	
Blood Typing	
Antigen	
Antibody	
Blood Donation	
Universal donor	
Universal recipient	

12

Lymphatic System Structures

Lymph capillary

Spleen

Thymus

Tonsil

Lymph node

Pre-Lab Exercise 12-2

Formed Elements

In this unit, we will identify the formed elements of blood on a peripheral blood smear. Each formed element has unique morphological characteristics and functions. Use Exercise 12-1 (p. 333) in this unit and your text to complete Table 12.1 with these functions and characteristics.

TABLE **12.1** Properties of Formed Elements

Formed Element	Nucleus Shape	Cytoplasm, Granule Color, or Both	Function	Prevalence
Erythrocyte				
Neutrophil				
Eosinophil				
Basophil				
Lymphocyte				
Monocyte				
Platelet				

12

Anatomy of the Lymphatic System

Color the structures of the lymphatic system in Figure 12.1, and label them with the following terms from Exercise 12-6 (pp. 346–348). Use Exercise 12-6 in this unit and your text for reference.

❑ Spleen
❑ Thymus

Lymph Vessels
❑ Thoracic duct
❑ Right lymphatic duct

Lymph Nodes
❑ Cervical lymph nodes
❑ Axillary lymph nodes
❑ Inguinal lymph nodes
❑ Mesenteric lymph nodes

Tonsils
❑ Palatine tonsil
❑ Pharyngeal tonsil
❑ Lingual tonsil

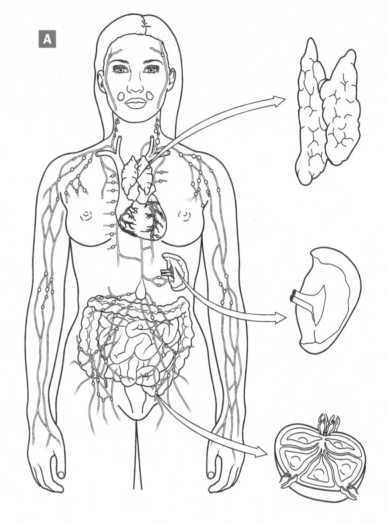

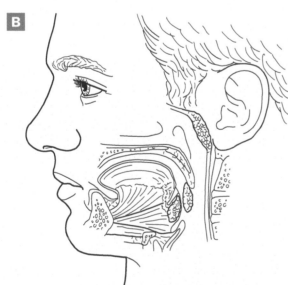

FIGURE **12.1** Lymphatic organs: (**A**) overview of the lymphatic organs and vessels; (**B**) the tonsils.

SEM of lymphatic nodules in the small intestine called Peyer's patches.

You have already examined the vessels that make up the cardiovascular system, and, in this unit, you will study the fluid tissue located within those vessels: **blood**. Safety concerns often preclude the use of real blood in the laboratory, but we can still demonstrate important principles by viewing prepared microscope slides of blood cells and by using simulated blood to learn about blood typing. There is no real concern over blood-borne diseases with simulated blood, but do keep in mind that the simulated blood contains chemicals that may be hazardous. So, as always, use appropriate safety protocols when handling all materials in this lab.

Your study of blood will begin with an examination of the formed elements of blood on microscope slides. After this, you will play a murder mystery game in which you use simulated blood to apply blood typing techniques in relation to blood donation. If your lab has the appropriate protocols for the use of real blood, you will type your own blood and determine its hemoglobin content.

This unit also examines a topic closely related to the cardiovascular system and blood: the **lymphatic system**, which contains vessels and organs that filter blood and another type of fluid known as **lymph** (LIMF). As you will see, this system works closely with the cardiovascular system to maintain fluid homeostasis. You will examine this and other functions of the lymphatic system in Exercise 12-6.

Exercise **12-1**

Formed Elements (Cells) of Blood

MATERIALS
- ❑ Blood slides
- ❑ Light microscope
- ❑ Colored pencils

Whole blood consists of two main components: **plasma**, the fluid portion of the blood, and **formed elements**, or the cellular portion of the blood. Plasma accounts for about 55 percent of the volume of whole blood and consists primarily of water, proteins, and other solutes such as nutrients and ions. Formed elements account for about 45 percent of the volume of whole blood. There are three classes of formed elements: erythrocytes, platelets, and leukocytes.

1. **Erythrocytes. Erythrocytes** (eh-RITH-roh-syt'z), also known as **red blood cells**, carry oxygen around the body on an iron-containing protein called **hemoglobin** (HEE-moh-glohb-in; Fig. 12.2). They are the most numerous blood cells, averaging about 44 percent of the total blood volume. This value, known as the **hematocrit** (heh-MAEH-toh-krit), is typically higher in males (40 to 50 percent) than in females (36 to 44 percent). Erythrocytes are easily distinguished from the other formed elements by their reddish-pink color and the fact that mature erythrocytes lack nuclei and most organelles.

2. **Platelets.** Note in Figure 12.2 that **platelets** (PLAYT-letz) aren't actually cells at all but are instead just small cellular fragments. As such, they lack nuclei and most organelles and are much smaller than the other formed elements. Platelets are involved in blood clotting and make up less than 1 percent of the total blood volume.

3. **Leukocytes. Leukocytes** (LOO-koh-syt'z), also known as **white blood cells**, play a role in the immune system and make up

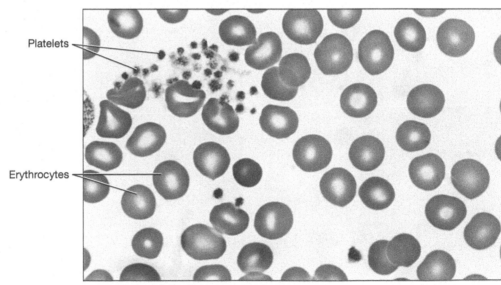

FIGURE **12.2** Photomicrograph of erythrocytes and platelets.

less than 1 percent of the total blood volume. The two subclasses of leukocytes are based upon the presence or absence of visible granules in their cytoplasm.

a. **Granulocytes.** As implied by their name, **granulocytes** (GRAN-yoo-loh-syt'z) are cells containing cytoplasmic granules that are visible when stained. The three types of granulocytes stain differently when treated with the dyes hematoxylin and eosin and are named for the type of stain with which they interact.

   i. **Neutrophils** (NOO-troh-filz) do not interact strongly with either type of dye, and their granules stain a light violet-pink color. They are the most numerous type of leukocyte, making up about 60 to 70 percent of the total leukocytes in the blood. They typically have multilobed nuclei, although their nuclei often vary in appearance. One common variation in nucleus shape, shown in Figure 12.3, is seen in immature neutrophils called *band cells*. As you can see, the nucleus of a band cell has a single lobe that is stretched into a "U"-shape (or a single "band"). Neutrophils are attracted to the site of any cellular injury, and are particularly active in ingesting and destroying bacteria.

   ii. **Eosinophils** (ee-oh-SIN-oh-filz) interact strongly with the red dye eosin, and so their granules stain bright red. They are far less numerous than neutrophils, accounting for only about 4 percent of the total leukocytes in the blood. As with neutrophils, their nuclei are segmented into lobes, although eosinophils' nuclei tend to be bilobed. Eosinophils play a role in the immune response to infection with parasitic worms and in the allergic response.

   iii. **Basophils** (BAY-zoh-filz) take up the dark purple stain hematoxylin (it is a basic dye, which is why they are named *baso*phils), and so their granules appear dark blue-purple. Like eosinophils, they tend to have bilobed nuclei, but their nuclei are often obscured by their dark granules. They are the least numerous of the leukocytes, making up fewer than 1 percent of the total leukocyte count, and will likely be the most difficult to find on your slide. Basophils are primarily involved in the allergic response.

b. **Agranulocytes.** The cells known as **agranulocytes** (AY-gran-yoo-loh-syt'z) lack visible cytoplasmic granules. There are two types of agranulocytes:

   i. **Lymphocytes** (LIMF-oh-syt'z) tend to be smaller than granulocytes and have large, spherical nuclei surrounded by a rim of light blue-purple cytoplasm. They are the second most numerous type of leukocyte, making up 20 to 25 percent of the total leukocyte count. There are two populations of lymphocytes. *B lymphocytes* produce proteins called *antibodies* that bind foreign substances called *antigens*. *T lymphocytes* play numerous roles, including enhancing other components of the immune response, destroying cancer cells, and destroying cells infected with viruses.

   ii. **Monocytes** (MAHN-oh-syt'z) are the largest of the leukocytes and have "U"-shaped or horseshoe-shaped nuclei with light blue or light purple cytoplasm. They are the third most numerous type of leukocyte, accounting for 3 to 8 percent of the total. Monocytes exit the blood to mature into cells called **macrophages** (MAK-roh-feyj-uhz), which are very active phagocytes.

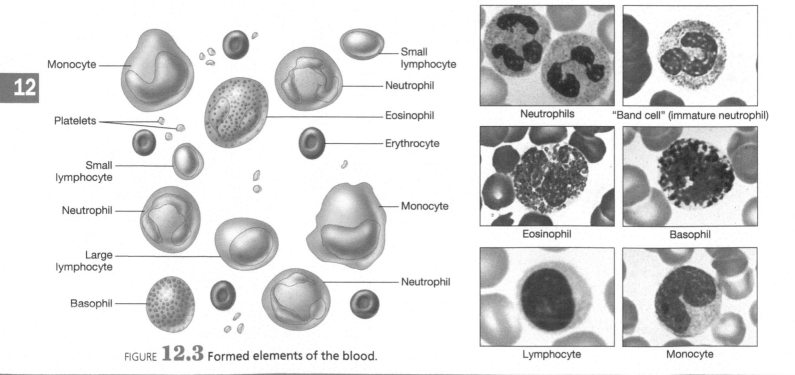

FIGURE **12.3** Formed elements of the blood.

In this procedure, you will examine a blood slide called a **peripheral blood smear**. Examine the peripheral blood smear on high power, and scroll through to find each of the formed elements. Note that you may have to find a second slide to locate certain cells, because some types are more difficult to find (in particular the eosinophils and basophils). In the spaces provided, use colored pencils to draw and describe each formed element you locate.

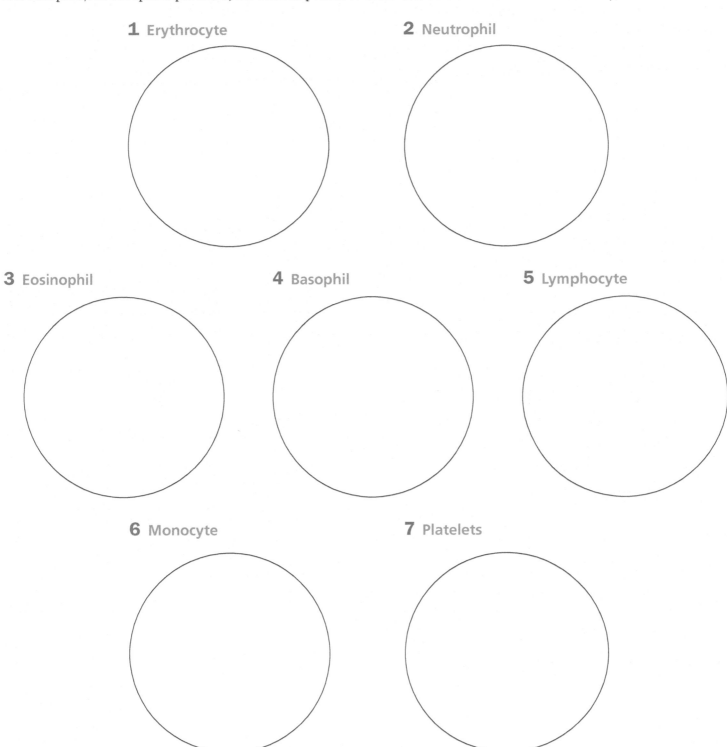

1 Erythrocyte

2 Neutrophil

3 Eosinophil

4 Basophil

5 Lymphocyte

6 Monocyte

7 Platelets

12

Exercise 12-2

ABO and Rh Blood Groups

MATERIALS

❑ Well plates

❑ Simulated blood types A–, B+, AB–, and O+

❑ Simulated antisera: anti-A, anti-B, anti-Rh

❑ Sharpie

Blood typing is done by checking the blood for the presence or absence of specific glycoproteins called **antigens** (AN-tih-jenz) that are found on the cell surface. Two clinically relevant antigens are the **A antigen** and the **B antigen**. The blood type is named based upon which of the antigens is present (Fig. 12.4).

▌ **Type A** blood has A antigens on the cell surface.

▌ **Type B** blood has B antigens on the cell surface.

▌ **Type AB** blood has both A and B antigens on the cell surface.

▌ **Type O** blood has neither A nor B antigens on the cell surface.

An additional clinically relevant antigen is the Rh antigen.

▌ Blood that has the Rh antigen is denoted as **Rh positive** (e.g., A+).

▌ Blood that lacks the Rh antigen is denoted as **Rh negative** (e.g., A–).

The prevalence of different blood types in the United States varies with different ethnic groups. In general, we can say that the most common type is O+, followed by A+, and then B+.

The antigens present on the surface of an erythrocyte can be determined by combining it with a solution called an **antiserum**. An antiserum is a solution that contains proteins produced by B lymphocytes called **antibodies** that bind to specific antigens. When antibodies bind to antigens on erythrocytes, they cause **agglutination** (uh-gloo-tin-AY-shun), or clumping of the erythrocytes, in the sample. In this exercise, we are using simulated blood, so you won't see agglutination unless your instructor sets up a demonstration or permits you to type your own blood (which is in a later, optional procedure). The antisera used to determine the blood type of a sample are named according to the antigen they bind:

▌ **Anti-A antiserum** contains anti-A antibodies that bind to erythrocytes with A antigens.

▌ **Anti-B antiserum** contains anti-B antibodies that bind to erythrocytes with B antigens.

▌ **Anti-Rh antiserum** contains anti-Rh antibodies that bind to erythrocytes with Rh antigens.

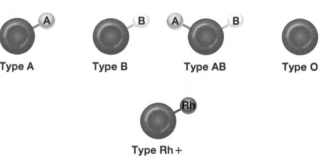

Type A   Type B   Type AB   Type O

Type Rh+

FIGURE **12.4** Basic ABO and Rh blood types.

Antigen-Antibody Reactions

This exercise allows you to examine the antigen-antibody reactions of known blood types. Each table should take one well plate and one set of dropper bottles. The bottles are labeled A–, B+, AB–, and O+ to represent each of those blood types, and anti-A, anti-B, and anti-Rh to represent the different antisera. You can see what a positive reaction looks like in Figure 12.5. Notice that a positive reaction is indicated by the formation of a white, cloudy precipitate. (**Note:** However, your sample may form either a granular or clumping precipitate, depending on the type of simulated blood used by your lab.) So, for type A+, a positive reaction is seen in the wells to which anti-A and anti-Rh antisera were added, but there is no reaction in the well containing anti-B antiserum.

12

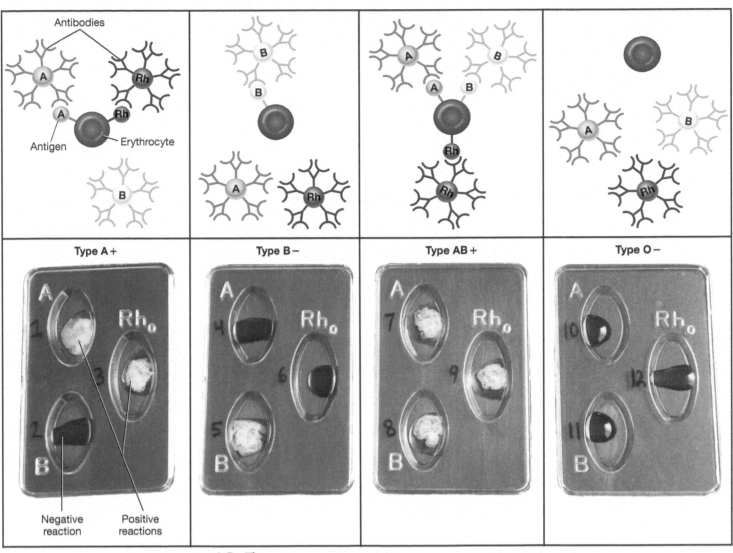

FIGURE **12.5** Reactions of simulated blood with simulated antisera.

Procedure **1** Testing Simulated Blood

Use Figure 12.6 as a guide for placement of samples in the wells.

1 Label wells on the well plate as Wells 1 through 12 with a Sharpie.

2 Drop two drops of type A− blood in Well 1, Well 2, and Well 3.

3 Drop two drops of type B+ blood in Well 4, Well 5, and Well 6.

4 Drop two drops of type AB− blood in Well 7, Well 8, and Well 9.

5 Drop two drops of type O+ blood in Well 10, Well 11, and Well 12.

6 Add two drops of the anti-A antiserum to Wells 1, 4, 7, and 10.

7 Add two drops of the anti-B antiserum to Wells 2, 5, 8, and 11.

8 Add two drops of the anti-Rh antiserum to Wells 3, 6, 9, and 12.

9 Observe the samples for changes in color symbolizing the agglutination, or clumping, that would normally occur between antisera and specific blood types. Record your results in Table 12.2.

Safety Note

Safety glasses and gloves are required!

12

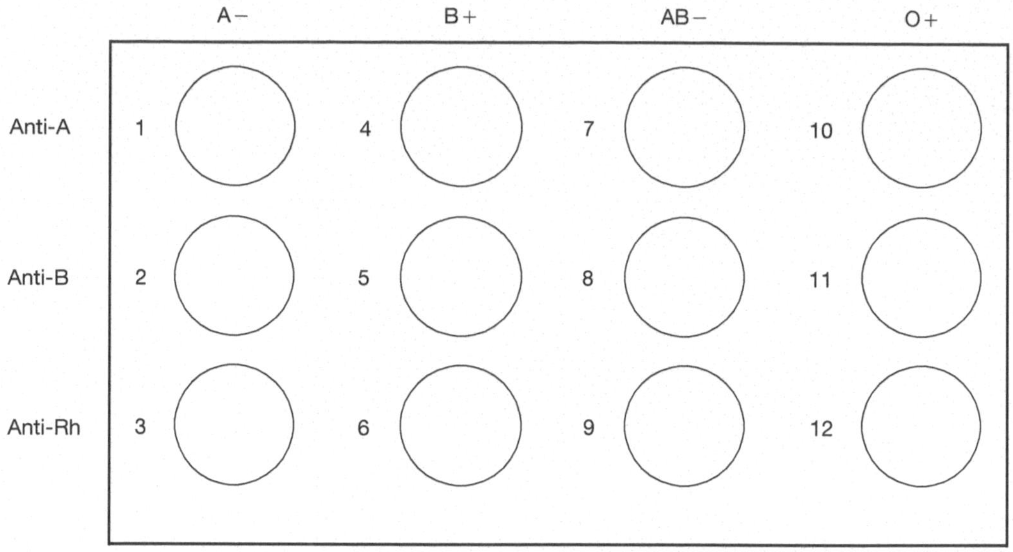

FIGURE **12.6** Well plate diagram.

TABLE **12.2** Blood Typing Results

Blood Type	Reacted with Anti-A? (Yes/No)	Reacted with Anti-B? (Yes/No)	Reacted with Anti-Rh? (Yes/No)	Antigens Present on Cell Surface
A–				
B+				
AB–				
O+				

Exercise 12-3

Murder Mystery Game

MATERIALS
❑ Well plate
❑ Simulated antisera: anti-A, anti-B, anti-Rh
❑ Murder mystery game

In this game, you will be applying the blood typing techniques you learned in Exercise 12-2 to solve a series of murder mysteries. Each of the following cases presents a victim, a murderer, three suspects, three possible murder rooms, and three possible murder weapons. Your job is to play the role of detective and determine the identity of the murderer, which weapon was used, and in which room the crime was committed.

Procedure 1 Solving the Murder Mysteries

For each case, a unique set of bottles is marked with a number that corresponds to the specific cases (i.e., the bottles are marked 1 for Case 1, 2 for Case 2, and 3 for Case 3). The murderer, rooms, and murder weapons are different for each case, but the cast of characters remains the same across the cases.

1   Assemble into groups of two or three students. Obtain a well plate, and choose one set of samples to test (e.g., the rooms from Case 1, the suspects from Case 2, or the weapons from Case 3).

2   To test the samples: Place two drops of the sample in each of three separate wells. Add two drops of anti-A antiserum to the first well, add two drops of anti-B antiserum to the second well, and add two drops of anti-Rh antiserum to the third well. After you have tested each of the samples, return them to their proper places in the front of the lab.

3   Watch for a reaction with the antisera. Remember that this is simulated blood, just as in Exercise 12-2. A positive reaction is denoted by the formation of a precipitate. Record your results in the spaces provided in your lab manual.

4   To determine the

   a   *Murderer:* Match the blood type of one of the **suspects** to that of the **murderer**.

   b   *Weapon and room:* Match the blood type of the **victim** to the blood types found in the **rooms** and on the **weapon**.

> **Note:** You are *not* trying to match the blood type of the victim to that of the murderer—that would mean the victim committed suicide!

Case 1: Ms. Magenta

We enter the scene to find the dearly departed Ms. Magenta. Forensic analysis determines that there are two types of blood on the body. One blood type is Ms. Magenta's, and the other blood type is a trace amount left behind from the murderer.

Ms. Magenta's blood type: _____   Murderer's blood type: _____

We have three suspects:

1. *Mrs. Blanc* was being blackmailed by Ms. Magenta and Col. Lemon. They had discovered that Mrs. Blanc had murdered her late husband. Mrs. Blanc knew this would ruin her reputation at the country club.

2. *Col. Lemon* wanted to keep the blackmail money for himself and wanted Ms. Magenta out of the way.

3. *Mr. Olive* had been secretly in love with Ms. Magenta for years, and when he told her of his feelings, she rejected him harshly.

Mrs. Blanc's blood type: _____   Col. Lemon's blood type: _____   Mr. Olive's blood type: _____

We have three possible murder rooms:

Ballroom blood type: _____   Library blood type: _____   Den blood type: _____

We have three possible murder weapons:

Candlestick blood type: _____   Noose blood type: _____   Knife blood type: _____

Case 1: Conclusion

Ms. Magenta was killed by _____, in the _____,

with the _____.

12

Case 2: Col. Lemon

Our next victim is poor Col. Lemon. On his body we find his blood and also trace amounts of the blood of another person, presumably the murderer.

Col. Lemon's blood type: _____   Murderer's blood type: _____

We have three potential suspects:

1. *Mrs. Blanc.* Now, with Ms. Magenta out of the way, Mrs. Blanc could easily rid herself of her problem by disposing of the only other person who knows her secret—Col. Lemon.
2. *Professor Purple* believed the colonel had stolen his groundbreaking FrozenForehead™ wrinkle-eraser formula.
3. *Mr. Olive* couldn't stand the colonel because of his close relationship with Ms. Magenta.

Mrs. Blanc's blood type: _____   Professor Purple's blood type: _____

Mr. Olive's blood type: _____

We have blood in three different rooms:

Hall blood type: _____   Kitchen blood type: _____   Billiards room blood type: _____

Forensics found blood on three different weapons:

Copper pipe blood type: _____   Hammer blood type: _____   Revolver blood type: _____

Case 2: Conclusion

Col. Lemon was killed by _____, in the _____,

with the _____.

Case 3: Mr. Olive

Our next (and hopefully last) victim is Mr. Olive. Analysis demonstrates two blood types: one belonging to Mr. Olive, and trace amounts of another belonging to the murderer.

Mr. Olive's blood type: _____   Murderer's blood type: _____

We have three potential suspects:

1. *Ms. Feather* simply didn't like Mr. Olive and thought he had a "bad aura."
2. *Mrs. Blanc* was worried that Mr. Olive knew her secret, and she wanted him out of the way.
3. *Professor Purple* discovered that Mr. Olive—not Col. Lemon—had actually stolen the FrozenForehead™ formula. Whoops!

Ms. Feather's blood type: _____   Mrs. Blanc's blood type: _____

Professor Purple's blood type: _____

Blood was found in three rooms:

Lounge blood type: _____   Dining room blood type: _____   Greenhouse blood type: _____

We have three potential murder weapons:

Noose blood type: _____   Hammer blood type: _____   Revolver blood type: _____

Case 3: Conclusion

Mr. Olive was killed by _____, in the _____,

with the _____.

Exercise 12-4

Blood Donation

Blood transfusion, the infusion of a recipient with a donor's blood cells, is a commonly performed medical procedure. Before a recipient is given a blood transfusion, the medical team must first learn the patient's blood type and then find a suitable, or "matching," donor. This is necessary because of the A, B, and Rh antigens on the surface of the donor erythrocytes and the presence of preformed antibodies in the recipient's blood. If a donor's erythrocytes have antigens that the recipient's immune system recognizes as foreign, the recipient's antibodies will agglutinate the foreign erythrocytes. The agglutinated erythrocytes are then destroyed by the immune system, a process known as **hemolysis** (heem-AW-lih-sis). This is called a **transfusion reaction**, and it is a medical emergency that can lead to kidney failure and death.

To ensure that a transfusion reaction does not occur, we must make sure the donor blood does not have antigens the recipient's immune system will recognize as foreign. For the ABO blood group, our immune systems produce antibodies to any antigen *not* present on the surface of our own cells.

- People with type A blood have A antigens and so produce anti-B antibodies.
- People with type B blood have B antigens and so produce anti-A antibodies.
- People with type O blood have neither A nor B antigens and so produce anti-A and anti-B antibodies.
- People with type AB blood have both A and B antigens and so produce neither anti-A nor anti-B antibodies.

If you're wondering about the Rh factor, wait just a moment—we're getting there. Let's do an example with the ABO blood group first: Felix has type B blood, which means that he has anti-A antibodies.

What will happen if we give him blood from a donor with:

- **Type A blood?** There are A antigens on these donor erythrocytes, and Felix's anti-A antibodies would agglutinate them. ✘
- **Type B blood?** Felix's anti-A antibodies would have no effect on the B antigens on these donor erythrocytes, so this blood is safe. ✔
- **Type O blood?** There are no antigens on these donor erythrocytes, so Felix's anti-A antibodies would have no effect on them, and this blood is safe. ✔
- **Type AB blood?** There are both A and B antigens on these donor erythrocytes, and Felix's anti-A antibodies would bind and agglutinate the A antigens. ✘

Now that's easy, isn't it?

Next let's address the Rh factor. The blood of an Rh-negative person does *not* contain preformed antibodies to the Rh antigen. However, anti-Rh antibodies are made if an Rh-negative person is exposed to the Rh antigen. In an emergency setting, it is generally not possible to determine if an Rh-negative person has been exposed to the Rh antigen, so healthcare professionals err on the side of caution and assume that the person has anti-Rh antibodies. For the sake of simplicity, we will assume the same thing in this exercise. So, for our purposes:

- People with Rh-positive blood *do not* produce anti-Rh antibodies.
- People with Rh-negative blood *do* produce anti-Rh antibodies.

Let's do one more example, taking into account the Rh factor this time: Lourdes has A − blood, which means that she has anti-B and anti-Rh antibodies.

What will happen if we give her blood from a donor with:

- **Type A + blood?** There are A and Rh antigens on these donor erythrocytes, and Lourdes' anti-Rh antibodies would agglutinate the Rh antigens. ✘
- **Type B − blood?** There are B antigens on these donor erythrocytes, and Lourdes' anti-B antibodies would agglutinate them. ✘
- **Type O − blood?** There are no antigens on these donor erythrocytes, so Lourdes' anti-B and anti-Rh antibodies would have no effect on them, and this blood is safe. ✔
- **Type AB + blood?** There are A, B, and Rh antigens on these donor erythrocytes, and Lourdes' anti-B and anti-Rh antibodies would agglutinate the B and Rh antigens. ✘

12

Procedure 1 Blood Type Matching Practice

Use the information you just read and your text to fill in Table 12.3.

TABLE **12.3** Blood Donation

Blood Type	Antigens Present	Antibodies Present	Can Donate Safely to Which Blood Types?	Can Receive Safely from Which Blood Types?
A+				
A−				
B+				
B−				
AB+				
AB−				
O+				
O−				

You should notice something from Table 12.3: People with AB+ blood can receive from any blood donor type, and people with O− blood can donate to any blood recipient type. This is because people with AB+ erythrocytes have all three antigens and so their blood contains no antibodies to bind to donor blood. Conversely, people with type O− erythrocytes have none of the three antigens and so there is nothing for any recipient's antibodies to bind. For these reasons, type AB+ blood is often called the **universal recipient** and type O− is often called the **universal donor.**

Procedure 2 Type Matching for Transfusions

Gasp! It turns out that Ms. Magenta, Col. Lemon, and Mr. Olive all survived their injuries! But they have lost blood and are in need of blood transfusions. All of the suspects have had a sudden change of heart and have offered to help the three victims by donating blood. When they arrived at the hospital, the blood of all victims and suspects was retested. This was fortunate, as it turned out the results obtained at the crime scenes were inaccurate due to a suspected mole in the police lab. The correct blood types are listed in Table 12.4. Your job is to determine who among the suspects could safely donate blood to whom.

TABLE **12.4** Corrected Blood Types of
the Suspects and Victims

Suspect/Victim	Blood Type
Ms. Magenta	B+
Ms. Feather	O+
Mrs. Blanc	A–
Prof. Purple	AB–
Col. Lemon	AB+
Mr. Olive	O–

HINTS & TIPS

Who Can Donate to Whom?

Remember—when trying to work out who can donate blood to whom, you are concerned with the recipient's antibodies and the donor's antigens. So first work out which antibodies the recipient has, then make sure the recipient's antibodies won't bind any antigens on the donor's erythrocytes.

Recipient 1: Ms. Magenta

Ms. Magenta's blood type:

Donors:

Ms. Feather's blood type:

Mrs. Blanc's blood type:

Professor Purple's blood type:

Who could safely donate blood to Ms. Magenta?

Who could not safely donate blood to Ms. Magenta?

Recipient 2: Col. Lemon

Col. Lemon's blood type:

Donors:

Ms. Feather's blood type:

Mrs. Blanc's blood type:

Professor Purple's blood type:

Who could safely donate blood to Col. Lemon?

Who could not safely donate blood to Col. Lemon?

Recipient 3: Mr. Olive

Mr. Olive's blood type:

Donors:

Ms. Feather's blood type:

Mrs. Blanc's blood type:

Professor Purple's blood type:

Who could safely donate blood to Mr. Olive?

Who could not safely donate blood to Mr. Olive?

12

Exercise 12-5

Typing and Examining Your Own Blood

MATERIALS

- ❏ Lancet
- ❏ Alcohol wipes
- ❏ Anti-A, anti-B, and anti-Rh antisera
- ❏ Blank microscope slide
- ❏ Sharpie
- ❏ Toothpicks
- ❏ Tallquist paper and scale
- ❏ Dissection microscope or magnifying glass

If your lab has the appropriate equipment and biohazard disposal means, your instructor may permit you to type and examine your own blood in this optional exercise. Working with blood is actually very safe provided you follow some basic procedures outlined here:

1. Wash your hands with soap and water before starting and after completing the procedures.

2. Prepare your work area by placing a disposable absorbent liner on the lab table.

3. Wear gloves and safety glasses when handling all materials for this lab.

4. Handle only your own lab materials to avoid coming into contact with blood other than your own.

5. Dispose of all lancets and microscope slides in the designated sharps container only.

6. Dispose of all nonsharp materials (disposable pads, gloves, alcohol wipes, and any other materials you have used) in the designated red biohazard bag.

7. When you are finished with the procedures, clean your work area with the disinfectant solution provided by your lab instructor.

In the following procedures, you will determine your blood type and measure the approximate hemoglobin content of your blood. Note that these exercises involve lancing your own finger to stimulate bleeding. If you have any medical conditions that render this activity unsafe, discuss this with your lab instructor.

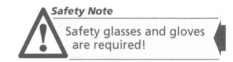

Safety Note

Safety glasses and gloves are required!

The process of determining the blood type of real blood is similar to the one you performed earlier with simulated blood: You use three blood samples, and apply three different antisera to look for a reaction; in this case, you are looking for agglutination of erythrocytes (Fig. 12.7). The reaction between real blood antigens and antibodies is more subtle than in the simulated blood; in particular, the reaction with Rh antigens and the anti-Rh antibodies can be difficult to see. You may wish to examine your samples under a magnifying glass to see the reaction more clearly.

Procedure 1 Determining Your Blood Type

1 Obtain a blank microscope slide, and draw three circles on it with a Sharpie. Label the circles A, B, and Rh.

2 Wash your hands, and prepare your work area with a disposable absorbent liner.

3 Place a glove on one hand, and prepare one finger of your ungloved hand by cleaning it with an alcohol wipe.

4 Use a fresh lancet to lance the finger you just cleaned. Dispose of the used lancet in the sharps bin.

5 Squeeze your finger to stimulate bleeding, and squeeze a small drop of blood onto each of the three circles on the slide.

6 Place a drop of anti-A antiserum in the circle marked A, anti-B in the circle marked B, and anti-Rh in the circle marked Rh.

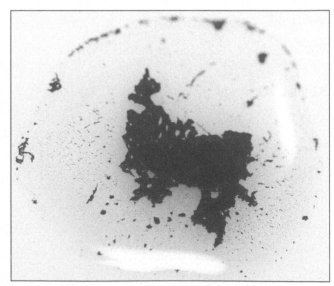

FIGURE **12.7** Erythrocyte agglutination.

12

7 Use three toothpicks, one for each circle, to gently mix the blood with the antisera. You must use a different toothpick for each circle to avoid cross-contamination. Watch carefully for a reaction, using a dissection microscope or magnifying glass if needed (be sure to hold a magnifying glass in your gloved hand to avoid coming into contact with blood from your classmates).

8 Determine your blood type based upon the reactions with the antisera. Remember, if the blood reacts with an antiserum, that antigen is present. For example, if your blood reacts with anti-B antiserum and anti-Rh antiserum, the B and Rh antigens are present, and the blood type is B+.

Record your blood type here: _____

9 Dispose of your microscope slide in the sharps bin when you have completed the blood typing procedure.

Procedure 2 Determining the Hemoglobin Content of Your Blood

As you learned in Exercise 12-1 (p. 333), erythrocytes are filled with the protein hemoglobin, which binds and transports oxygen through the blood. The amount of hemoglobin in erythrocytes averages about 12–16 g/dl in females and about 14–18 g/dl in males. Medically, this is an important value, because a decreased amount of hemoglobin can indicate conditions such as *iron-deficiency anemia*. In a hospital lab, the amount of hemoglobin in the blood is measured with an instrument known as a *hemoglobinometer*. However, in this lab we will employ an older (and less expensive) method using *Tallquist paper*, which provides an estimate of the blood's hemoglobin content.

1 Squeeze your finger to continue to stimulate bleeding. If you need to lance your finger again, obtain a new lancet, and follow the same steps from the first procedure.

2 Obtain a piece of Tallquist paper. Roll your bleeding finger on the paper.

3 Compare the color the paper turns with the scale on the container while the paper is still slightly wet—do not let the paper dry, or you will need to repeat the procedure.

Estimated hemoglobin value: _____

4 Dispose of any remaining sharps in the sharps container and all other blood-containing materials, including the Tallquist paper, in the red biohazard bag. Clean your work area with the disinfectant provided by your lab instructor.

Exercise 12-6

Lymphatic System

MATERIALS
❑ Human torso models
❑ Head and neck models
❑ Intestinal villus model
❑ Laminated outline of the human body
❑ Water-soluble marking pens
❑ Colored pencils

Now let's turn our attention to the lymphatic system (limf-AEH-tik), which consists of organs that have three primary functions:

1. **Transporting excess interstitial fluid back to the heart.** Approximately 1.5 mL/min. of fluid is lost from the circulation in the blood capillaries to the interstitial fluid. This may not sound like a lot, but if this fluid were not returned to the blood vessels, we could lose our entire plasma volume in about a day. Fortunately, the lymphatic system picks up this lost fluid, carries it through lymphatic vessels, and returns it to the cardiovascular system.

The water in the interstitial fluid first enters small, blind-ended **lymph capillaries** that surround blood capillary beds. After it is inside the lymph capillaries, the fluid is called **lymph** (LIMF). Next, the lymph is delivered to larger **lymph-collecting vessels**, which drain into larger **lymph trunks**. There are nine main lymph trunks that drain lymph from major body regions (Fig. 12.8A): the **jugular trunks**, which drain the head and the neck; the **subclavian trunks**, which drain the upper limbs; the **bronchomediastinal trunks** (brongk-oh-mee-dee-ah-STYN-uhl), which drain the thorax; the **intestinal trunk**, which drains the abdomen; and the **lumbar trunks**, which drain the pelvis and lower limbs.

The lymph trunks then drain into the final vessels of the lymphatic circuit: **lymph ducts**. Note in Figure 12.8A that there are two lymph ducts: the **right lymphatic duct**, which drains the right upper limb and the right side of the head, neck, and thorax, and the **thoracic duct**, which drains lymph from the remainder of the body. The right lymphatic duct delivers lymph to the blood at the junction of the right subclavian and internal jugular veins. Similarly, the thoracic duct drains lymph into the blood at the junction of the left subclavian and internal jugular veins (Fig. 12.8B).

2. **Activating the immune system.** Several of the lymphatic organs activate the *immune system*, a group of cells and proteins that protects the body from cellular injury such as trauma or pathogens (disease-causing organisms, cells, or chemicals). The lymphatic organs include the following:

a. **Thymus.** The **thymus** is an organ composed of two lobes and many small lobules that is located in the anterior mediastinum (Fig. 12.8C). Recall from Unit 10, the endocrine unit, that the thymus secretes hormones that stimulate T lymphocyte maturation. The thymus is largest and most active in infants and young children. In adults, it atrophies (shrinks) and becomes filled with adipose and other connective tissue. (*Note:* The adult thymus is smaller and less distinct than what is shown in Figure 12.8C; the thymus is shown larger and more distinct here for clarity.)

b. **Lymph nodes.** The lymphatic organs called **lymph nodes** are arrangements of lymphatic tissue surrounded by a connective tissue capsule. Lymph nodes are often called "lymph glands," but this is a misnomer—they are not glands because they do not secrete any products. Instead, they act as filters that remove potential pathogens from the lymph before it is delivered to the blood. Lymph nodes are found along lymphatic vessels, where lymph is delivered by *afferent lymphatic vessels* and drained by an *efferent lymphatic vessel*. All lymph nodes are individual, but there are clusters

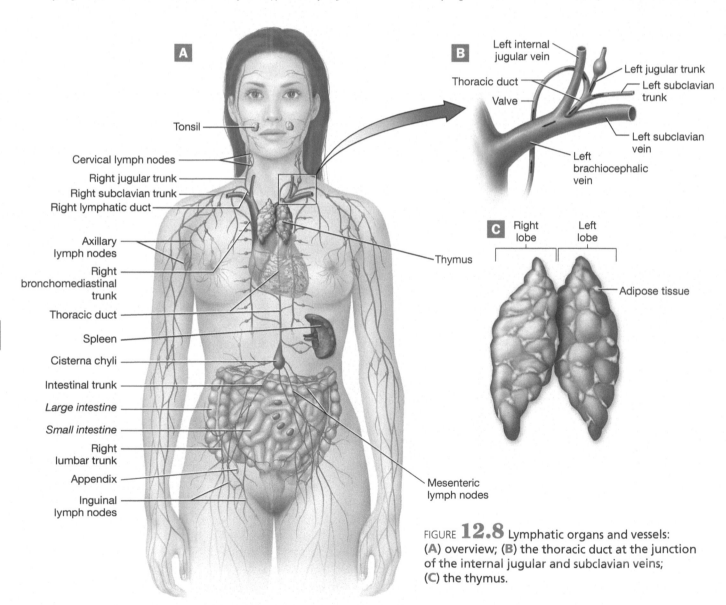

FIGURE **12.8** Lymphatic organs and vessels: (**A**) overview; (**B**) the thoracic duct at the junction of the internal jugular and subclavian veins; (**C**) the thymus.

of them that are named as a group; these groupings are referred to as the **cervical, axillary, inguinal** (IN-gwin-uhl), and mesenteric (or intestinal) **lymph nodes.** The internal structure of a lymph node, which consists of an outer **cortex** and an inner **medulla,** is shown in Figure 12.9.

c. **Spleen.** The **spleen** is a lymphatic organ that resides in the upper left quadrant of the abdominopelvic cavity. In the immune system, the spleen filters the blood and houses phagocytes. As you can see in Figure 12.10, it has two histologically distinct regions: red pulp and white pulp. **Red pulp** contains macrophages that destroy old or damaged erythrocytes. **White pulp** contains leukocytes that destroy pathogens that filter out of arterial blood.

d. **Mucosa-associated lymphatic tissue and tonsils.** Clusters of loosely organized lymphatic tissue are scattered throughout mucous membranes in locations such as the gastrointestinal tract. These clusters are known as **mucosa-associated lymphatic tissue,** or **MALT.** Most MALT lacks a connective tissue capsule; however, some types of MALT, known as specialized MALT, are partially encapsulated. The most prominent examples of specialized MALT are the **tonsils,** which are found in the posterior oropharynx and nasopharynx (Fig. 12.11). Notice in the figure the three named tonsils: the **pharyngeal tonsil** (or **adenoid**), located in the posterior nasopharynx; the paired **palatine tonsils** (PAL-uh-ty'n), located in the posterior oropharynx; and the **lingual tonsil** (LING-yoo-uhl), located at the base of the tongue.

3. **Absorbing dietary fats.** Fats are not absorbed from the small intestine directly into the blood capillaries because they are too large to enter these small vessels. Instead, fats enter a lymphatic capillary called a **lacteal** (lak-TEEL). The lacteal delivers the fats to the lymph in a large lymphatic vessel called the **cisterna chyli** (sis-TUR-nuh KY-lee), which then drains into the thoracic duct.

In the following activities, you will identify structures of the lymphatic system, and then trace the flow of lymph through the vessels on its way back to the cardiovascular system. Your instructor may also wish you to dissect a preserved small mammal to identify some of the structures difficult to see on models, such as the thymus. If so, follow the procedure outlined in Unit 1 (p. 14) to open the animal and identify the required structures.

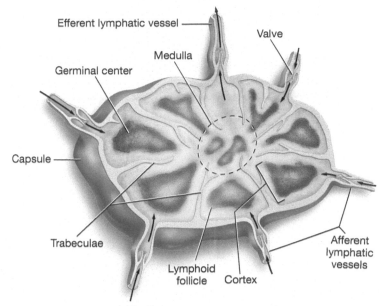

FIGURE **12.9** Structure of a lymph node.

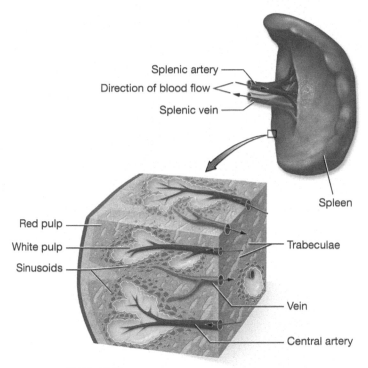

FIGURE **12.10** Anatomy and microanatomy of the spleen.

12

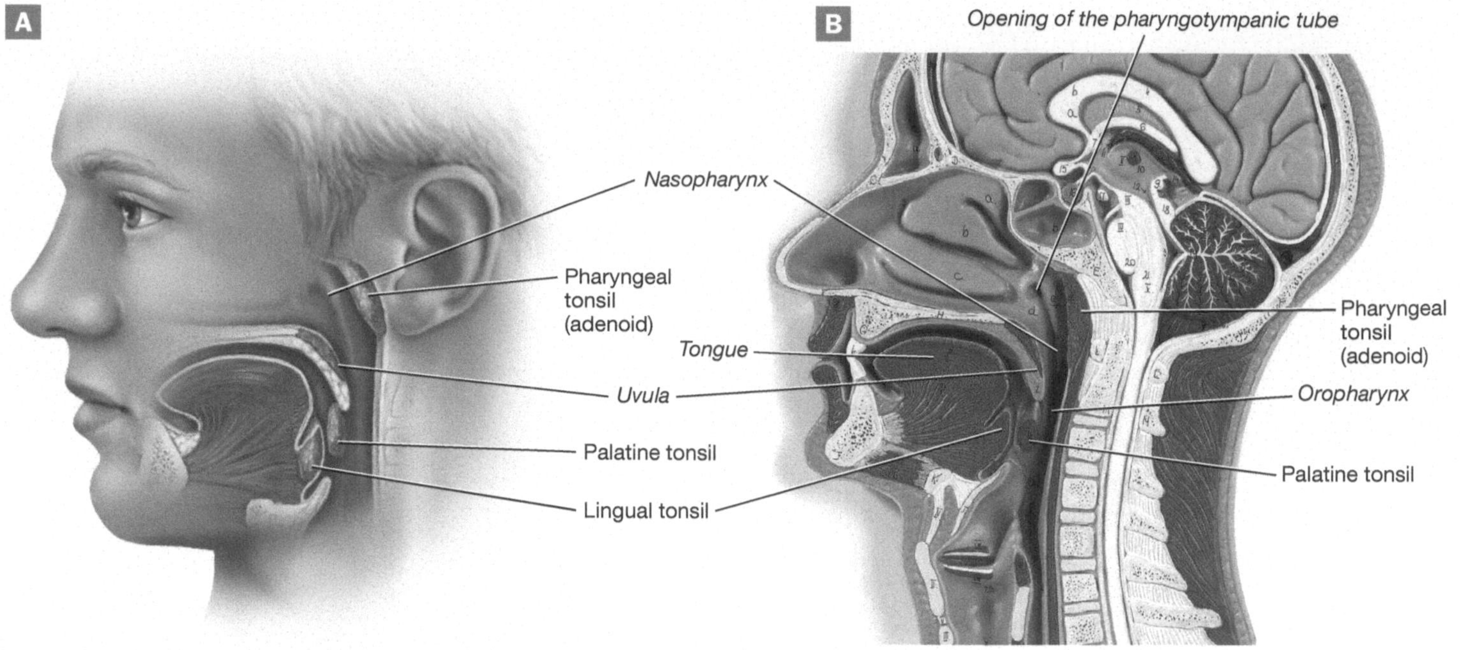

A

Nasopharynx

Pharyngeal
tonsil
(adenoid)

Tongue

Uvula

Palatine tonsil

Lingual tonsil

B

Opening of the pharyngotympanic tube

Pharyngeal
tonsil
(adenoid)

Oropharynx

Palatine tonsil

FIGURE **12.11** Anatomy of the tonsils: (**A**) illustration of a midsagittal section of
the oral cavity; (**B**) anatomical model photo, midsagittal section of the head.

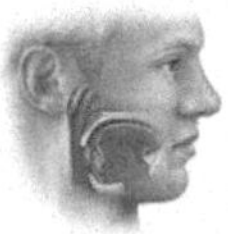

Procedure **1** Model Inventory for the Lymphatic System

Identify the following structures of the lymphatic system on models and diagrams, using this unit and your
textbook for reference. As you examine the anatomical models and diagrams, record the name of the model
and the structures you were able to identify on the model inventory in Table 12.5.

1. Lymph vessels
   a. Thoracic duct
   b. Right lymphatic duct
   c. Lymph trunks
   d. Lacteal
   e. Cisterna chyli
2. Lymph nodes
   a. Cervical lymph nodes
   b. Axillary lymph nodes
   c. Inguinal lymph nodes
   d. Mesenteric lymph nodes

3. Spleen
4. Thymus (this is best viewed on a fetal pig)
5. Mucosa-associated lymphatic tissue (MALT)
6. Tonsils
   a. Palatine tonsil
   b. Pharyngeal tonsil
   c. Lingual tonsil

12

TABLE **12.5** Model Inventory for the Lymphatic System

Model	Structures Identified

12

Procedure **2** Tracing the Flow of Lymph through the Body

In this procedure, you will trace the pathway of lymph flow from the starting location to the point at which the lymph is delivered to the cardiovascular system. You will trace the flow through the major lymph-collecting vessels, trunks, and ducts, and highlight clusters of lymph nodes through which the lymph passes as it travels.

1 Write the sequence of the flow.

2 Then use differently colored water-soluble markers to draw the pathway on a laminated outline of the human body. If no outline is available, use colored pencils and Figure 12.12.

3 Trace the flow from the following locations:

Start: Right foot

Start: Right arm

Start: Intestines (with fat)

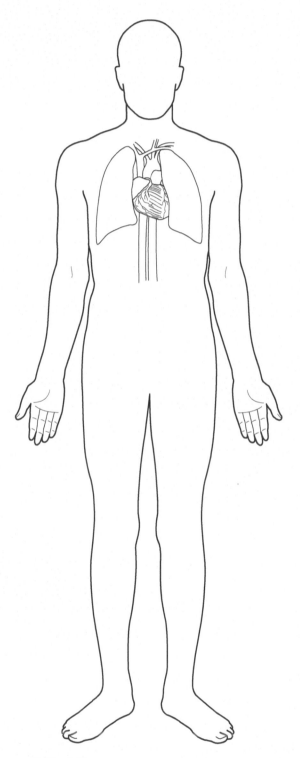

FIGURE **12.12** Outline of human body, anterior view.

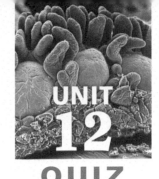

1 *True/False:* Mark the following statements as true (T) or false (F). If the statement is false, correct it to make it a true statement.

_____ a. The liquid portion of blood is known as plasma.

_____ b. Erythrocytes are generally involved in some aspect of the immune system.

_____ c. The granulocytes include the neutrophils, eosinophils, and basophils.

_____ d. Eosinophils are involved in mediating responses to parasitic worms and allergies.

_____ e. Lymphocytes leave the blood and become macrophages, very active phagocytes.

2 Label the formed elements on the peripheral blood smear in **Figure 12.13**.

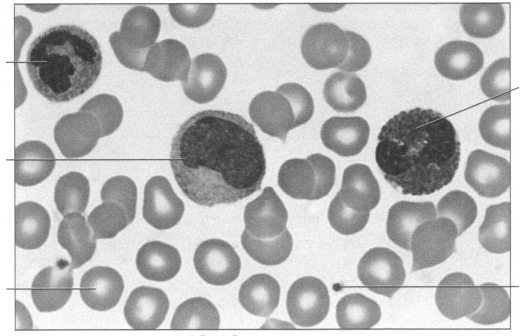

FIGURE **12.13** Peripheral blood smear.

3 Which of the following is *not* an antigen that may be found on the surface of an erythrocyte?

a. A antigen.

b. B antigen.

c. O antigen.

d. Rh antigen.

4 *Circle all that apply:* A person with type O blood has

a. anti-A antibodies.

b. anti-B antibodies.

c. anti-O antibodies.

d. no antibodies.

12

5 *Circle all that apply:* A person with type A– blood has which of the following antibodies? (Assume the person has been exposed to Rh antigens.)

    a. Anti-A antibodies.

    b. Anti-B antibodies.

    c. Anti-Rh antibodies.

    d. No antibodies.

6 List all of the blood types to which the following people could donate:

    a Person 1: Type A+ _____

    b Person 2: Type O– _____

    c Person 3: Type AB+ _____

    d Person 4: Type B– _____

7 The disease *erythroblastosis fetalis* (also called *hemolytic disease of the newborn*) develops in a fetus or a newborn infant with Rh-positive blood and an Rh-negative mother. Symptoms result when maternal anti-Rh antibodies cross the placenta and interact with the fetus' erythrocytes. Why are the children of Rh-positive mothers not at risk for this disease? Why are Rh-negative fetuses not at risk for this disease?

_____

_____

_____

_____

_____

8 Explain why a person who is blood type O– can donate blood to any blood type but can receive only from individuals who are also blood type O–.

_____

_____

_____

_____

9 A common consequence of many cancer chemotherapy drugs is destruction of bone marrow cells. This leads to a decrease in the number of circulating erythrocytes and leukocytes, resulting in conditions called *anemia* and *leukopenia*, respectively. Predict the consequences of anemia and leukopenia, considering the function of these types of cells.

_____

_____

_____

_____

12

10 Label **Figure 12.14** with the terms below.

Figure 12.14A

❏ Lymph nodes:
   ☐ Axillary
   ☐ Cervical
   ☐ Inguinal
   ☐ Mesenteric
❏ Spleen
❏ Thoracic duct
❏ Thymus
❏ Cisterna chyli

Figure 12.14B

❏ Lingual tonsil
❏ Palatine tonsil
❏ Pharyngeal tonsil

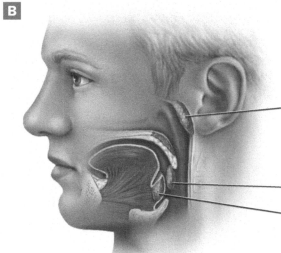

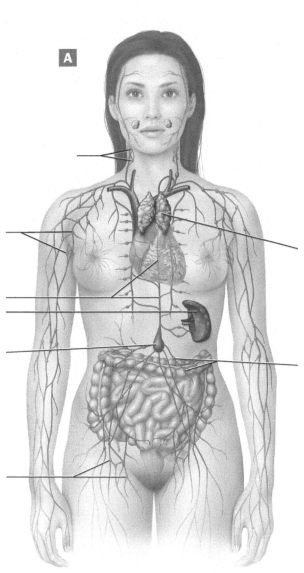

FIGURE **12.14** Lymphatic organs: (**A**) overview of the lymphatic organs and vessels; (**B**) the tonsils.

12

11 Which of the following is not a function of the lymphatic system?

    a. Maintaining blood pressure.

    b. Absorbing dietary fats.

    c. Activating the immune system.

    d. Transporting excess interstitial fluid back to the heart.

12 What is the function of the thymus?

    a. Traps pathogens in lymph.

    b. Filters the blood.

    c. Mediates inflammation.

    d. Serves as the site of T lymphocyte maturation.

13 Nilesh has *Hodgkin's lymphoma*, a cancer that primarily involves lymphocytes. As part of his treatment plan, he is undergoing surgery to remove his spleen, a procedure called a *splenectomy*, and several affected cervical lymph nodes.

    a How does the role of the spleen differ from the role of lymph nodes?

    b What potential complications might Nilesh face without a spleen? Explain.

    c How would the complications in part b differ from potential issues resulting from removal of multiple cervical lymph nodes?

12

14 Juliet is a 42-year-old patient who is preparing to undergo surgery to remove her thymus gland, which has a tumor (a *thymoma*). She has read about the thymus and its functions and is concerned that her immune system will be much weaker after the surgery. What do you tell her, and why?

Respiratory System

When you have completed this unit, you should be able to:

1 Describe and identify structures of the respiratory system.

2 Trace the pathway of gases through the respiratory system.

3 Describe the anatomical changes associated with inflation of the lungs.

4 Describe the pressure-volume relationships in the lungs.

5 Measure and define respiratory volumes and capacities.

PRE-LAB EXERCISES

Complete the following exercises prior to coming to lab, using your lab manual and textbook for reference.

Pre-Lab Exercise **13-1**

✎ Key Terms

You should be familiar with the following terms before coming to lab.

Term	Definition

General Structures of the Respiratory System

Respiratory tract

Pleural cavity

Lungs

Structures of the Respiratory Tract

Nasal cavity

Pharynx

Larynx

Trachea

13 Bronchi

Bronchioles

Alveoli

Pulmonary Ventilation Terms

Inspiration

Expiration

Boyle's law

Respiratory volume

Respiratory capacity

13

Respiratory System Anatomy

Color the diagrams of the structures of the respiratory system in Figure 13.1, and label them with the following terms from Exercise 13-1 (pp. 361 and 365). Use Exercise 13-1 in this unit and your text for reference.

❑ Nasal cavity

❑ Larynx

❑ Trachea

Right Lung

☐ Upper, middle, lower lobes

☐ Horizontal fissure

☐ Oblique fissure

Left Lung

☐ Upper and lower lobes

☐ Oblique fissure

Pharynx

☐ Nasopharynx

☐ Oropharynx

☐ Laryngopharynx

Bronchi

☐ Primary bronchi

☐ Secondary bronchi

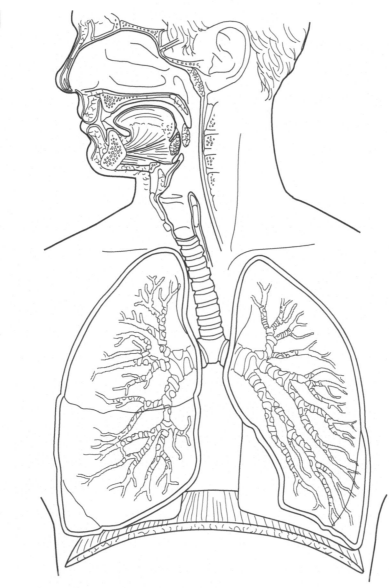

A

FIGURE **13.1** Structures of the respiratory system: **(A)** anatomy of the lungs and respiratory tract; *(continues)*

13

Bronchioles

- ☐ Terminal bronchiole
- ☐ Respiratory bronchiole
- ☐ Alveolar duct
- ☐ Alveoli
- ☐ Pulmonary capillaries
- ☐ Pulmonary arteriole
- ☐ Pulmonary venule

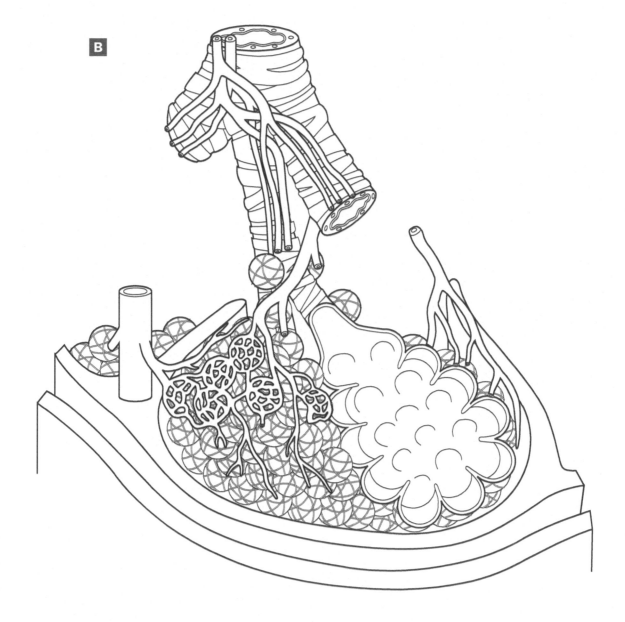

FIGURE **13.1** Structures of the respiratory system *(cont.)*: **(B)** bronchiole and alveolar sac.

Pre-Lab Exercise 13-3
Defining Pulmonary Volumes and Capacities

Define and give the normal values for males and females for each of the following respiratory volumes and capacities listed in Table 13.1.

TABLE **13.1** Respiratory Volumes and Capacities

Volume/Capacity	Definition	Normal Value
Tidal volume		Male: _____ mL Female: _____ mL
Inspiratory reserve volume		Male: _____ mL Female: _____ mL
Expiratory reserve volume		Male: _____ mL Female: _____ mL
Residual volume		Male: _____ mL Female: _____ mL
Inspiratory capacity		Male: _____ mL Female: _____ mL
Functional residual capacity		Male: _____ mL Female: _____ mL
Vital capacity		Male: _____ mL Female: _____ mL
Total lung capacity		Male: _____ mL Female: _____ mL

EXERCISES

Cells require oxygen in the reactions that synthesize ATP, and these reactions produce carbon dioxide as a waste product. The **respiratory system** and the cardiovascular system work together to supply the cells with the oxygen they need and to rid them of carbon dioxide.

The first exercise in this unit will familiarize you with the anatomy of the respiratory system, including the paired *lungs* and the collection of airway passages known as the *respiratory tract*. Exercises 13-2 and 13-3 focus on the physiology of the respiratory system and the process known as *ventilation*.

Ciliated epithelium of the respiratory tract.

Exercise **13-1**

Respiratory System Anatomy

MATERIALS
- ❏ Lung models
- ❏ Larynx models
- ❏ Alveolar sac model
- ❏ Head and neck model
- ❏ Colored pencils

The **lungs** are composed of elastic connective tissue and tiny air sacs called **alveoli** (al-vee-OH-lye), where gas exchange takes place. Each lung is divided into smaller structures called **lobes**. The right lung has three lobes (upper, middle, and lower), and the left lung has two lobes (upper and lower) (Fig. 13.2). The lobes are separated from one another by deep indentations called **fissures**. The **horizontal fissure** separates the right upper and right middle lobes; the **right oblique fissure** separates the right middle and right lower lobes; and the **left oblique fissure** separates the left upper and left lower lobes. The left upper lobe has a groove, called the **cardiac notch**, on its medial surface where it comes into contact with the heart. Each lung has an **apex** that sits just above the clavicle, and a **base** that rests on the **diaphragm muscle**, the main muscle for ventilation.

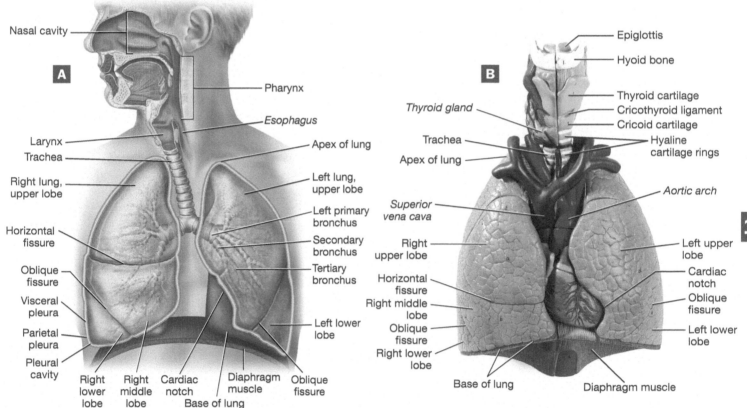

FIGURE **13.2** Anatomy of the lungs and respiratory tract: (**A**) illustration; (**B**) model photo.

Each lung is surrounded by serous membranes, similar in structure and function to the pericardial membranes, known as the **pleural membranes** (PLOO-ruhl). There are two layers of the pleural membranes:

1. **Parietal pleura.** The outer **parietal pleura** lines the interior of the thoracic cavity and the superior surface of the diaphragm.

2. **Visceral pleura.** When the parietal pleura reaches the structures of the mediastinum, it folds inward to become the **visceral pleura.** The visceral pleura adheres tightly to the surface of the lung.

The very thin potential space between the parietal and visceral pleurae is the **pleural cavity.** The space is only a "potential" space because it is filled with a thin layer of serous fluid that reduces friction as the lungs change in shape and size during ventilation.

If we flip a lung on its side and examine its medial or mediastinal surface (the surface that faces the mediastinum), as we do in Figure 13.3, we can see an area called the **hilum** (HY-lum) of the lung. This is an indentation where the pulmonary vessels, nerves, lymphatic vessels, and airway passages called *primary bronchi* enter and exit the lungs.

Air is delivered to the lungs' alveoli through the passageways of the **respiratory tract.** The respiratory tract may be divided into two structural regions: (1) the *upper respiratory tract*, which consists of the passages from the nasal cavity to the larynx; and (2) the *lower respiratory tract*, which consists of the passages from the trachea to the alveoli.

The respiratory tract begins with the **nasal cavity** (Fig. 13.4), which warms, filters, and humidifies the incoming air. Air first enters through the **nares** (or nostrils), then enters the actual nasal cavity. Lining the walls of the nasal cavity are three projections known as the **superior, middle,** and **inferior nasal conchae** (KAHN-kee). These projections make airflow turbulent, which removes dust and other debris from the inhaled air. The nasal cavity is lined with pseudostratified ciliated columnar epithelium with copious mucus-secreting **goblet cells,** a type of tissue known as **respiratory epithelium.**

Air flow is continuous between the nasal cavity and the

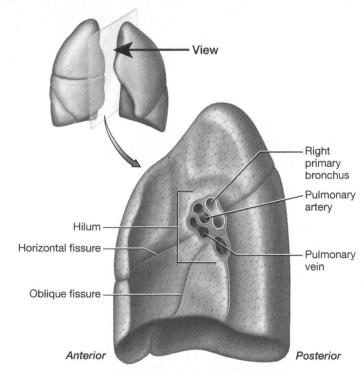

FIGURE **13.3** Mediastinal surface of the right lung.

paranasal sinuses, which assist in filtering, warming, and humidifying the inhaled air. Recall from Unit 6 that there are four paranasal sinuses: the frontal, maxillary, ethmoid, and sphenoid sinuses (see Figure 6.17, p. 137, for a review).

From the nasal cavity, air next enters the **pharynx** (FEHR-inks), also known as the throat, which has the following three divisions:

1. **Nasopharynx.** The **nasopharynx** (nayz-oh-FEHR-inks) is the region posterior to the nasal cavity, color-coded green in Figure 13.4A. The muscles of the soft palate and *uvula* move superiorly to close off the nasopharynx during swallowing to prevent food from entering the passage. Sometimes this mechanism fails (such as when a person is laughing and swallowing simultaneously), and the unfortunate result is that food or liquid comes out of the nose. Like the nasal cavity, the nasopharynx is lined with respiratory epithelium.

2. **Oropharynx.** The **oropharynx** (ohr-oh-FEHR-inks) is the region posterior to the oral cavity, color-coded yellow in Figure 13.4A. Both food and air pass through the oropharynx, and it is therefore lined with stratified squamous epithelium. This tissue provides more resistance to mechanical and thermal stresses.

3. **Laryngopharynx.** The **laryngopharynx** (lah-ring-oh-FEHR-inks) is the intermediate region between the larynx and the esophagus, color-coded light blue in Figure 13.4A. As with the oropharynx, both food and air pass through the laryngopharynx, and it is lined with stratified squamous epithelium.

13

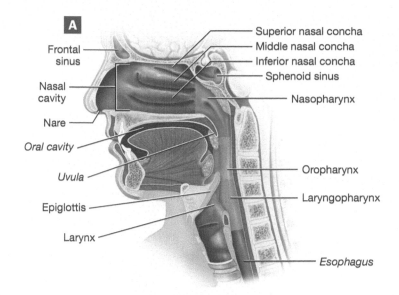

A

Frontal sinus
Nasal cavity
Nare
Oral cavity
Uvula
Epiglottis
Larynx

Superior nasal concha
Middle nasal concha
Inferior nasal concha
Sphenoid sinus
Nasopharynx
Oropharynx
Laryngopharynx
Esophagus

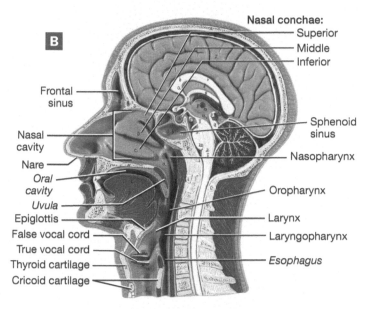

B

Nasal conchae:
Superior
Middle
Inferior

Frontal sinus
Nasal cavity
Nare
Oral cavity
Uvula
Epiglottis
False vocal cord
True vocal cord
Thyroid cartilage
Cricoid cartilage

Sphenoid sinus
Nasopharynx
Oropharynx
Larynx
Laryngopharynx
Esophagus

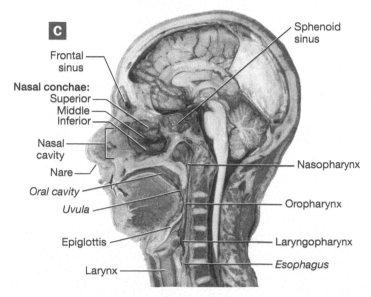

C

Sphenoid sinus
Frontal sinus

Nasal conchae:
Superior
Middle
Inferior

Nasal cavity
Nare
Oral cavity
Uvula

Epiglottis

Larynx

Nasopharynx
Oropharynx
Laryngopharynx
Esophagus

FIGURE **13.4** Midsagittal section of the head and neck: (**A**) illustration; (**B**) model photo; (**C**) cadaver dissection.

Next, air passes from the pharynx to the **larynx** (LEHR-inks), a short passage framed by the hyoid bone and nine cartilages (Fig. 13.5). The "lid" of the larynx is a piece of elastic cartilage called the **epiglottis** (ep-ih-GLAW-tiss). During swallowing, muscles of the pharynx and larynx move the larynx superiorly, and the epiglottis seals off the larynx from food and liquids. The largest cartilage of the larynx is the shieldlike **thyroid cartilage**, which is inferior to the hyoid bone. It forms the larynx's anterior and lateral walls. Inferior to the thyroid cartilage is the smaller **cricoid cartilage** (KRY-koyd), which is the only cartilage in the respiratory tract that is a complete ring—the remaining cartilages are "C"-shaped. The thyroid and cricoid cartilages are united by a connective tissue membrane called the **cricothyroid ligament** (kry-koh-THY-royd). The smaller cartilages are labeled in Figure 13.5 for your reference.

As its common name "voice box" implies, the larynx is the structure where sound is produced. As you can see in Figure 13.5C, it contains two sets of elastic ligaments known as the **vocal folds** or **vocal cords**. The superior set of vocal folds, the **false vocal cords** (also called the *vestibular folds*), plays no role in sound production. They do, however, serve an important sphincter function and can constrict to close off the larynx. The inferior set of vocal folds, called the **true vocal cords**, vibrates as air passes over it to produce sound. Superior to the vocal cords, the larynx is lined with stratified squamous epithelium; inferior to the vocal cords, it is lined with respiratory epithelium, as is the remainder of the respiratory tract until we reach much smaller passageways.

Inspired air passes from the larynx into the **trachea** (TRAY-kee-uh), a passage supported by "C"-shaped rings of hyaline cartilage (Fig. 13.6). The trachea bifurcates in the mediastinum at its final cartilage ring, called the **carina** (kuh-RY-nuh), into two **primary bronchi** (BRONG-kye) that begin the large and branching **bronchial tree** (BRONG-kee-uhl). The right primary bronchus is short, fairly straight, and wide, and the left primary bronchus is long, more horizontal, and narrow because of the position of the heart.

Each primary bronchus divides into smaller **secondary bronchi**, each of which serves one lobe of the lung. The two left secondary bronchi serve the two lobes of the left lung, while the three right secondary bronchi serve the three lobes of the right lung. The bronchi continue to branch, becoming tertiary bronchi, quaternary bronchi, and so on until the air reaches tiny air passages smaller than 1 millimeter in diameter called **bronchioles** (BRONG-kee-ohlz; Fig. 13.7). At this level, the epithelium changes to simple cuboidal epithelium.

13

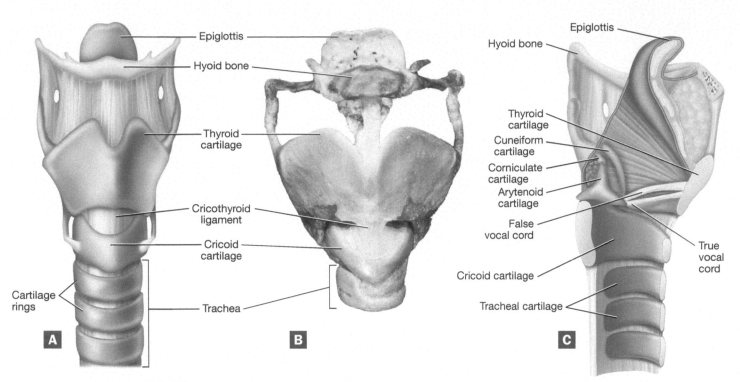

FIGURE **13.5** Larynx: (**A**) anterior view; (**B**) anterior view from a cadaver; (**C**) midsagittal section.

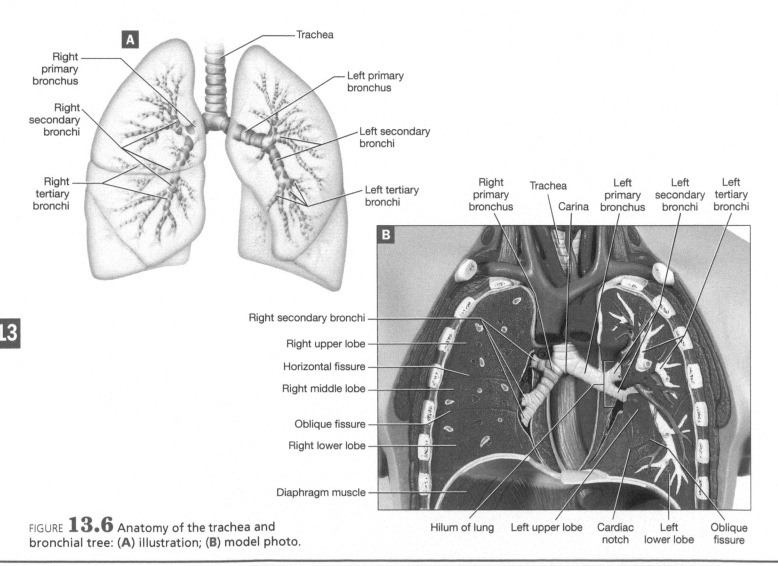

FIGURE **13.6** Anatomy of the trachea and bronchial tree: (**A**) illustration; (**B**) model photo.

Bronchioles smaller than 0.5 mm in diameter are called the **terminal bronchioles**, and these branch into **respiratory bronchioles**, which have alveoli in their walls. As the respiratory bronchioles progressively branch, the number of alveoli in their walls increases until the walls are made up exclusively of alveoli, at which point they are termed **alveolar ducts** (al-vee-OH-lahr). The terminal portions of the respiratory zone, called **alveolar sacs**, are grapelike clusters of alveoli.

The alveoli are surrounded by **pulmonary capillaries**, which are fed by **pulmonary arterioles** and drained by **pulmonary venules**. The capillary-alveolus junction, along with their fused basal laminae, forms a structure known as the **respiratory membrane**, which is where pulmonary gas exchange takes place. During pulmonary gas exchange, oxygen from the air in the alveoli diffuses into the blood, and carbon dioxide from the blood diffuses into the air in the alveoli to be exhaled. Alveolar walls and capillary walls are both composed of simple squamous epithelium, so the gases have only a short distance to diffuse, which is critical to effective gas exchange. In addition, the structure of the alveolar sacs creates a huge surface area (around 1,000 square feet on average), another factor that allows gas exchange to take place rapidly and efficiently.

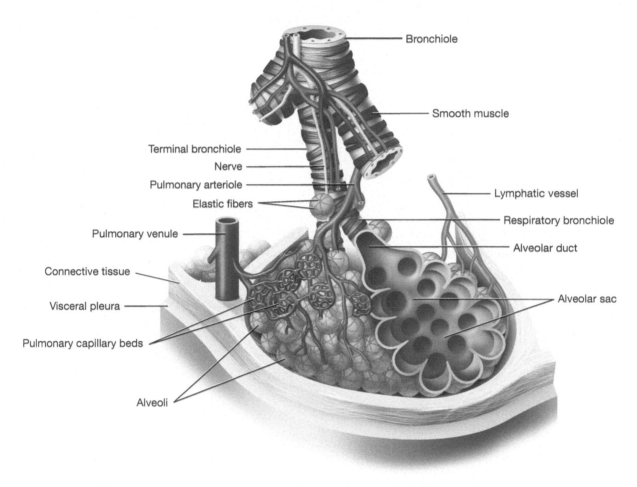

FIGURE **13.7** Bronchiole and alveolar sac.

13

Procedure 1 Model Inventory for the Respiratory System

Identify the following structures of the respiratory system on models and diagrams, using this unit and your textbook for reference. As you examine the anatomical models and diagrams, record the name of the model and the structures you were able to identify on the model inventory in Table 13.2.

1. Right lung
   a. Upper, middle, lower lobes
   b. Horizontal fissure
   c. Oblique fissure
2. Left lung
   a. Upper and lower lobes
   b. Oblique fissure
   c. Cardiac notch
3. Lungs, general
   a. Apex
   b. Base
   c. Diaphragm muscle
   d. Pleurae
      (1) Parietal pleura
      (2) Visceral pleura
      (3) Pleural cavity
   e. Hilum

4. Nasal cavity
   a. Nares
   b. Nasal conchae
   c. Paranasal sinuses
5. Pharynx
   a. Nasopharynx
   b. Oropharynx
   c. Laryngopharynx
6. Larynx
   a. Hyoid bone
   b. Epiglottis
   c. Thyroid cartilage
   d. Cricoid cartilage
   e. Cricothyroid ligament
   f. False vocal cords
   g. True vocal cords

7. Trachea
   a. Hyaline cartilage rings
   b. Carina
8. Bronchi
   a. Right and left primary bronchi
   b. Secondary bronchi
9. Bronchioles
   a. Terminal bronchioles
   b. Respiratory bronchioles
   c. Alveolar duct
10. Alveoli and alveolar sacs
11. Vascular structures
    a. Pulmonary capillaries
    b. Pulmonary arterioles
    c. Pulmonary venules

TABLE **13.2** Model Inventory for the Respiratory System

Model/Diagram	Structures Identified

13

Procedure 2 Time to Draw

In the space provided, draw, color, and label one of the respiratory system models that you examined. In addition, write the function or definition of each structure that you label.

Procedure 3 Time to Trace!

In this next procedure, you will trace the pathway taken by oxygen and carbon dioxide molecules through the respiratory tract and through parts of the systemic and pulmonary vascular circuits.

You are a molecule of oxygen floating happily through the atmosphere when all of a sudden you are inhaled by Ms. Magenta.

1 Trace your pathway through Ms. Magenta's respiratory tract, beginning in her nasal cavity to the point where you enter the pulmonary capillaries.

Start:

End

13

The cardiovascular system delivers you to a cell in Ms. Magenta's liver, where you are used as the final electron acceptor in the process of cellular respiration as the cell generates ATP. While in her cell, you notice that a molecule of carbon dioxide has just been produced.

2 Trace the carbon dioxide's pathway from Ms. Magenta's inferior vena cava through her heart and to her pulmonary capillaries. You may want to review Unit 11 (p. 281) for some help with this.

Start: _____

_____

_____

_____

_____

_____ End

3 Trace the pathway of the carbon dioxide from the pulmonary capillaries through the respiratory tract to the point where it exits from Ms. Magenta's body through her nares.

Start: _____

_____

_____

_____

_____

_____ End

13

Exercise 13-2

Respiratory System Physiology: Pressure-Volume Relationships in the Lungs

MATERIALS
- ❏ Bell-jar model of the lungs
- ❏ Air hose
- ❏ Air compressor
- ❏ Fresh lungs

Respiration consists of four basic physiological processes:

1. **Pulmonary ventilation** is the physical movement of air into and out of the lungs.

2. **Pulmonary gas exchange** is the movement of gases across the respiratory membrane.

3. **Gas transport** is the movement of gases through the blood.

4. **Tissue gas exchange** is the exchange of gases between the blood in the systemic capillaries and the tissues.

Of these four processes, the easiest one to examine in the lab is pulmonary ventilation, which consists of two phases: (1) **inspiration**, during which air is brought into the lungs, and (2) **expiration**, during which air is expelled from the lungs. The movement of air during inspiration and expiration is driven by changes in the lungs' volume and pressure (Fig. 13.8). The relationship of gas pressure and volume is expressed in what is known as **Boyle's law**, expressed mathematically as:

$$P_1V_1 = P_2V_2 \text{ or } P = \frac{1}{V}$$

Stated simply, this means that pressure (P) and volume (V) are inversely proportional: As the volume of a container increases, the pressure inside the container decreases, and as the volume of a container decreases, the pressure inside the container increases.

The changes in volume during the phases of ventilation are driven by the **inspiratory muscles**. The main inspiratory muscle is the **diaphragm muscle**, and it is assisted by the **external intercostal muscles** (see Figures 7.10 and 7.11, p. 180, for a review). During forced inspiration, several other muscles, termed *accessory muscles of inspiration*, assist the diaphragm and external intercostal muscles. When the inspiratory muscles contract, they increase both the height and the diameter of the thoracic cavity, which increases its volume. Recall that the lungs are attached to the thoracic cavity directly by the pleural membranes. For this reason, the lungs' height and diameter increase as well, as do their volume. As the lungs' volume increases, the pressure within the alveoli, called the **intrapulmonary pressure**, decreases. When intrapulmonary pressure is lower than the **atmospheric pressure**, inspiration occurs, and air rushes into the lungs.

Expiration is achieved primarily by the elastic recoil of the lungs. As the inspiratory muscles relax, the lungs' elastic tissue causes them to recoil to their original, smaller size. This decreases the volume of the lungs and increases the intrapulmonary pressure. When the intrapulmonary pressure is higher than the atmospheric pressure, air exits the lungs, and expiration occurs. In the event of forced expiration, several accessory muscles of expiration, including the **internal intercostal muscles**, will further decrease the height and diameter of the thoracic cavity.

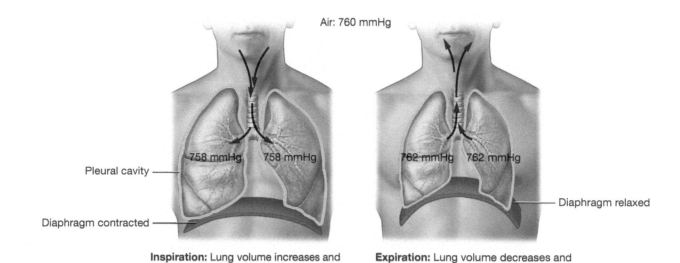

Air: 760 mmHg

Pleural cavity

758 mmHg   758 mmHg

762 mmHg   762 mmHg

Diaphragm relaxed

Diaphragm contracted

Inspiration: Lung volume increases and intrapulmonary pressure decreases.

Expiration: Lung volume decreases and intrapulmonary pressure increases.

FIGURE **13.8** Pressure-volume relationships in the lungs during inspiration and expiration.

13

Procedure 1 Model Ventilation with the Bell-Jar Model

In this procedure, we will use a bell-jar model of the lungs, shown in Figure 13.9, to view the effects of pressure and volume on ventilation. The bell-jar model has two balloons, each representing one lung, and a flexible membrane on the bottom that represents the diaphragm muscle.

1 Apply upward pressure to the membrane. This represents how the diaphragm muscle looks when it is relaxed. What has happened to the pressure of the system (has it increased or decreased)? What happened to the volume of the lungs?

FIGURE **13.9** Bell-jar model of the lungs, simulated inspiration.

2 Now slowly release the membrane. This represents the diaphragm muscle flattening out as it contracts. What is happening to the pressure as you release the membrane? What happened to the volume of the lungs?

3 If your bell-jar model has a rubber stopper in the top, you can use it to demonstrate the effects of a **pneumothorax** (noo-moh-THOHR-ax) on lung tissue. A pneumothorax generally is caused by a tear in the pleural membranes that allows air to enter the pleural cavity. With the membrane flat and the lungs (balloons) inflated, loosen the rubber stopper. What happens to the lungs? Why?

13

Procedure 2 Lung Inflation

Now let's see what the lungs look like as they inflate and deflate during inspiration and expiration. In this procedure, you will obtain fresh lungs from a sheep, a pig, or a cow, and inflate them with an air hose. As you perform the procedure, note the difference in the textures and appearances of the lungs as they inflate and deflate.

1 Obtain a fresh specimen and a large air hose.

2 Examine the specimen for structures covered in Exercise 13-1 (p. 361), particularly the epiglottis and vocal folds, the trachea and its hyaline cartilage rings, and the pleural membranes.

Safety Note

⚠ Safety glasses and gloves are required!

3 Squeeze the deflated lungs between your fingertips, and record their texture below. Deflated lungs are shown in Figure 13.10A.

4 Insert the air hose into the larynx, and feed it down into the trachea. Take care not to get the hose stuck in one of the primary bronchi.

5 Attach the hose to the air outlet, and turn it on slowly. You may have to squeeze the trachea and the hose to prevent air leakage.

6 Observe the lungs as they inflate, shown in Figure 13.10B. You may inflate the lungs quite fully. Don't worry—they're very unlikely to pop.

7 Squeeze the inflated lungs between your fingers, and note their texture.

8 Crimp the air hose, and watch the lungs deflate. Again feel the lungs, and note changes in texture.

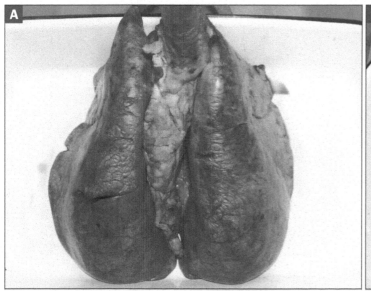
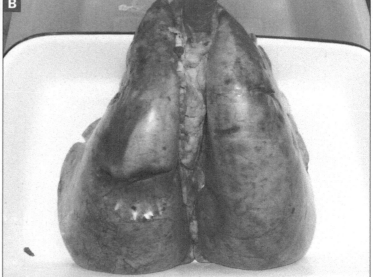

FIGURE **13.10** Fresh lungs: (**A**) deflated; (**B**) inflated.

Exercise 13-3

Measuring Pulmonary Volumes and Capacities

MATERIALS
- ❏ Wet or handheld spirometer
- ❏ Disposable mouthpiece

Respiratory volumes are the volumes of air that are exchanged with each breath. These volumes of air are measured with an instrument called a **spirometer** (spih-RAH-meh-ter) and include the following (Fig. 13.11):

1. **Tidal volume (TV).** The **tidal volume**, or **TV**, is the amount of air exchanged with each breath during normal, quiet breathing. It measures about 500 mL in a healthy adult.

2. **Expiratory reserve volume (ERV).** The **expiratory reserve volume**, or the **ERV**, is the volume of air that may be expired after a tidal expiration. It averages between 700 and 1,200 mL of air.

3. **Inspiratory reserve volume (IRV).** The **inspiratory reserve volume**, or the **IRV**, is the amount of air that may be inspired after a tidal inspiration. It averages between 1,900 and 3,100 mL of air.

Note that there is also a fourth volume, called the **residual volume (RV)**, which cannot be measured with a spirometer. It is defined as the amount of air that remains in the lungs after maximal expiration, and is generally equal to about 1,100 to 1,200 mL of air. This amount of air accounts for the difference between the IRV and the ERV.

Two or more respiratory volumes may be combined to give **respiratory capacities**. As you can see in Figure 13.11, there are four respiratory capacities:

1. **Inspiratory capacity (IC).** The **inspiratory capacity**, or **IC**, is equal to the TV plus the IRV and is the amount of air that a person can maximally inspire after a tidal expiration. It averages between 2,400 and 3,600 mL of air.

2. **Functional residual capacity (FRC).** The **functional residual capacity**, or the **FRC**, is the amount of air that is normally left in the lungs after a tidal expiration. It is the sum of the ERV and the RV and averages between 1,800 and 2,400 mL of air. It is not measurable with general spirometry.

3. **Vital capacity (VC).** The **vital capacity**, or **VC**, represents the total amount of exchangeable air that moves in and out of the lungs. It averages between 3,100 and 4,800 mL of air and is equal to the sum of the TV, IRV, and the ERV.

4. **Total lung capacity (TLC).** The **total lung capacity**, or **TLC**, represents the total amount of exchangeable and nonexchangeable air in the lungs. It is the total of all four respiratory volumes and is not measurable with general spirometry. It averages between 4,200 and 6,000 mL of air.

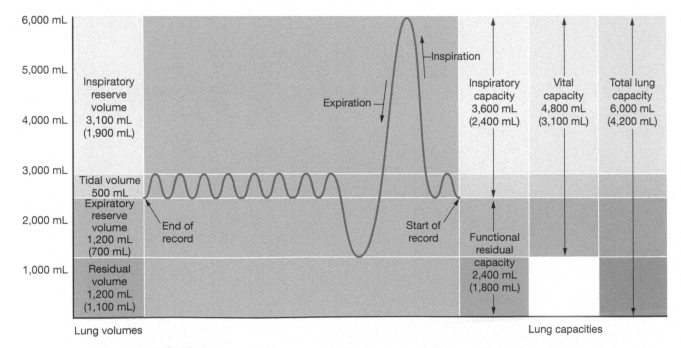

FIGURE **13.11** Respiratory volumes and capacities (note that the top value in the figure is the average value for males, and the value in parentheses is the average value for females).

Spirometry is a useful tool with which to assess pulmonary function. The respiratory volumes and capacities are especially helpful in differentiating the two primary types of respiratory disorders: restrictive diseases and obstructive diseases.

- **Restrictive diseases**, such as pneumoconiosis (coal-miner's lung) and tuberculosis, are characterized by a loss of lung or chest distensibility, meaning that the lungs or chest are unable to stretch and increase in size. As a result, restrictive diseases cause difficulty with *inspiration*. This decreases the IRV, IC, VC, and TLC.

- **Obstructive diseases**, such as chronic obstructive pulmonary disease (COPD) and asthma, are characterized by increased airway resistance caused by narrowing of the bronchioles, increased mucus secretion, and/or an obstructing body such as a tumor. It may seem counterintuitive, but obstructive diseases make *expiration* difficult. This is because the increased intrapulmonary pressure during expiration naturally tends to shrink the diameter of the bronchioles. When the bronchioles are already narrowed, as in an obstructive disease, the increased intrapulmonary pressure can actually collapse the bronchioles and trap oxygen-poor air in the distal respiratory passages. Therefore, patients with obstructive diseases often exhale slowly and through pursed lips to minimize the pressure changes and maximize the amount of air exhaled. Obstructive diseases decrease the ERV and VC and increase the RV and FRC.

Procedure 1 Measuring Respiratory Volumes with a Wet Spirometer

Three types of spirometers are commonly found in anatomy and physiology labs: (1) a bell, or "wet" spirometer; (2) a computerized spirometer; and (3) a handheld spirometer (shown in Figure 13.12). Many types of wet and handheld spirometers allow you to assess only expiratory volumes, whereas computerized spirometers typically allow you to assess both inspiratory and expiratory volumes. All types of devices allow measurement of one or more respiratory capacities.

The following procedure is intended for either a wet or handheld spirometer. If your lab has a computerized spirometer, however, you can either follow the procedure below or follow the prompts by the computer program.

Safety Note

You should not be the subject of this experiment if you have cardiovascular or pulmonary disease or are prone to dizziness or fainting.

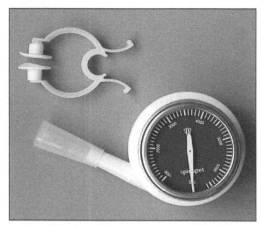

FIGURE **13.12** Handheld spirometer.

1 Obtain a disposable mouthpiece, and attach it to the end of the tube.

2 Before you begin, practice exhaling through the tube several times. Note that you are supposed to only *exhale* into the tube, as the tube likely has no filter.

3 *Measure the tidal volume*: Sit in a chair with your back straight and your eyes closed. Inhale a normal tidal inspiration, and exhale this breath through the tube. Getting a true representation of the tidal volume is often difficult, because people have a tendency to force the expiration. To get the most accurate tidal volume, take several measurements and average the numbers. Note that some handheld spirometers are not terribly sensitive, so they give a value for the tidal volume that is too low. If you find that this is the case, you may need to take extra readings to get an approximately accurate measurement.

Measurement 1: _____   Measurement 2: _____   Measurement 3: _____

Average Tidal Volume: _____

4 *Measure the expiratory reserve volume*: Before taking this measurement, inhale and exhale a series of tidal volumes. Then inspire a normal tidal inspiration, breathe out a normal tidal expiration, put the mouthpiece to your mouth, and exhale as forcibly as possible. Don't cheat by taking in a large breath first! As before, perform several measurements, and average the numbers.

Measurement 1: _____   Measurement 2: _____   Measurement 3: _____

Average Expiratory Reserve Volume: _____

5 *Measure the vital capacity*: As before, inhale and exhale a series of tidal volumes. Then bend over, and exhale maximally. Once you have exhaled as much air as you can, raise yourself upright, and inhale as much air as you possibly can (until you feel like you are about to "pop"). Quickly place the mouthpiece to your mouth, and exhale as forcibly and as long as possible. Take several measurements (you may want to give yourself a minute to rest between measurements), and record the data below.

Measurement 1:   Measurement 2:   Measurement 3:

Average Vital Capacity:

6 *Calculate the inspiratory reserve volume*: Even though the handheld or bell spirometer cannot measure inspiratory volumes, you can calculate this volume now that you have the vital capacity (VC), tidal volume (TV), and expiratory reserve volume (ERV). Recall that VC = TV + IRV + ERV. Rearrange the equation: IRV = VC − (TV + ERV).

Average IRV:

7 How do your values compare with the average values? What factors, if any, do you think may have affected your results?

8 Pool the results for your class, and divide the results into four categories: female cigarette smokers, female nonsmokers, male cigarette smokers, and male nonsmokers. Calculate the average TV, ERV, IRV, and VC for each group, and record these data in Table 13.3.

TABLE **13.3** Respiratory Volumes and Capacities for Smokers and Nonsmokers

Group	Tidal Volume	Expiratory Reserve Volume	Inspiratory Reserve Volume	Vital Capacity
Female smokers				
Female nonsmokers				
Male smokers				
Male nonsmokers				

9 Interpret your results:

   a How did the average values differ for males and females?

   b How did the average values differ for smokers and nonsmokers?

   c What disease pattern would you expect to see with the smokers? Did your results follow this expectation?

13

1 Label **Figure 13.13** with the terms below.

- ❏ Epiglottis
- ❏ Laryngopharynx
- ❏ Larynx
- ❏ Left primary bronchus
- ❏ Nasal cavity

- ❏ Nasopharynx
- ❏ Pleural cavity
- ❏ Right primary bronchus
- ❏ Secondary bronchi

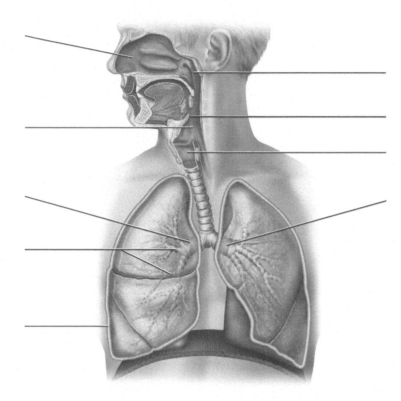

FIGURE **13.13** Anatomy of the lungs and respiratory tract.

13

2 Label **Figure 13.14** with the terms below.

- ❏ Alveolar duct
- ❏ Alveoli
- ❏ Pulmonary arteriole
- ❏ Pulmonary capillaries
- ❏ Pulmonary venule
- ❏ Respiratory bronchiole

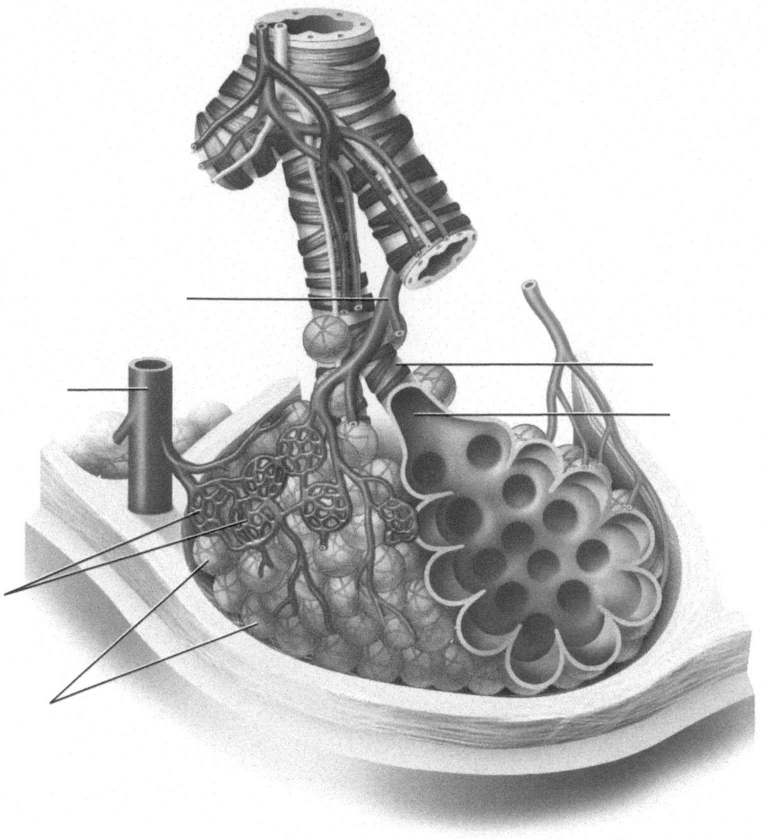

FIGURE **13.14** Bronchiole and alveolar sac.

3 *Fill in the blanks:* The _____ is the lungs' outer serous membrane, which adheres to the inner wall of the thoracic cavity. At the lungs' root, it folds inward on itself to become the inner serous membrane called the _____, which adheres to the lungs' surface. Between the two layers of membrane is the _____.

4 Number the following structures of the respiratory tract in the proper order. The structure that comes into contact with oxygenated air first should be number 1, and the structures where gas exchange takes place should be number 11.

_____ Oropharynx		_____ Nasopharynx	
_____ Trachea		_____ Bronchi	
_____ Nasal cavity		_____ Terminal bronchiole	
_____ Bronchiole		_____ Laryngopharynx	
_____ Alveolar sac		_____ Respiratory bronchiole	
_____ Larynx			

5 The piece of elastic cartilage that seals off the larynx during swallowing is called the

a. uvula.

b. false vocal cord.

c. true vocal cord.

d. epiglottis.

13

6 Passages in the respiratory tract smaller than 1 mm in diameter are called
   a. bronchi.
   b. bronchioles.
   c. alveolar ducts.
   d. paranasal sinuses.

7 *Fill in the blanks:* According to Boyle's law, as the volume of a container increases, the

pressure _____. Conversely, as the volume of a container decreases,

the pressure _____.

8 Air moves out of the lungs when
   a. intrapulmonary pressure is less than atmospheric pressure.
   b. intrapulmonary pressure is greater than atmospheric pressure.
   c. blood pressure is greater than intrapulmonary pressure.
   d. blood pressure is less than intrapulmonary pressure.

9 Pulmonary ventilation is best defined as
   a. the movement of gases across the respiratory membrane.
   b. the exchange of gases between the blood in the systemic capillaries and the tissues.
   c. the physical movement of air into and out of the lungs.
   d. the movement of gases through the blood.

10 A *pulmonary embolus* is a piece of a blood clot, adipose tissue, or other substance that lodges somewhere in the pulmonary circuit.

   a Henry has a pulmonary embolus lodged in one of his pulmonary arterioles. Trace the pathway of blood flow from his inferior vena cava to the pulmonary arteriole where the embolus is lodged.

   _____

   _____

   _____

   b How will the embolus affect gas exchange in Henry's pulmonary capillaries? Be specific.

   _____

   _____

   _____

   _____

13

11 The condition *pulmonary edema*, in which fluid collects around alveoli, increases the thickness of the respiratory membrane. Predict the effect this would have on the efficiency of gas exchange. Explain your reasoning.

_____

_____

_____

_____

12 Certain viral or bacterial infections can cause inflammation of the epiglottis, a condition known as *epiglottitis*, which is considered a medical or surgical emergency. Why do you think epiglottitis is so dangerous?

_____

_____

_____

_____

13 The condition *emphysema* results in loss of elastic recoil of the lung tissue. Would this make inspiration or expiration difficult? Explain.

_____

_____

_____

_____

14 A male patient presents with the following respiratory volumes and capacities: TV = 500 mL, ERV = 600 mL, IRV = 2,700 mL. What is this patient's VC? Are these values normal? If not, are they more consistent with an obstructive or restrictive disease pattern? Explain.

_____

_____

_____

_____

13

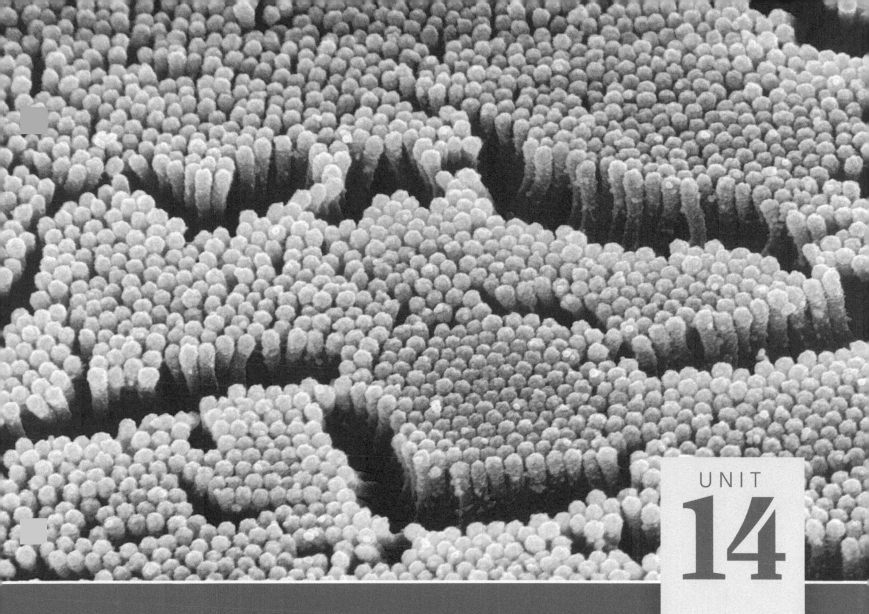

Digestive System

When you have completed this unit, you should be able to:

1 Describe and identify structures of the digestive system.

2 Demonstrate and describe the action of emulsifying agents on lipids.

3 Demonstrate and describe the functions of digestive enzymes.

4 Trace the pathway of physical and chemical digestion of carbohydrates, proteins, and lipids.

PRE-LAB EXERCISES

Complete the following exercises prior to coming to lab, using your lab manual and textbook for reference.

Pre-Lab Exercise **14-1**

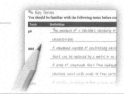

✎ Key Terms

You should be familiar with the following terms before coming to lab. Please note that this list is not all-inclusive, because the terminology for the digestive system is extensive.

Term	Definition
Digestive System Structures	
Alimentary canal	
Accessory organ	
Peritoneal cavity	
Esophagus	
Stomach	
Small intestine	
Large intestine	
Salivary glands	
Pancreas	
Liver	
Gallbladder	

14

Digestive Physiology

Digestive enzyme

Chemical digestion

Salivary amylase

Emulsification

Bile

Pre-Lab Exercise **14-2**

Anatomy of the Digestive System

Color the structures of the digestive system in Figures 14.1 and 14.2, and label them with the following terms from Exercise 14-1 (pp. 384 and 390). Use Exercise 14-1 in this unit and your text for reference.

- ❏ Pancreas
  - ☐ Pancreatic duct
- ❏ Liver
  - ☐ Common hepatic duct
- ❏ Gallbladder
  - ☐ Cystic duct
  - ☐ Common bile duct
- ❏ Duodenum

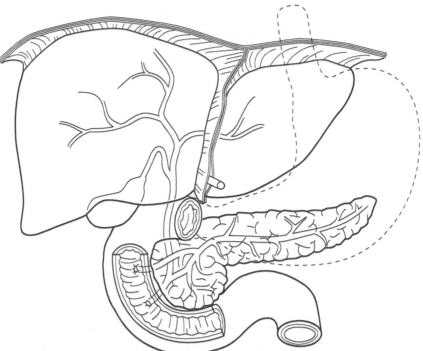

FIGURE **14.1** Liver, gallbladder, pancreas, and duodenum.

14

- ❏ Esophagus
- ❏ Stomach
- ❏ Small intestine

Salivary Glands
- ☐ Parotid gland
- ☐ Submandibular gland
- ☐ Sublingual gland

Pharynx
- ☐ Oropharynx
- ☐ Laryngopharynx

Large Intestine
- ☐ Cecum
- ☐ Vermiform appendix
- ☐ Ascending colon
- ☐ Transverse colon
- ☐ Descending colon
- ☐ Sigmoid colon
- ☐ Rectum
- ☐ Anal canal

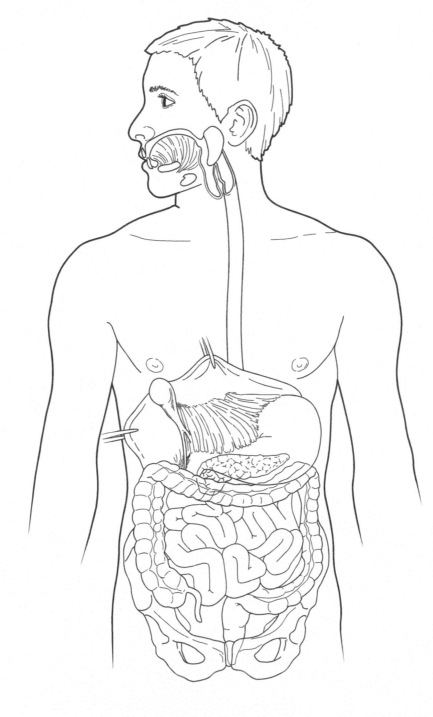

FIGURE **14.2** Anatomy of the digestive system.

14

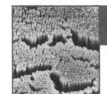

Colored SEM of microvilli in the duodenum.

The food we eat contains nutrients our cells use to build and repair body tissues and to make ATP. Food macromolecules, however, are typically too large for the body to absorb and use, so they must be broken down into smaller compounds. This process of breaking down foods into smaller substances that can enter cells is called **digestion** and is carried out by the **digestive system.** In general, the functions of the digestive system include taking in food, breaking down food mechanically and chemically into nutrients, absorbing these nutrients into the bloodstream, and eliminating indigestible substances.

We begin this unit with an introduction to the anatomy of the organs of the digestive system. Next we examine the physiological processes of chemical digestion and emulsification. We conclude with a "big picture" view of digestion through a tracing exercise.

Exercise 14-1

Digestive System Anatomy

MATERIALS

- ❏ Digestive system models
- ❏ Head and neck models
- ❏ Human torso models
- ❏ Human skulls with teeth
- ❏ Digestive organ models (stomach, pancreas, liver, and duodenum)
- ❏ Colored pencils

The digestive system is composed of two groups of organs: (1) the organs of the **alimentary canal,** also known as the **gastrointestinal** or **GI tract,** through which food travels, and (2) the **accessory organs,** which assist in mechanical or chemical digestion (Fig. 14.3). The organs of the alimentary canal include the *oral cavity, pharynx, esophagus, stomach, small intestine,* and *large intestine.* The accessory organs include the *teeth, tongue, salivary glands, liver, gallbladder,* and *pancreas.* We discuss the structure of each in greater detail shortly.

Much of the alimentary canal and many of the accessory organs reside inside a cavity known as the **peritoneal cavity** (pehr-ih-toh-NEE-uhl; Fig. 14.4). Like the pleural and pericardial cavities, the peritoneal cavity is found between a double-layered serous membrane. The membrane secretes serous fluid to enable the organs within the cavity to slide over one another without friction. Organs that are within the peritoneal cavity are known as *intraperitoneal,* while those that are behind the cavity are *retroperitoneal.* You can see in Figure 14.4 how widespread the peritoneal cavity is, and how many of the abdominal organs are located either partially or totally within it.

The two layers of the peritoneal membranes are as follows:

1. **Parietal peritoneum.** The outer **parietal peritoneum** is a thin membrane functionally fused to the abdominal wall and certain organs.

2. **Visceral peritoneum.** The inner **visceral peritoneum** adheres to the surface of many abdominal organs. The visceral peritoneum around the intestines folds over on itself to form a thick membrane known as the **mesentery** (MEZ-en-tehr-ee; Fig. 14.5). The mesentery houses blood vessels, nerves, and lymphatic vessels, anchoring these structures and the intestines in place. Mesenteries in different regions have specific names. One example is the **greater omentum** (oh-MEN-tum), which covers the abdominal organs like an apron. Another example is the smaller **lesser omentum,** which runs from the liver to the lesser curvature of the stomach (visible in Figures 14.3 and 14.4).

Let's now turn to the alimentary canal, which is a hollow passageway that extends from the mouth to the anus. Ingested food within the alimentary canal is broken down mechanically and chemically into nutrient molecules. The nutrients, along with water, vitamins, ions, and other substances, then cross the epithelium of the canal to enter the body and be absorbed into the blood or lymph.

14

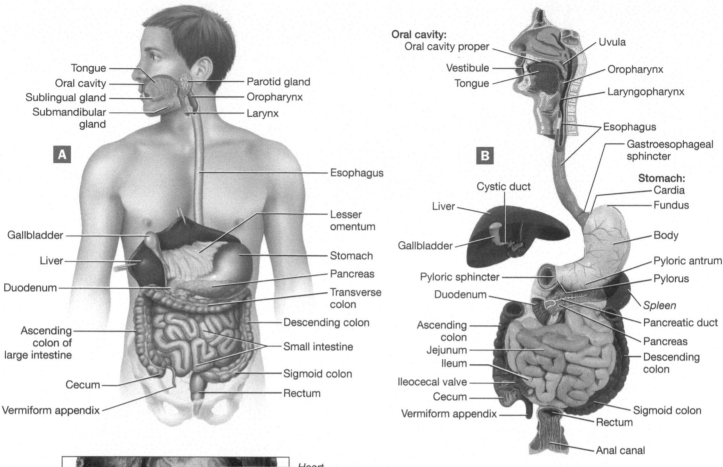

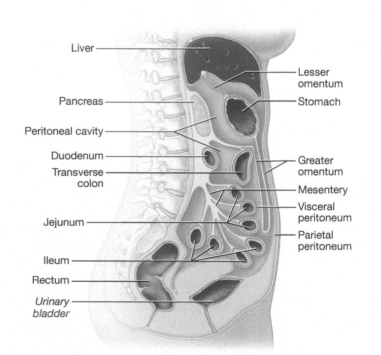

FIGURE 14.3 Overview of the digestive system: **(A)** illustration; **(B)** anatomical model photo; **(C)** dissected abdominopelvic cavity.

FIGURE 14.4 Structure of the peritoneal membranes and cavity.

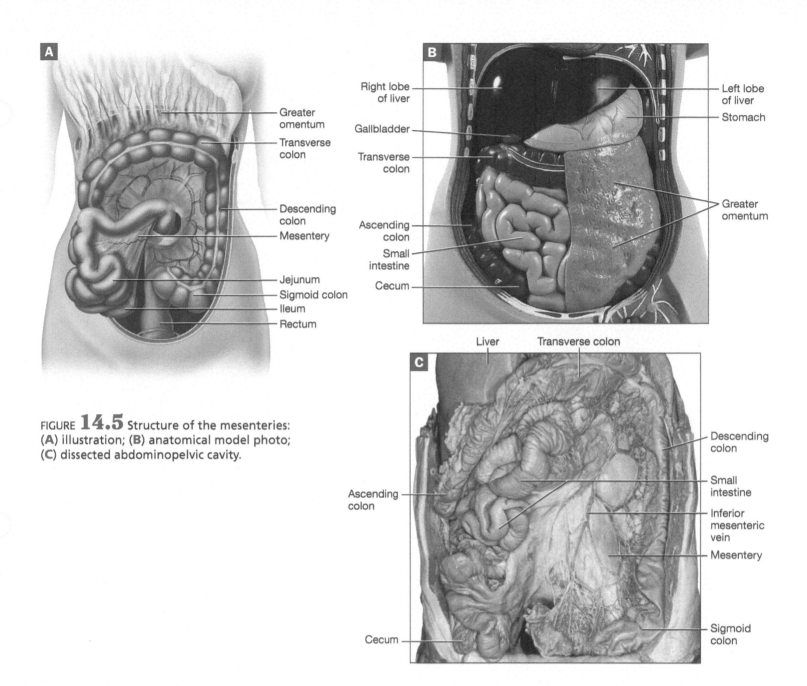

FIGURE **14.5** Structure of the mesenteries: (**A**) illustration; (**B**) anatomical model photo; (**C**) dissected abdominopelvic cavity.

The alimentary canal consists of the following organs:

1. **Mouth**. The alimentary canal begins with the **mouth**, which has two internal regions: the vestibule and the oral cavity (Fig. 14.6). The **vestibule** is the space between the lips and the teeth. Within the vestibule we find the **gums**, or *gingivae* (JIN-jih-vay), and the maxilla and mandible, where the teeth are housed.

    Posterior to the vestibule we find the **oral cavity**, which is defined as the area posterior to the teeth and bounded by the palate, cheeks, and tongue. The "roof" of the mouth is the **palate** (PAL-it), which consists of two portions. The first, which makes up the anterior two-thirds of the palate, is the bony **hard palate**. The second, which makes up its posterior one-third, is the muscular, arch-shaped **soft palate** (you can see this shape in Figure 14.6A). Extending inferiorly from the soft palate is the **uvula** (YOO-vyoo-luh)—both move posteriorly to prevent food from entering the nasopharynx and nasal cavity when we swallow.

2. **Pharynx**. The food next enters the **pharynx** (FEHR-inks), also known as the throat. Food passes through two divisions of the pharynx: the **oropharynx** (ohr-oh-FEHR-inks) and the **laryngopharynx** (lah-ring-oh-FEHR-inks; Fig. 14.6B). The muscles surrounding the pharynx propel swallowed food, which is known as a *bolus*, into the next portion of the alimentary canal. Recall from the lymphatic exercises that two sets of tonsils are located in the oropharynx—the lingual tonsil and the palatine tonsil—which protect the alimentary canal from ingested pathogens.

14

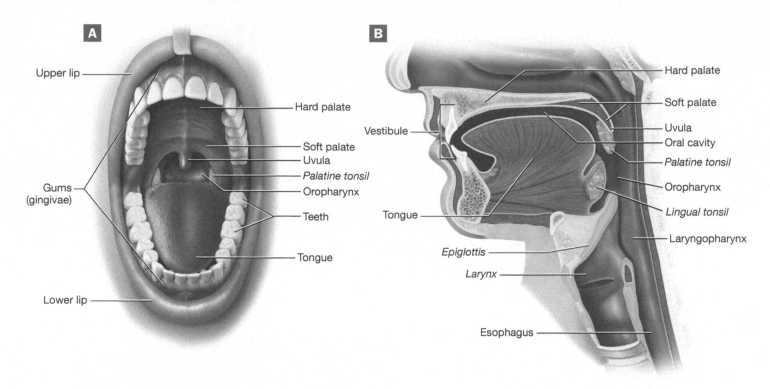

FIGURE **14.6** Structure of the mouth and oral cavity: (**A**) anterior view; (**B**) midsagittal section.

3. **Esophagus.** The bolus next enters the **esophagus** (eh-SOF-ah-gus), a narrow tubular organ posterior to the heart and the trachea in the thoracic cavity. Smooth muscle and skeletal muscle in the wall of the esophagus propel the bolus toward the stomach via rhythmic contractions called **peristalsis** (pehr-ih-STAHL-sis). At the inferior end of the esophagus is a ring of smooth muscle called the **gastroesophageal sphincter** (also known as the *lower esophageal sphincter* or *cardiac sphincter*). The gastroesophageal sphincter normally remains closed, which prevents the contents of the stomach from regurgitating into the esophagus.

4. **Stomach.** The **stomach**, shown in Figure 14.7, has five regions: the **cardia** near the gastroesophageal sphincter, the dome-shaped **fundus**, the middle **body**, the inferior **pyloric antrum**, and the terminal portion of the stomach, the **pylorus** (py-LOHR-us). At the pylorus we find another sphincter, the **pyloric sphincter**, which regulates the flow of material from the stomach to the initial portion of the small intestine.

   Note in Figure 14.7 that the stomach has interior folds called **rugae** (ROO-ghee) that allow it to expand considerably when it is filled with food and liquid. In addition, notice that the wall of the stomach has three layers of smooth muscle (inner oblique, middle circular, and outer longitudinal layers) that work together to pummel the bolus into a liquid material called **chyme** (KY'M).

5. **Small intestine.** Chyme passes through the pyloric sphincter to enter the **small intestine**, the portion of the alimentary canal where most chemical digestion and absorption take place. It has three divisions: the initial, short **duodenum** (doo-AW-den-um); the middle, approximately 8-foot-long **jejunum** (jeh-JOO-num); and the terminal, 12-foot-long **ileum** (ILL-ee-um; Fig. 14.8). At the terminal ileum is a sphincter called the **ileocecal valve** (ill-ee-oh-SEE-kuhl) that abuts the first portion of the large intestine, the *cecum*. This valve helps to prevent bacteria in the large intestine from reaching the normally sterile small intestine.

6. **Large intestine.** Material from the small intestine that was not digested or absorbed moves next to the terminal part of the alimentary canal, the **large intestine**, which is named for its large diameter rather than its length (about 5.5 feet). The large intestine may be divided into four regions (Fig. 14.9):

   a. The **cecum** (SEE-kum) is the blind pouch that receives contents from the ileum, from which it is separated by the ileocecal valve. It features an extension called the **vermiform appendix**, which is a blind-ended sac that contains lymphoid follicles.

14

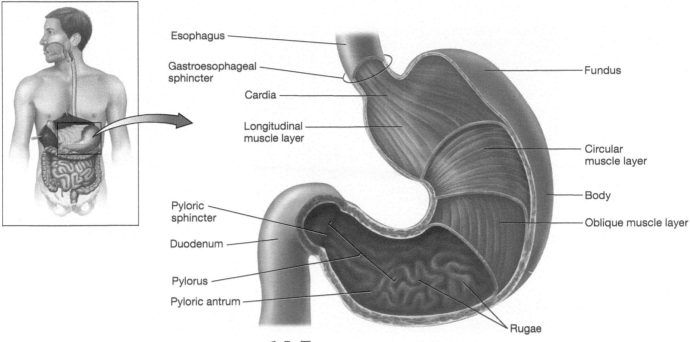

FIGURE **14.7** Anatomy of the stomach.

b. The next segment is the **colon,** which itself has four divisions. It begins as the **ascending colon,** which makes a left-hand turn at the **hepatic flexure** and crosses the superior abdomen as the **transverse colon.** At the spleen, the colon turns inferiorly at the **splenic flexure** to become the **descending colon,** which curves into an "S"-shape as it approaches the sacrum to become the **sigmoid colon.**

c. The **rectum** is the straight part of the large intestine that runs anterior to the sacrum.

d. The terminal portion of the large intestine is the **anal canal.** It has two sphincters: the involuntary **internal anal sphincter** and the voluntary **external anal sphincter.**

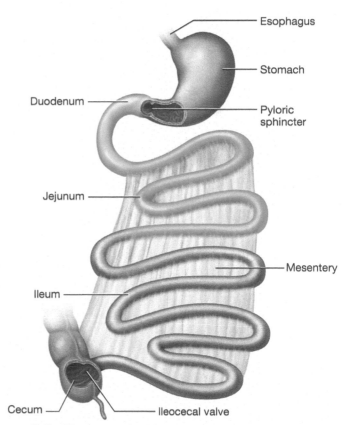

FIGURE **14.8** Small intestine, spread out to show the divisions.

14

The accessory digestive organs generally do not come into direct contact with the ingested food (the teeth and tongue are exceptions). Most of them instead secrete substances such as bile salts and enzymes that travel through a duct to the alimentary canal. The accessory digestive organs are as follows:

1. **Teeth and tongue.** The **teeth** and the **tongue** are accessory organs located in the mouth that assist in mechanical digestion of ingested food. The tongue's surface features keratinized papillae called **filiform papillae** that provide a rough surface to help break down food physically. Note that filiform papillae do not contain taste buds, which are located only on circumvallate, foliate, and fungiform papillae (for a review of tongue anatomy, see Unit 9, p. 254).

   The **teeth** are located in bony sockets known as *alveoli* within the mandible and maxilla. There are three classes of teeth (Fig. 14.10):

   ■ **Incisors.** The **incisors** are the broad and flat central teeth. There are four incisors: the two *central incisors*, and the two *lateral incisors*.

   ■ **Canines.** The **canines** (KAY-nynz), also known as *cuspids*, are the pointed teeth located to the sides of the lateral incisors.

   ■ **Molars.** There are two sets of molars. First are the **premolars**, located lateral to the canines, and second are the larger **molars**, which are the posterior teeth.

2. **Salivary glands.** The **salivary glands** are exocrine glands located around the mouth (Fig. 14.11). There are three sets of salivary glands:

   a. The largest salivary glands are the paired **parotid glands** (puh-RAWT-id), which are located superficial to the masseter muscles.

   b. The smaller **submandibular glands** are located just medial to the mandible.

   c. The **sublingual glands** (sub-LING-gwuhl) are located under the tongue, as their name implies.

All three types of glands secrete **saliva,** which contains water, mucus, an enzyme called *salivary amylase,* and antimicrobial proteins such as *lysozyme.* Saliva is secreted from the glands into the mouth through ducts.

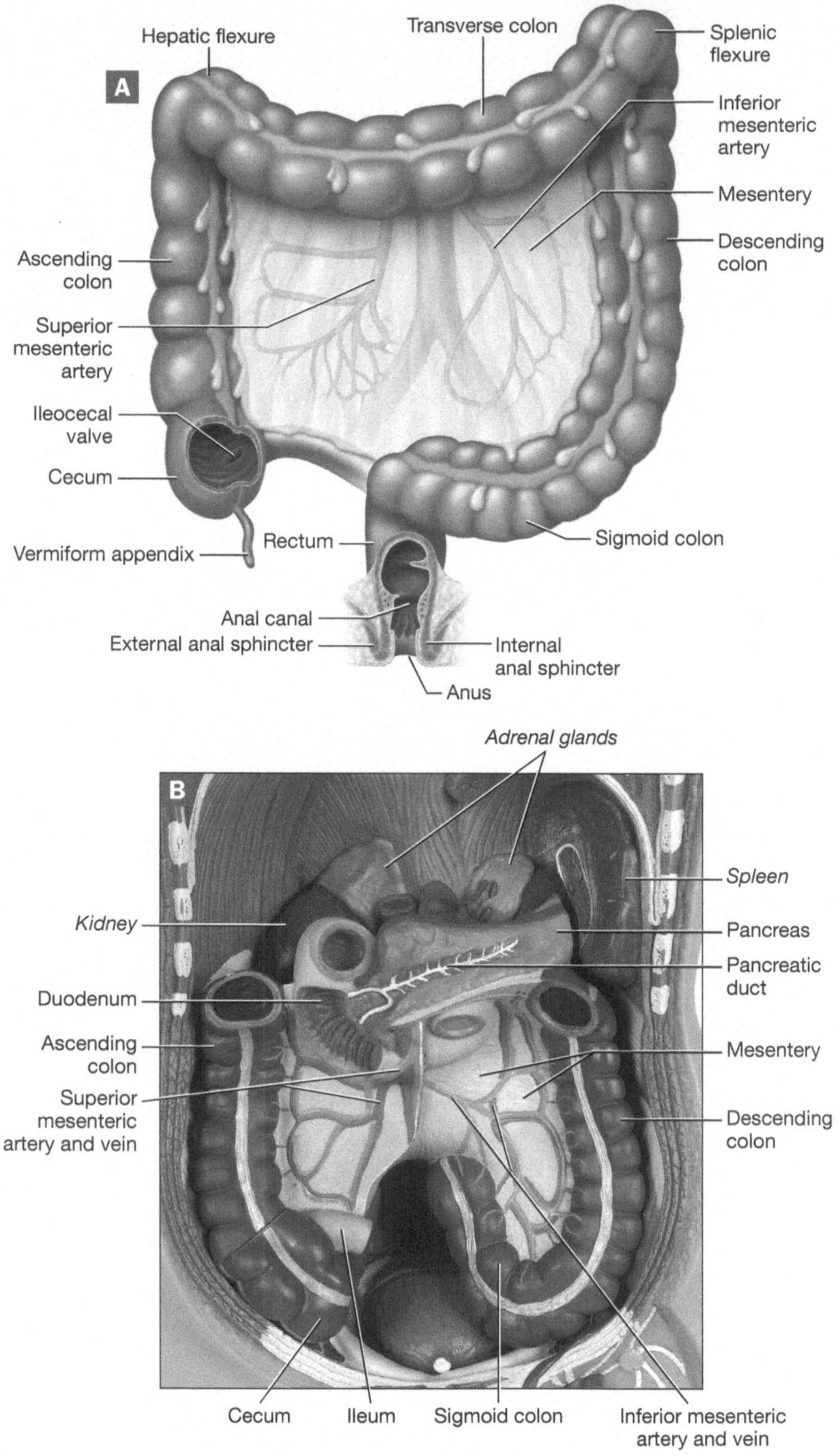

FIGURE **14.9** Structure of the large intestine: (**A**) illustration; (**B**) anatomical model photo.

3. **Liver and gallbladder.** The liver and gallbladder are organs located in the right upper quadrant of the abdominal cavity.

a. The **liver** consists of four lobes: the large **right** and **left lobes,** and the small **caudate** and **quadrate lobes,** which are located on the posterior side of the right lobe (Fig. 14.12). The liver is wrapped in a thin connective tissue capsule, and most of it is covered by the visceral peritoneum.

On the liver's posterior side near the quadrate lobe, shown in Figure 14.12B, is an area called the **porta hepatis** (POHR-tuh heh-PAEH-tis). This area serves as a main "gateway" into and out of the liver, as the hepatic artery, the hepatic portal vein, and the common hepatic duct all enter and exit via the porta hepatis. Note the exception is the hepatic veins, which exit at the superior side of the liver, where they drain into the inferior vena cava.

The liver has multiple functions in the body, most of which are metabolic in nature. For example, recall from Unit 11 that all blood from the digestive organs and the spleen travels via the hepatic portal vein to the hepatic portal system of the liver. There, the absorbed nutrients and other substances are processed before they enter the general circulation. One of the liver's main digestive functions is to produce a chemical called *bile*, required for the digestion and absorption of fats.

b. The **gallbladder** is a small, saclike organ on the posterior side of the liver that stores and concentrates bile. Bile leaves the right and left lobes of the liver by the *right* and *left hepatic ducts*, which merge to form the **common hepatic duct.** Some bile may be released into the duodenum depending on hormonal signals, but most of it enters the gallbladder for storage (Fig. 14.13). When stimulated by certain hormones, the gallbladder contracts, and the bile contained within it is ejected through the **cystic duct** (SIS-tik). Notice that the cystic duct joins with the common hepatic duct to form the **common bile duct,** which empties into the duodenum at the **hepatopancreatic ampulla** (heh-PAEH-toh-payn-kree-at-ik am-POOL-ah). A sphincter controls the release of bile from the ampulla to the duodenum.

4. **Pancreas.** The **pancreas** (PAYN-kree-us) is an exocrine and endocrine gland that sits posterior and inferior to the stomach. Its exocrine functions are digestive, whereas its endocrine functions are metabolic. The exocrine portion of the pancreas produces a fluid called **pancreatic juice** that contains water, bicarbonate ions to neutralize the acid produced by the stomach, and multiple digestive enzymes. Pancreatic juice is released through the **pancreatic duct** and enters the duodenum at the hepatopancreatic ampulla.

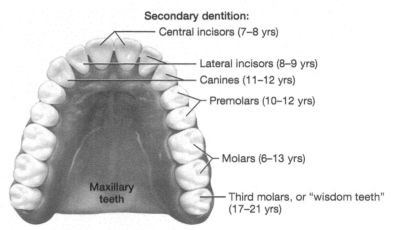

Secondary dentition:
- Central incisors (7–8 yrs)
- Lateral incisors (8–9 yrs)
- Canines (11–12 yrs)
- Premolars (10–12 yrs)
- Molars (6–13 yrs)
- Third molars, or "wisdom teeth" (17–21 yrs)

Maxillary teeth

FIGURE **14.10** Dentition of the maxilla.

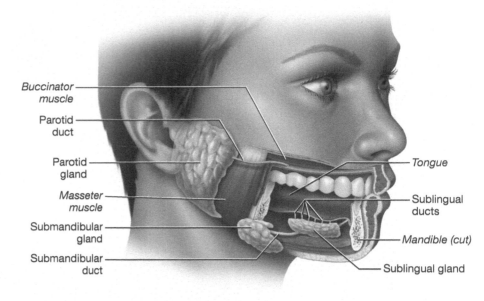

Buccinator muscle
Parotid duct
Parotid gland
Masseter muscle
Submandibular gland
Submandibular duct
Tongue
Sublingual ducts
Mandible (cut)
Sublingual gland

FIGURE **14.11** Salivary glands.

14

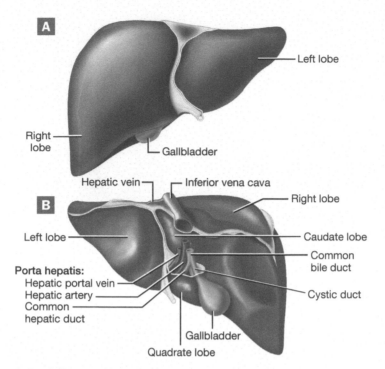

FIGURE **14.12** Anatomy of liver: **(A)** anterior view; **(B)** posteroinferior view.

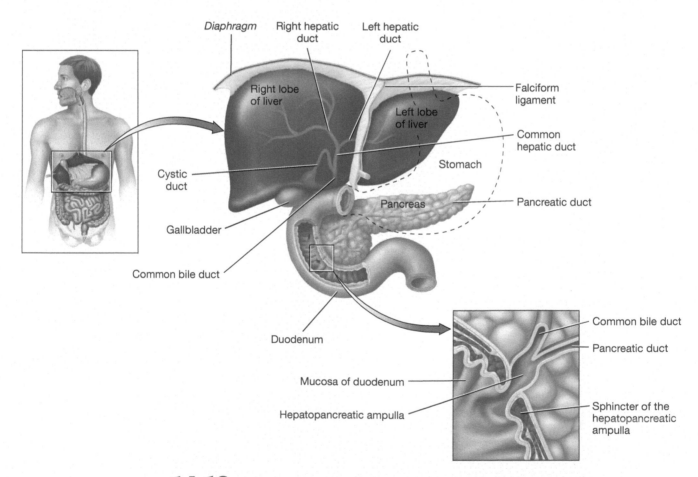

FIGURE **14.13** Structure of the liver, gallbladder, pancreas, and duodenum.

Procedure 1 Model Inventory for the Digestive System

Identify the following structures of the digestive system on models and diagrams, using this unit and your textbook for reference. As you examine the anatomical models and diagrams, record the name of the model and the structures you were able to identify on the model inventory in Table 14.1.

Peritoneum
1. Peritoneal cavity
2. Parietal peritoneum
3. Visceral peritoneum
4. Mesentery
5. Greater omentum
6. Lesser omentum

Alimentary Canal
1. Mouth
   a. Vestibule
   b. Gums (gingivae)
   c. Oral cavity
   d. Hard palate
   e. Soft palate
      (1) Uvula
2. Pharynx
   a. Oropharynx
   b. Laryngopharynx
3. Esophagus
   a. Gastroesophageal sphincter
      (or lower esophageal sphincter)
4. Stomach
   a. Cardia
   b. Fundus
   c. Body
   d. Pyloric antrum
   e. Pylorus
   f. Pyloric sphincter
   g. Rugae

5. Small intestine
   a. Duodenum
   b. Jejunum
   c. Ileum
   d. Ileocecal valve
6. Large intestine
   a. Cecum
      (1) Vermiform appendix
   b. Colon
      (1) Ascending colon
      (2) Hepatic flexure
      (3) Transverse colon
      (4) Splenic flexure
      (5) Descending colon
      (6) Sigmoid colon
   c. Rectum
   d. Anal canal
      (1) Internal anal sphincter
      (2) External anal sphincter

Accessory Organs
1. Tongue
   a. Filiform papillae
2. Teeth
   a. Incisors
   b. Canines
   c. Premolars
   d. Molars

3. Salivary glands
   a. Parotid gland
   b. Submandibular gland
   c. Sublingual gland
4. Liver
   a. Right lobe, left lobe, caudate lobe, and quadrate lobe
   b. Porta hepatis
   c. Hepatic portal vein
   d. Common hepatic duct
5. Gallbladder
   a. Cystic duct
   b. Common bile duct
   c. Hepatopancreatic ampulla
6. Pancreas
   a. Pancreatic duct

TABLE **14.1** Model Inventory for the Digestive System

Model	Digestive Structures Identified

14

(continues)

TABLE **14.1** *(cont.)* Model Inventory for the Digestive System

Procedure 2 Time to Draw

In the space below, draw, color, and label the arrangement of the liver, gallbladder, pancreas, and duodenum. In addition, write the function or definition of each structure that you label.

14

Exercise 14-2

Digestion

MATERIALS

- ❏ 7 glass test tubes
- ❏ Graduated cylinder
- ❏ Starch solution
- ❏ Amylase enzyme
- ❏ Distilled water
- ❏ Water bath set to 37°C
- ❏ Boiling water bath
- ❏ Glass stirring rod
- ❏ Lugol's iodine solution
- ❏ Benedict's reagent
- ❏ Test tube rack
- ❏ Vegetable oil
- ❏ Rubber stopper
- ❏ Sudan red stain
- ❏ Liquid detergent
- ❏ Sharpie
- ❏ Laminated outline of the human body
- ❏ Water-soluble marking pens

We now turn our attention to the physiology of food breakdown and absorption. Three major types of nutrients are broken down and absorbed in the alimentary canal: *carbohydrates*, *proteins*, and *lipids*. Carbohydrate digestion begins in the mouth with the enzyme **salivary amylase** (AM-uh-layz), which catalyzes the reactions that digest polysaccharides into smaller oligosaccharides. The remaining polysaccharides and oligosaccharides are digested in the small intestine. These reactions are catalyzed by *pancreatic amylase* and enzymes associated with the enterocytes of the small intestine called **brush border enzymes** (including lactase, maltase, and sucrase).

Protein digestion begins in the stomach with the enzyme **pepsin,** secreted from chief cells in the gastric glands as the inactive pre-enzyme **pepsinogen.** Pepsinogen becomes the active enzyme pepsin when it encounters an acidic environment. This acidic environment is provided by cells in the stomach known as **parietal cells** that secrete hydrochloric acid. Pepsin begins digesting proteins into polypeptides and some free amino acids. The remainder of protein digestion occurs in the small intestine with pancreatic enzymes, such as *trypsin* and *chymotrypsin*, and also brush border enzymes, such as *carboxypeptidase* and *aminopeptidase*.

Most lipid digestion begins when lipids reach the small intestine. The process is more complicated than it is for protein or carbohydrate digestion because lipids are nonpolar molecules that do not dissolve in the water-based environment of the small intestine. This causes the lipids to clump together and form fat "globules" in the small intestine. In this form, it is extremely difficult for the enzyme **pancreatic lipase** (LY-payz) to catalyze the breakdown of lipids into monoglycerides and free fatty acids because it has only a small surface area on which to work. Therefore, the first step in lipid digestion is to break up fat globules into smaller pieces, a process called **emulsification** (Fig. 14.14).

Emulsification is accomplished by chemicals called **bile salts,** which are produced by the liver and stored in the gallbladder. Bile salts are compounds that have both nonpolar and polar parts. Notice in Figure 14.14 what happens to the fat globule when bile salts come into contact with it: The bile salts' nonpolar parts interact with the lipids, but their polar parts repel other lipids. The overall effect is that the fat globule is physically broken down into smaller fat pieces. This gives pancreatic lipase much more surface area on which to work, and the fats are digested into monoglycerides and free fatty acids.

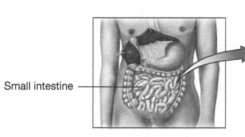

FIGURE **14.14** Emulsification of fats by bile salts.

14

After chemical digestion is complete, all three nutrients are absorbed through the plasma membranes of the enterocytes lining the small intestine. Both monosaccharides and amino acids are absorbed into the enterocytes via facilitated diffusion or secondary active transport mechanisms. These nutrients then exit the other side of the enterocytes and enter the capillaries in the small intestine.

Lipid absorption, like lipid digestion, is more complicated because lipids are nonpolar and cannot be transported as free fatty acids through the water-filled plasma of the blood. In the small intestinal lumen, digested lipids remain associated with bile salts to form clusters of lipids and other nonpolar chemicals called **micelles** (my-SELLZ). Micelles escort the lipids to the enterocytes' plasma membranes; they can do so because of their polar outer surface. The lipids then leave the micelles and enter the enterocytes by simple diffusion. Note that in the absence of bile salts, nearly all the lipids will simply pass through the small intestine unabsorbed and be excreted in the feces.

In the enterocytes, lipids associate with other nonpolar substances and proteins to form protein-coated droplets called **chylomicrons** (ky-loh-MY-krahnz). Chylomicrons exit the enterocytes by exocytosis, but they are too large to enter the blood capillaries. Instead, they enter a lymphatic vessel called a *lacteal* and join the lymph in the lymphatic system. The chylomicrons travel within the lymphatic vessels until they enter the blood at the junction of the left subclavian vein and the left internal jugular vein.

In this exercise, you will be examining the processes of chemical digestion with the example of carbohydrate digestion. You will also examine the process of emulsification, and perform a tracing exercise to tie together the anatomy and physiology that you've learned.

Procedure 1 Test Carbohydrate Digestion

In the following procedure, you will test for carbohydrate digestion using a starch solution and the enzyme amylase. You will check for digestion using two reagents: Lugol's iodine, which turns dark blue in the presence of starch, and Benedict's reagent, which forms a precipitate (solid) in the presence of simple sugars. You may interpret the results of your tests as follows:

▮ If the solution turns dark blue with Lugol's iodine, *no or limited* carbohydrate digestion occurred because starch is still present (Fig. 14.15).

▮ If the solution develops a greenish-brownish red precipitate with Benedict's reagent, monosaccharides are present and carbohydrate digestion *did* occur (Fig. 14.16).

1 Obtain six glass test tubes and number them 1 through 6 with a Sharpie.

2 Place 3 mL of starch solution into Tube 1 and Tube 2.

3 Add 3 mL of amylase to Tube 1.

4 Add 3 mL of distilled water to Tube 2.

5 Set Tubes 1 and 2 into a 37°C water bath and leave them in there for 30 minutes.

6 Remove the tubes from the water bath and mix the contents of each tube thoroughly with stirring rods (be certain to use two separate rods so you don't contaminate the contents of the tubes).

7 Divide the mixture in Tube 1 equally into Tubes 3 and 4 (each tube should contain about 3 mL of the mixture).

8 Divide the mixture in Tube 2 equally into Tubes 5 and 6 (each tube should contain about 3 mL of the mixture).

9 Add two drops of Lugol's iodine solution to Tubes 3 and 5. Record the results in Table 14.2.

10 Add 10 drops of Benedict's reagent to Tubes 4 and 6. Swirl the tubes gently to mix the contents.

11 Place Tubes 4 and 6 into a boiling water bath for 3 minutes. *Use caution* to prevent the mixture from splattering and burning you or your classmates.

12 Remove the tubes from the boiling water, and record the results of the test in Table 14.2.

Reagent	Negative Reaction	Positive Reaction
Lugol's iodine	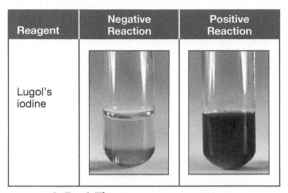	

FIGURE **14.15** Positive results for Lugol's iodine, which turns dark purple in the presence of starch.

Reagent	Negative Reaction	Positive Reaction
Benedict's reagent	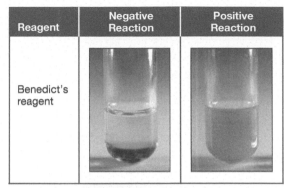	

FIGURE **14.16** Possible results for Benedict's reagent, in which a light blue color indicates no monosaccharides are present and a greenish-red color indicates monosaccharides are present.

13 Interpret your results:

a In which tube(s) did carbohydrate digestion occur? How do you know?

b In which tube(s) did no carbohydrate digestion occur? How do you know?

c What conclusions can we draw?

TABLE **14.2** Results of Carbohydrate Digestion Experiment

Tube	Reaction with Lugol's Iodine
Tube 3 (starch + amylase)	
Tube 5 (starch + water)	
Tube	**Reaction with Benedict's Reagent**
Tube 4 (starch + amylase)	
Tube 6 (starch + water)	

14

Procedure 2 Demonstrate Lipid Emulsification

Let's now examine what happens during the process of emulsification. In this procedure, you will use four substances to observe emulsification in action:

▌ *Lipids*: The source of lipids for this procedure is vegetable oil.

▌ *Emulsifying agent*: Detergents are considered to be emulsifiers because they have both polar parts and nonpolar parts, similar to bile salts. The emulsifying agent in this procedure will be a liquid detergent.

▌ *Distilled water.*

▌ *Sudan red stain*: This is a stain that binds only to lipids. The sole purpose of this stain is to make the lipids more visible during the procedure.

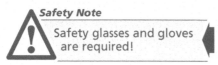

Safety Note

Safety glasses and gloves are required!

1 Obtain a glass test tube and add approximately 2 mL of distilled water to it.

2 Add about 2 mL of vegetable oil to the water.

3 Place a rubber stopper into the tube and shake it vigorously for 15 seconds. Allow it to stand for 2 minutes. What happens to the oil and water?

4 Add three or four drops of Sudan red stain, and shake the tube again for 15 seconds. What color is the oil? What color is the water?

5 Add about 1 mL of liquid detergent to the mixture and shake the tube vigorously for 15 seconds. Allow the tube to stand for 2 minutes. What has happened to the solution? Is it still two distinct colors? Explain your results.

14

Now it's time to put all the digestive anatomy and physiology together to get a "big picture" view of the digestive system. In this procedure, you will trace the pathways that three different nutrients take from their ingestion at the mouth to their arrival in the blood or lymph. You will trace a cookie (primarily carbohydrates), an egg (primarily protein), and greasy fried food (primarily lipids).

Along the way, detail the following for each:

1. The *anatomical pathway* each takes, from ingestion, through its passage through the alimentary canal, to its absorption into the blood or lymph.

2. The *physical* and *chemical processes* that break down each substance, including enzyme-catalyzed chemical digestion and processes of mechanical digestion, such as churning, chewing, and emulsification.

Some hints:

▮ Remember that digestion and absorption are quite different for lipids. For example, fats are not absorbed into the intestinal blood capillaries.

▮ Use the text in this exercise for reference about the enzymes involved in the chemical digestion of each nutrient.

▮ You may find it helpful to physically trace the pathway on a laminated outline of the human body to better visualize the processes.

Tracing Steps

1 Cookie: **Start:** mouth →

→ blood **End**

2 Egg: **Start:** mouth →

→ blood **End**

14

3 Greasy fried food: **Start:** mouth →

→ lymph **End**

14

UNIT 14

QUIZ

1 Label the following structures on Figure 14.17.

❑ Cecum
❑ Esophagus
❑ Gallbladder
❑ Liver
❑ Parotid gland

❑ Sigmoid colon
❑ Sublingual gland
❑ Submandibular gland
❑ Transverse colon

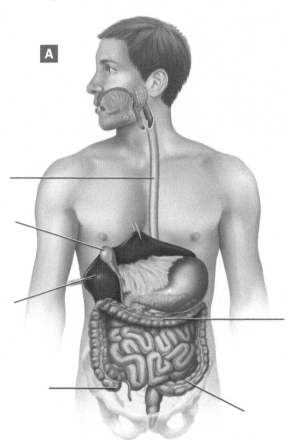

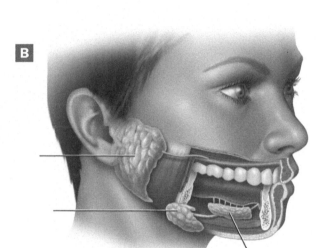

FIGURE **14.17** Organs of the digestive system: (**A**) overview of the digestive system; (**B**) salivary glands.

14

2 Label the following structures on Figure 14.18.

❏ Body
❏ Cardia
❏ Fundus
❏ Gastroesophageal sphincter

❏ Pyloric sphincter
❏ Pylorus
❏ Rugae

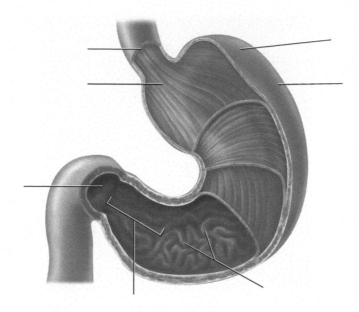

FIGURE **14.18** Anatomy of the stomach.

3 Label the following structures on Figure 14.19.

❏ Common bile duct
❏ Common hepatic duct
❏ Cystic duct

❏ Duodenum
❏ Pancreas
❏ Pancreatic duct

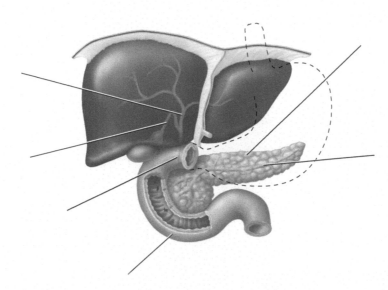

FIGURE **14.19** Structure of the liver, gallbladder, pancreas, and duodenum.

14

4 Which of the following organs is *not* an accessory organ of digestion?

a. Salivary glands.

b. Gallbladder.

c. Cecum.

d. Pancreas.

5 Which of the following digestive functions is performed by the organs in the alimentary canal, but not by accessory organs?

a. Absorption.

b. Mechanical digestion.

c. Chemical digestion.

d. Ingestion.

6 *True/False:* Mark the following statements as true (T) or false (F). If the statement is false, correct it so it becomes a true statement.

_____ a. The peritoneal cavity is located between the visceral peritoneum and the mesentery.

_____ b. The longest segment of the small intestine is the duodenum.

_____ c. The stomach has three layers of smooth muscle that contract to churn food into chyme.

_____ d. The gallbladder produces and stores bile.

_____ e. Bile leaves the liver through the cystic duct.

7 *Matching:* Match the following with the correct definition.

_____ Salivary amylase   A. Clusters of bile salts and digested lipids

_____ Bile salts   B. Enzyme(s) that digest(s) lipids into free fatty acids and monoglycerides

_____ Micelles   C. Chemical required to activate pepsinogen

_____ Hydrochloric acid   D. Protein-digesting enzyme(s) produced by the stomach

_____ Pancreatic lipase   E. Emulsifies/emulsify fats

_____ Pepsin   F. Begin(s) carbohydrate digestion in the mouth

8 How does the absorption of lipids differ from the absorption of carbohydrates and proteins?

a. Lipids are absorbed into capillaries; carbohydrates and proteins are absorbed into lacteals.

b. Lipids are absorbed by bile salts; carbohydrates and proteins are absorbed by micelles.

c. Lipids, carbohydrates, and proteins are absorbed the same way.

d. Lipids are absorbed into lacteals; carbohydrates and proteins are absorbed into capillaries.

14

9 If the ileocecal valve fails to close properly, the contents of the cecum can reflux back into the ileum. Why might this cause problems? (*Hint:* What is found in the large intestine that is not normally found in the small intestine?)

---

10 The condition known as *heartburn* is most often caused by acid regurgitating from the stomach into the esophagus. Some of the drugs that treat heartburn work by decreasing the secretion of acid by the cells of the stomach. Could this affect the chemical digestion of certain nutrients? Explain.

---

11 Many dietary supplements contain digestive enzymes the manufacturers claim are necessary to digest food properly. What will happen to these enzymes in the stomach? (*Hint:* Enzymes are proteins.) Will the enzymes continue to function once they have reached the small intestine? Why or why not?

---

12 Eva has been diagnosed with *gallstones*, which are lumps of cholesterol and other components of bile. One of the gallstones is blocking her cystic duct, preventing the release of bile from the gallbladder.

a Will the gallstones prevent bile from being produced and released into the duodenum? Why or why not?

---

b Predict what might happen if a gallstone were to block Eva's pancreatic duct, preventing the release of pancreatic juice into the duodenum.

---

13 Sadaf has an ulcer of her large intestine that has *perforated*, meaning that it has developed a hole that has gone through all tissue layers. A major concern with large intestine perforation is *peritonitis*, or infection of the peritoneal fluid. Why could peritonitis have wide-ranging effects on Sadaf's other organs of the abdomen and digestive system? Are there any organs that would be unlikely to be directly affected by peritonitis? Explain.

14

Urinary System

When you have completed this unit, you should be able to:

1 Describe and identify gross and microscopic structures of the urinary system.

2 Model the physiology of the kidney, and test for chemicals in the filtrate.

3 Perform and interpret urinalysis on simulated urine specimens.

4 Trace an erythrocyte, a glucose molecule, and a urea molecule through the gross and microscopic anatomy of the kidney.

PRE-LAB EXERCISES

Complete the following exercises prior to coming to lab, using your lab manual and textbook for reference.

Pre-Lab Exercise **15-1**

✎ Key Terms

You should be familiar with the following terms before coming to lab.

Term	Definition

Gross Structures of the Kidney

Renal cortex _____

Renal medulla _____

Renal pelvis _____

Major and minor calyces _____

Other Structures of the Urinary System

Ureter _____

Urinary bladder _____

Urethra _____

Microscopic Anatomy of the Kidney

Nephron _____

15 Glomerulus _____

Peritubular capillaries _____

Renal tubule _____

Collecting duct _____

Urinary Physiology Terms

Glomerular filtration _____

Tubular reabsorption _____

Tubular secretion _____

Filtrate _____

Urine _____

Urinalysis _____

15

Pre-Lab Exercise **15-2**

Structures of the Urinary System

Color the structures of the urinary system in **Figures 15.1** and **15.2,** and label them with the following terms from Exercise 15-1 (pp. 411 and 413). Use Exercise 15-1 in this unit and your text for reference.

❑ Ureter
❑ Urinary bladder
❑ Urethra

Regions of the Kidney

☐ Renal cortex
☐ Renal medulla
  ▪ Renal columns
  ▪ Renal pyramids
☐ Minor calyces
☐ Major calyces
☐ Renal pelvis

Blood Supply

☐ Renal artery
☐ Renal vein

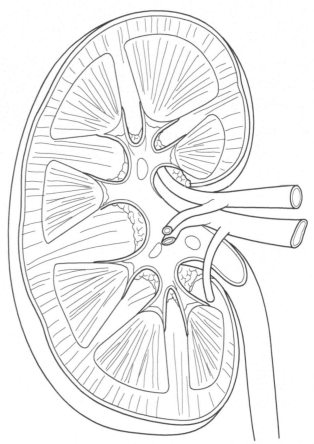

FIGURE **15.1** Right kidney, frontal section.

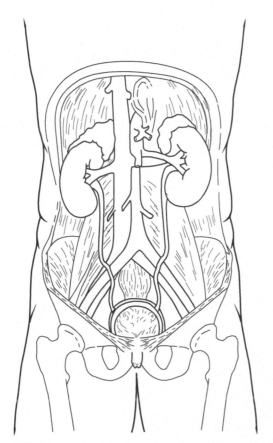

FIGURE **15.2** Organs of the urinary system, anterior view.

15

Pre-Lab Exercise **15-3**

Structures of the Nephron

Color the structures of the nephron in Figure 15.3, and label them with the following terms from Exercise 15-1 (pp. 411 and 413). Use Exercise 15-1 in this unit and your text for reference.

- ❑ Renal corpuscle
- ❑ Glomerulus
  - ☐ Afferent arteriole
  - ☐ Efferent arteriole
  - ☐ Peritubular capillaries
- ❑ Glomerular capsule
  - ☐ Parietal layer
  - ☐ Visceral layer
- ❑ Renal tubule
  - ☐ Proximal tubule
  - ☐ Nephron loop
  - ☐ Distal tubule
- ❑ Collecting duct
  - ☐ Cortical collecting duct
  - ☐ Medullary collecting duct

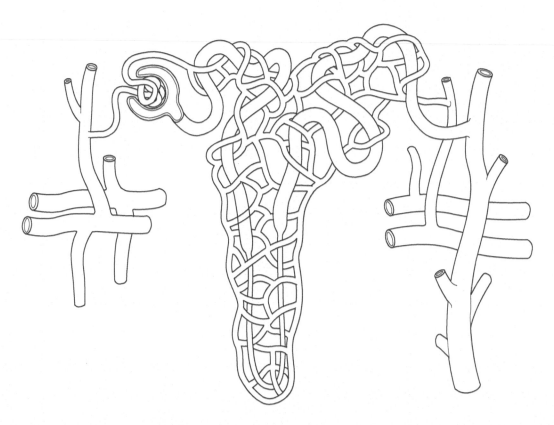

FIGURE **15.3** Structure of the nephron and collecting system.

15

Pre-Lab Exercise **15-4**

Glomerular Filtration and Tubular Reabsorption

Table 15.1 lists chemicals and cells found in the blood. Do the following for each chemical, using your textbook for reference:

1 Determine whether each listed substance is filtered at the glomerulus (i.e., whether it is able to leave the blood and enter the nephron to become part of the filtrate under normal conditions).

2 If the substance is found in the filtrate, determine whether the substance is reabsorbed into the blood or whether it is found in the urine.

3 If the substance is reabsorbed into the blood, determine where this substance is reabsorbed in the nephron (e.g., the proximal tubule, the nephron loop, etc.).

TABLE **15.1** Substances Found in the Blood and Their Filtration and Reabsorption by the Nephron

Substance	Filtered in the Glomerulus? (Yes/No)	Reabsorbed? (Yes/No)	Location Where Reabsorption Takes Place
Cells			
Water			
Glucose			
Proteins			
Urea			
Creatinine			
Electrolytes (sodium, potassium, chloride ions)			
Uric acid			

15

EXERCISES

Colored SEM of kidney stone crystals.

The **urinary system** is the group of organs that consists of the *kidneys* and the *urinary tract*—the *ureters*, *urinary bladder*, and *urethra*—and performs many functions critical to the maintenance of homeostasis. Most such functions are directly performed by the kidneys. For example, the kidneys filter the blood to remove *metabolic wastes*, which are chemicals produced by the body that the body cannot use for any purpose. The kidneys also regulate the body's fluid, electrolyte, and acid-base balance, as kidneys can conserve or excrete water, specific electrolytes, and bicarbonate and hydrogen ions. Finally, the kidneys help the liver detoxify certain compounds, make glucose during times of starvation, and produce the hormone *erythropoietin*, which regulates blood cell formation.

In this unit, you will become acquainted with the anatomy and the physiology of the urinary system. You will first examine the gross and microscopic structures of its organs, after which you will construct a model kidney and test samples of simulated urine. In the final exercise, you will trace the pathway of different substances through the general circulation and the microanatomy of the kidney.

Exercise 15-1

Urinary System Anatomy

MATERIALS
- ❏ Urinary system models
- ❏ Kidney models
- ❏ Nephron models
- ❏ Preserved kidney
- ❏ Dissection equipment
- ❏ Dissecting tray
- ❏ Colored pencils

The paired **kidneys** are situated against the posterior body wall posterior to the peritoneal membranes, meaning they are *retroperitoneal* (Fig. 15.4). Externally, they are encased within three layers of connective tissue: the superficial **renal fascia** (REE-nuhl), a layer of dense connective tissue that anchors the kidneys to the posterior abdominal wall and the peritoneum; the **adipose capsule**, a thick layer of adipose tissue that wedges the kidneys in place; and the **renal capsule**, a very thin layer of dense connective tissue that encases each kidney like plastic wrap. With the layers of connective tissue removed, the kidneys are bean-shaped and have a medial indentation where blood vessels and the ureter enter and exit, which is known as the **hilum** (HY-lum).

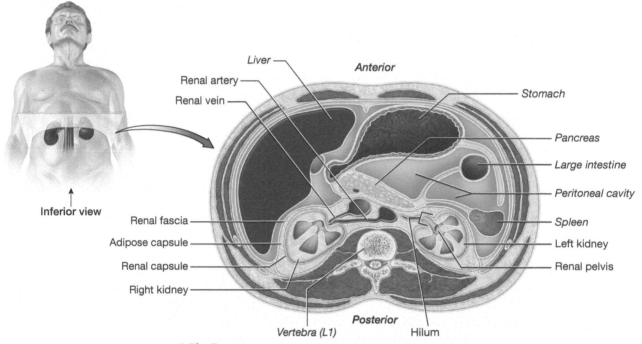

FIGURE **15.4** Transverse section of the abdomen and kidneys.

15

Internally, each kidney has three distinct regions that can be seen in a frontal section (Fig. 15.5):

1. **Renal cortex.** The most superficial region is known as the **renal cortex**. It is dark brown because it consists of many blood vessels that serve the tiny blood-filtering structures of the kidney, the *nephrons*.

2. **Renal medulla.** The kidney's middle region is the **renal medulla**, which consists of triangular **renal pyramids**. The renal pyramids are separated from one another by inward extensions of the renal cortex called **renal columns**. Like the renal cortex, the renal columns contain many blood vessels. Each pyramid contains looping tubules of nephrons as well as structures that drain fluid from nephrons. These tubes give the pyramids a striped (or *striated*) appearance.

3. **Renal pelvis.** Fluid from the renal papilla drains into tubes called **minor calyces** (KAY-lih-seez) that in turn drain into even larger **major calyces**. The major calyces drain into the kidney's innermost region, the **renal pelvis**, which serves as a basin for collecting urine.

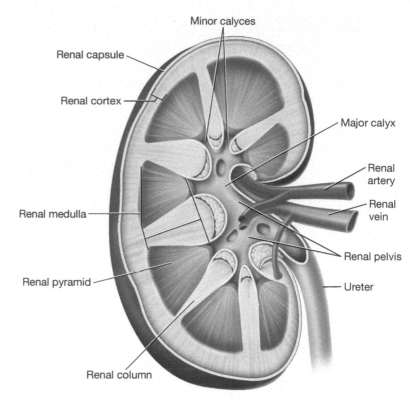

FIGURE **15.5** Right kidney, frontal section.

The blood flow through the kidney follows a unique pattern that allows it to carry out its function of maintaining the homeostasis of the blood. The large **renal arteries** deliver about 1,200 milliliters of blood per minute to the kidney to be filtered (Fig. 15.6). The renal arteries branch into progressively smaller arteries as they pass through the medulla to the cortex, including the **segmental arteries** in the renal pelvis, the **interlobar arteries** between the medullary pyramids, the **arcuate arteries** (ARK-yoo-it) that curve around the top of the pyramids, and, finally, the small **interlobular arteries** (also known as *cortical radiate arteries*) in the renal cortex. The interlobular arteries branch into tiny **afferent arterioles**, each of which supplies a ball of capillaries known as the **glomerulus** (gloh-MEHR-yoo-luhs). This is where the blood is filtered. Note that the glomerulus is not the primary site for gas and nutrient exchange for the tissues of the kidneys.

You learned in Unit 12 (p. 329) that a capillary bed generally drains into a venule, but notice in Figure 15.6B that the capillaries of the glomerulus drain into a second *arteriole* called the **efferent arteriole**. The efferent arteriole then branches to form a second capillary bed known as the **peritubular capillaries**. These capillaries surround the tubules of the nephron, where they provide them with oxygen and nutrients and also take substances reabsorbed by the tubules back into the blood. The peritubular capillaries then drain out through the small **interlobular veins** (also known as *cortical radiate veins*), which drain into **arcuate veins**, then into **interlobar veins**, and, finally, into the large **renal vein**. This pattern of blood flow allows the kidneys both to filter blood and to reclaim most of the fluid and solutes filtered.

Microscopically, each kidney is composed of more than a million tiny units called **nephrons** (NEF-rahnz). The nephron can be divided into two parts: the *renal corpuscle* and the *renal tubule*. The **renal corpuscle** itself consists of two parts: the glomerulus and the glomerular capsule (Fig. 15.7). As you've seen, the glomerulus is a ball of looping capillaries. The capillaries themselves are *fenestrated*, meaning that they have large slits or pores, which allows large volumes of fluid to rapidly exit them. The second portion of the renal corpuscle is the **glomerular capsule**, which has two layers: (1) the outer **parietal layer**, which is simple squamous epithelium; and (2) the inner **visceral layer**, which consists of cells called **podocytes** (POH-doh-syt'z) that surround the capillaries of the glomerulus. The space between the parietal and visceral layers is known as the **capsular space**. Fluid forced out of the glomerular capillaries enters this space; once in the capsular space, the fluid is called *filtrate*.

From the capsular space, the filtrate next enters the **renal tubule**, which can be likened to the "plumbing" of the kidney. The renal tubule consists of three parts, shown in Figure 15.8: (1) the **proximal tubule** (sometimes called the *proximal convoluted tubule*), (2) the descending and ascending limbs of the **nephron loop**, and (3) the **distal tubule** (sometimes called the *distal convoluted tubule*). Note that the majority of the renal tubule is confined to the renal cortex; only the nephron loops of certain nephrons dip down into the renal medulla.

15

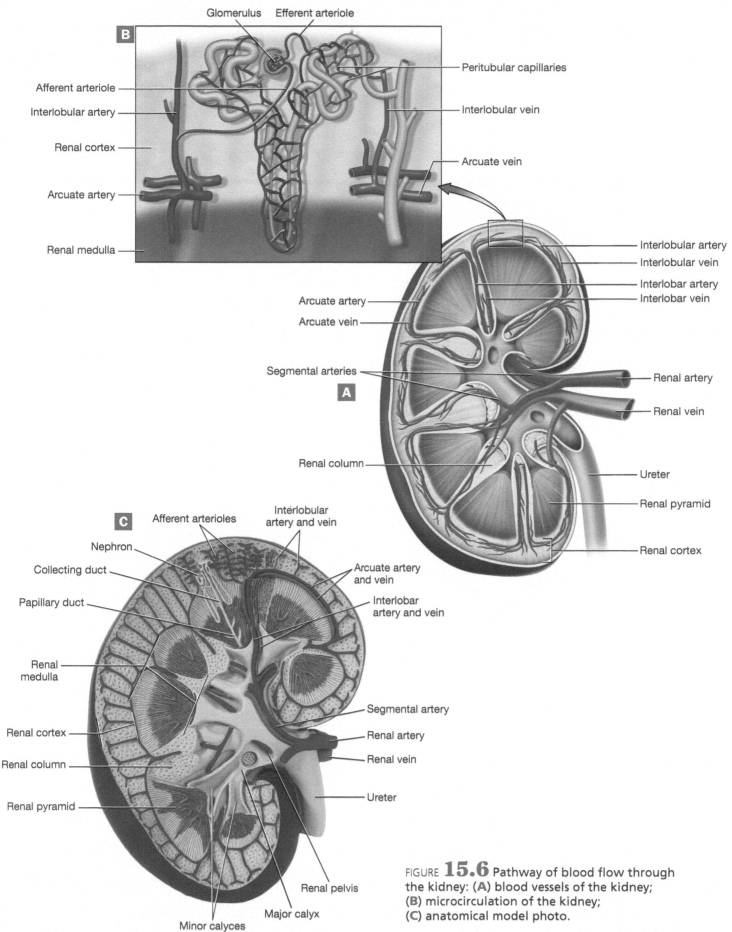

Glomerulus Efferent arteriole

B

Afferent arteriole

Interlobular artery

Renal cortex

Arcuate artery

Renal medulla

Peritubular capillaries

Interlobular vein

Arcuate vein

Interlobular artery
Interlobular vein
Interlobar artery
Interlobar vein

Arcuate artery
Arcuate vein

Segmental arteries

A

Renal column

Renal artery

Renal vein

Ureter

Renal pyramid

Renal cortex

C Afferent arterioles

Nephron

Collecting duct

Papillary duct

Renal medulla

Renal cortex

Renal column

Renal pyramid

Interlobular artery and vein

Arcuate artery and vein

Interlobar artery and vein

Segmental artery

Renal artery

Renal vein

Ureter

Renal pelvis

Major calyx

Minor calyces

FIGURE **15.6** Pathway of blood flow through the kidney: (**A**) blood vessels of the kidney; (**B**) microcirculation of the kidney; (**C**) anatomical model photo.

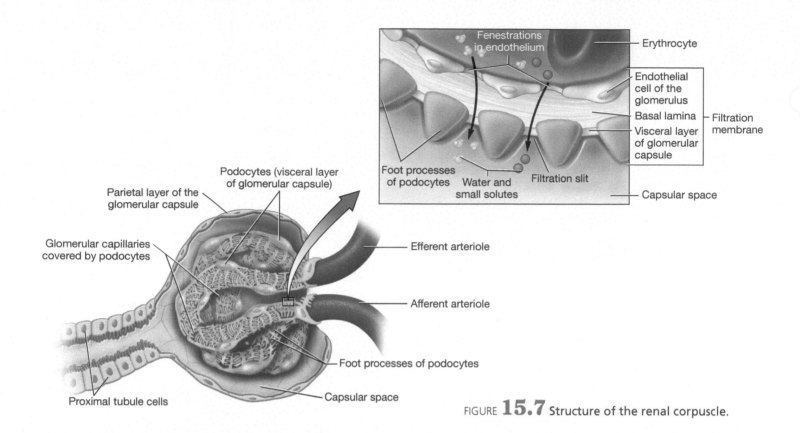

Podocytes (visceral layer of glomerular capsule)

Parietal layer of the glomerular capsule

Glomerular capillaries covered by podocytes

Proximal tubule cells

Foot processes of podocytes

Capsular space

Efferent arteriole

Afferent arteriole

Fenestrations in endothelium

Erythrocyte

Endothelial cell of the glomerulus

Basal lamina

Visceral layer of glomerular capsule

Filtration membrane

Foot processes of podocytes

Water and small solutes

Filtration slit

Capsular space

FIGURE **15.7** Structure of the renal corpuscle.

Several distal tubules drain into one **collecting duct**, which is part of a larger collecting system that is not part of the nephron. Collecting ducts in the renal cortex are **cortical collecting ducts**; those in the renal medulla are **medullary collecting ducts**. In the deep renal medulla are **papillary ducts**, which empty into minor calyces.

As filtrate passes through the renal tubule and collecting system, it is heavily modified, and most of the water and solutes are reclaimed and returned to the blood. By the time the filtrate leaves the papillary ducts to enter the minor calyces, it is known as **urine**, which then drains into the major calyces and, finally, into the renal pelvis.

From the renal pelvis, urine enters the next organs of the urinary system, which are collectively called the **urinary tract**. The first portion of the urinary tract consists of the tubes called the **ureters** (YOOR-eh-terz; Figs. 15.9 and 15.10). Ureters are lined by a type of epithelium called **transitional epithelium**, and their walls contain smooth muscle that massages the urine inferiorly via peristalsis.

The ureters drain urine into the posteroinferior wall of the organ known as the **urinary bladder** at the **ureteral orifices** (yoo-REE-ter-uhl; Fig. 15.10). Like the ureters, the urinary bladder is lined with transitional epithelium and contains smooth muscle, sometimes called the **detrusor muscle** (dee-TROO-sur), in its wall. The majority of the urinary bladder contains folds called **rugae** (ROO-ghee) that allow it to expand when it is filled with urine. The smooth inferior portion of the urinary bladder wall features a triangular-shaped area known as the **trigone** (TRY-gohn). This opens into the final organ of the urinary system—the **urethra** (yoo-REETH-ruh). The urethra is surrounded by two rings of muscle: the involuntary **internal urethral sphincter**, composed of smooth muscle, and the voluntary **external urethral sphincter**, composed of skeletal muscle. The external urethral sphincter is an extension of the *levator ani muscle,* also known as the **pelvic** or **urogenital diaphragm.** When both of these sphincters relax, urine is expelled from the body through the **external urethral orifice** via a process called **micturition** (mik-choo-RISH-un).

Notice in Figure 15.11 that the male and female urethras are quite different. In females, the urethra is short, measuring only about 4 cm. In males, the urethra is much longer, about 20 cm, and has three divisions: the **prostatic urethra**, which passes through the *prostate gland*; the short **membranous urethra**, which passes through the pelvic diaphragm; and the **spongy urethra**, which passes through the penis. Another difference you may notice is the position of the urinary bladder. The urinary bladder of the male is anterior to the rectum. The urinary bladder of the female sits anterior to the vagina and inferior to the uterus. This position is one of the reasons why pregnant women have to urinate so frequently—the enlarged uterus sits right on top of the urinary bladder.

15

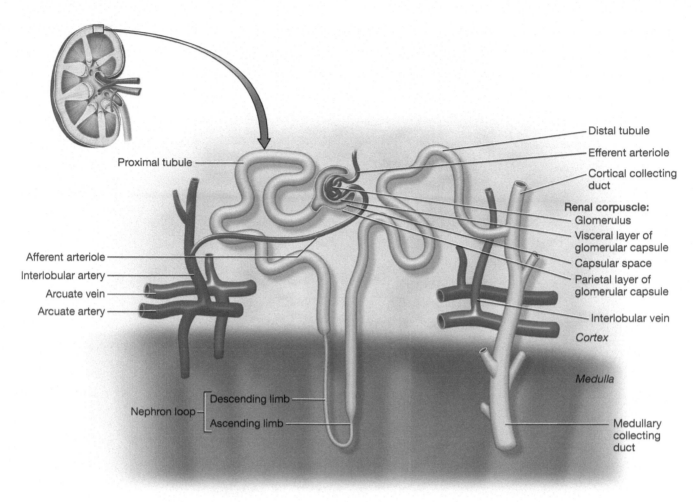

FIGURE **15.8** Structure of the nephron and collecting system.

Labels in figure 15.8:
- Proximal tubule
- Afferent arteriole
- Interlobular artery
- Arcuate vein
- Arcuate artery
- Nephron loop
  - Descending limb
  - Ascending limb
- Distal tubule
- Efferent arteriole
- Cortical collecting duct
- Renal corpuscle:
  - Glomerulus
  - Visceral layer of glomerular capsule
  - Capsular space
  - Parietal layer of glomerular capsule
- Interlobular vein
- Cortex
- Medulla
- Medullary collecting duct

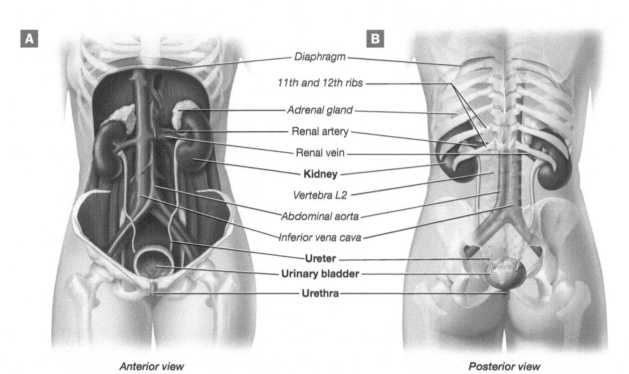

A   **B**

Labels in figure 15.9:
- Diaphragm
- 11th and 12th ribs
- Adrenal gland
- Renal artery
- Renal vein
- **Kidney**
- Vertebra L2
- Abdominal aorta
- Inferior vena cava
- **Ureter**
- **Urinary bladder**
- **Urethra**

Anterior view   *Posterior view*

FIGURE **15.9** Organs of the female urinary system: (**A**) anterior view; (**B**) posterior view; *(continues)*

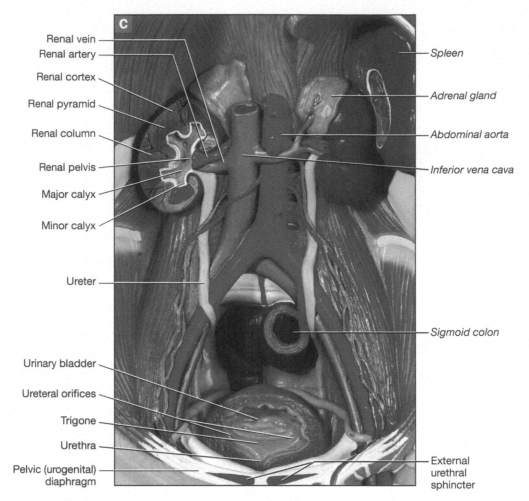

Renal vein
Renal artery
Renal cortex
Renal pyramid
Renal column
Renal pelvis
Major calyx
Minor calyx

Ureter

Urinary bladder
Ureteral orifices
Trigone
Urethra
Pelvic (urogenital) diaphragm

Spleen
Adrenal gland
Abdominal aorta
Inferior vena cava

Sigmoid colon

External urethral sphincter

FIGURE **15.9** Organs of the female urinary system (*cont.*): (**C**) anatomical model photo.

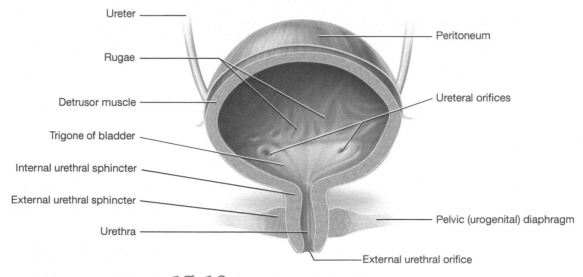

Ureter
Rugae
Detrusor muscle
Trigone of bladder
Internal urethral sphincter
External urethral sphincter
Urethra

Peritoneum
Ureteral orifices
Pelvic (urogenital) diaphragm
External urethral orifice

FIGURE **15.10** Female urinary bladder, frontal section.

15

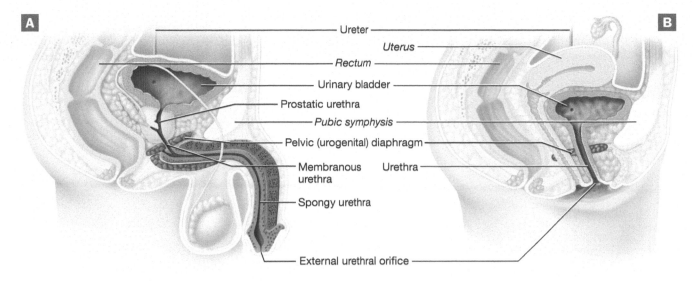

FIGURE **15.11** Organs of the male and female urinary systems, sagittal sections: (**A**) male; (**B**) female.

Procedure **1** Model Inventory for the Urinary System

Identify the following structures of the urinary system on models and diagrams, using this unit and your textbook for reference. As you examine the anatomical models and diagrams, record the name of the model and the structures you were able to identify on the model inventory in Table 15.2.

Kidney Anatomy
1. Surrounding connective tissue:
   a. Renal fascia
   b. Adipose capsule
   c. Renal capsule
2. Hilum
3. Regions:
   a. Renal cortex
   b. Renal medulla
      (1) Renal pyramids
      (2) Renal columns
   c. Renal pelvis
      (1) Minor calyces
      (2) Major calyces
4. Blood supply:
   a. Renal artery
      (1) Segmental artery
      (2) Interlobar artery
      (3) Arcuate artery
      (4) Interlobular artery
   b. Renal vein
      (1) Interlobular vein
      (2) Arcuate vein
      (3) Interlobar vein

Structures of the Nephron and Collecting System
1. Renal corpuscle
   a. Glomerulus
      (1) Afferent arteriole
      (2) Efferent arteriole
      (3) Peritubular capillaries
   b. Glomerular capsule
      (1) Parietal layer
      (2) Visceral layer
      (3) Podocytes
      (4) Capsular space
2. Renal tubule
   a. Proximal tubule
   b. Nephron loop
   c. Distal tubule
3. Collecting system
   a. Cortical collecting duct
   b. Medullary collecting duct
   c. Papillary duct

Urinary Tract
1. Ureter
2. Urinary bladder
   a. Ureteral orifices
   b. Detrusor muscle
   c. Rugae
   d. Trigone
3. Urethra
   a. Internal urethral sphincter
   b. External urethral sphincter
   c. External urethral orifice
   d. Prostatic urethra
   e. Membranous urethra
   f. Spongy urethra

15

TABLE **15.2** Model Inventory for Urinary Anatomy

Model/Diagram	Structures Identified

Procedure **2** Time to Draw the Urinary System

In the space provided, draw, color, and label one of the overview of the urinary system models that you examined. In addition, write the function of each structure that you label.

15

Procedure 3 Time to Draw a Nephron

In the space provided, draw, color, and label a nephron. In addition, write the function of each structure that you label.

Procedure 4 Kidney Dissection

In this procedure, you will now dissect a preserved kidney and locate several of the structures you just identified on models and diagrams.

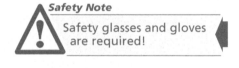

Safety Note

Safety glasses and gloves are required!

1 Obtain a fresh or preserved kidney specimen and dissection supplies.

2 If the thick surrounding connective tissue coverings are intact, note their thickness and amount of adipose tissue.

3 Use scissors to cut through the connective tissue coverings, and remove the kidney.

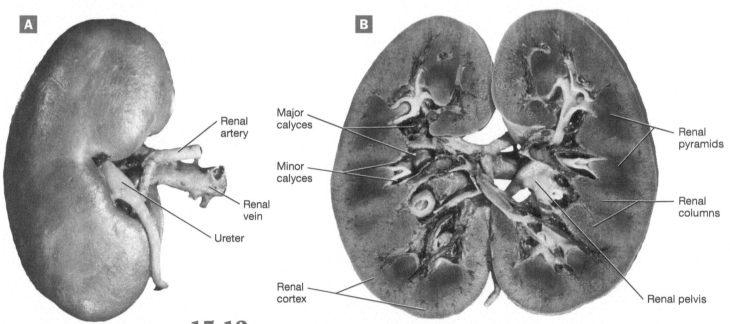

FIGURE **15.12** Preserved kidney: (**A**) anterior view; (**B**) frontal section.

15

4 List surface structures you are able to identify (see Figure 15.12A for reference):

_____

_____

5 Distinguishing between the ureter, the renal artery, and the renal vein is often difficult. Following are some hints to aid you:

■ The renal artery typically has the thickest and most muscular wall, and it branches into several segmental arteries prior to entering the kidney.

■ The renal vein is thinner, flimsier, and often larger in diameter than the renal artery.

■ The ureter has a thick, muscular wall, but it does not branch after it leaves the kidney. Also, its diameter is usually smaller than either the renal artery or the renal vein.

Keeping these points in mind, determine the location of the renal artery, the renal vein, and the ureter on your specimen. Sketch the arrangement of the three structures in the space provided.

6 Use a scalpel to make a frontal section of the kidney. Draw the kidney in the space provided below and label the structures you are able to identify (see Figure 15.12B).

15

Exercise 15-2

Renal Physiology: The Model Kidney

MATERIALS
- ❏ 4-inch piece of dialysis tubing
- ❏ 2 pieces of string
- ❏ Animal blood or simulated blood
- ❏ 200 mL beaker
- ❏ Deionized water
- ❏ URS-10 and bottle with key

The kidneys filter the blood in order to remove metabolic wastes and regulate the body's fluid, electrolyte, and acid-base balance. Recall that the kidneys' filters are located within the glomerular capillaries. However, the glomerular capillaries can't filter the blood unless there is a force to push fluid and solutes through them and into the nephron tubule. This force is supplied by a pressure called the **net filtration pressure (NFP)**.

The NFP is the combination of three forces (Fig. 15.13):

1. **Glomerular hydrostatic pressure (GHP)**. The **GHP** is equal to the blood pressure in the glomerulus and averages about 50 mmHg. This pressure pushes fluid and solutes out of the glomerulus and into the capsular space.

2. **Colloid osmotic pressure (COP)**. The glomerular **COP** (or **GCOP**) is the pressure created by the proteins and other solutes in the blood and averages about 30 mmHg. This pressure tends to oppose filtration and pulls water *into* the glomerular capillaries by osmosis.

3. **Capsular hydrostatic pressure (CHP)**. The final pressure, the **CHP**, is the force of the fluid in the capsular space. It averages about 10 mmHg, and, like the COP, tends to oppose filtration, pushing water back into the blood in the glomerular capillaries.

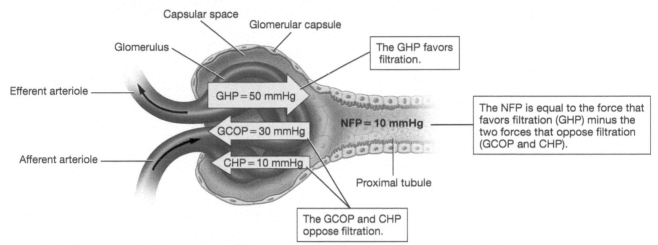

FIGURE **15.13** Net filtration pressure in the renal corpuscle.

As you can see in Figure 15.13, to get the NFP, simply subtract the two forces that oppose filtration from the one force that favors filtration:

$$GHP - (COP + CHP) = NFP$$
$$50 - (30 + 10) = 10 \text{ mmHg}$$

So there is a net pressure of 10 mmHg in the glomerulus that drives fluid from it into the capsular space. When the fluid enters the capsular space, it becomes a liquid called **filtrate**.

Filtrate is similar in composition to blood, but it lacks most of the proteins and cells we find in blood. This selectivity is possible because of the **filtration membrane**, which consists of the glomerular endothelial cells, the basal lamina, and the visceral layer of the glomerular capsule (Fig. 15.14). Notice in the figure that the cells of the visceral layer, known as **podocytes**, have extensions called **foot processes** that interlock to form narrow **filtration slits**. The filtration membrane acts in a similar manner to the filter in your coffeemaker. Just as a coffee filter holds back the coffee grounds while allowing water and other solutes to pass through into your coffee, the filtration membrane prevents large items, such as proteins and cells, from leaving the blood while allowing small substances, such as water, glucose, amino acids, electrolytes, and metabolic wastes, to leave the blood and enter the filtrate. This selectivity is vitally important, as the nephron tubules and collecting system cannot reabsorb large proteins or cells; for this reason, any that entered the filtrate would be lost to the urine.

15

The rate of filtrate production averages about 120 mL/min, a value called the **glomerular filtration rate (GFR)**. The GFR is determined largely by the three pressures that factor into the NFP. To put it simply, if the NFP increases, then the GFR generally also increases. The opposite is also true—if the NFP decreases, the GFR also generally decreases. Note that this is only true up to a point because there are many compensatory mechanisms that work to maintain a stable GFR. These compensatory mechanisms exist because the GFR is a critical determinant of kidney function; if the GFR is too low or too high, the kidneys will not be able to carry out their many functions.

The filtrate contains a number of substances our bodies need to reclaim, including water, glucose, amino acids, and many different ions. This reclamation occurs by a process known as **tubular reabsorption** (Fig. 15.15). About 99 percent of the water in the filtrate and most of the solutes are reabsorbed through the epithelium of the nephron tubules and returned to the blood. The importance of this function cannot be overstated. If the water were not reabsorbed in the renal tubules and the collecting ducts, we would lose our entire plasma volume in fewer than 30 minutes. This demonstrates another important reason to tightly regulate the GFR—if the GFR is too high, the nephron tubules will not have time to reabsorb all of the necessary substances in the filtrate, and we will lose much of our water and important solutes to the urine.

There are certain substances that aren't filtered at the glomerulus that need to be excreted in the urine, as well as substances that the body needs to excrete in greater amounts than are filtered. Notice in Figure 15.15 that such substances can move from the blood in the peritubular capillaries to the filtrate in the nephron tubules by a process known as **tubular secretion**. Examples of secreted substances include potassium ions, hydrogen ions, and uric acid.

In this exercise, you will examine the process of glomerular filtration and the selectivity of the filtration membrane by constructing a model

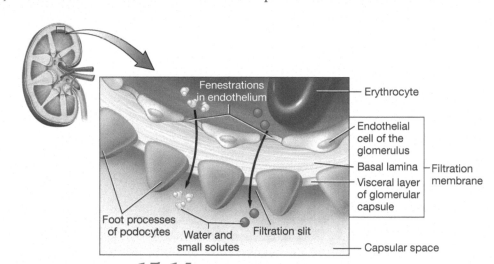

FIGURE **15.14** Structure of the filtration membrane.

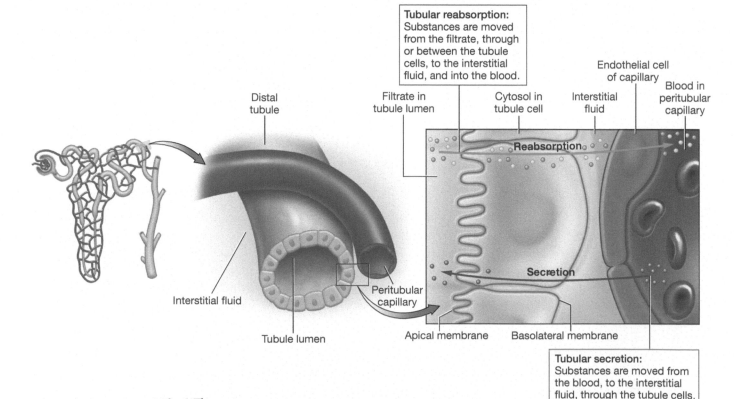

FIGURE **15.15** Reabsorption and secretion in the renal tubule.

15

kidney and by testing for substances that appear in the filtrate. A kidney is easily modeled with either animal blood or simulated blood and simple dialysis tubing, which has a permeability similar to that of the filtration membrane. Glomerular filtration is mimicked by immersing the blood-filled tubing in water. Substances to which the "filtration membrane" is permeable will enter the surrounding water (the "filtrate"), whereas those to which the membrane is not permeable will remain in the tubing.

The results of this test are analyzed using **urinalysis reagent test strips** (URS-10; Fig. 15.16). Each strip consists of 10 small, colored pads that change color in the presence of certain chemicals, such as glucose or hemoglobin. The strip is interpreted by watching the pads for color changes and comparing the color changes to a color-coded key on the side of the bottle. The color that is closest to the color on the strip is recorded as your result. Please note that for this exercise you will read only four of the 10 boxes: blood, leukocytes, protein, and glucose.

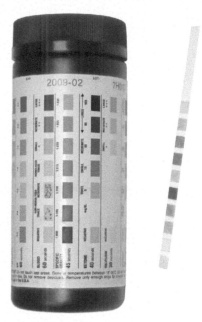

FIGURE **15.16** URS-10 strip and vial.

Procedure **1** Making a Model Kidney and Testing Glomerular Filtration

Now let's construct our model kidneys. Note that the model kidney must sit in water for 25 minutes while you wait for your results.

Safety Note
Safety glasses and gloves are required!

1 Cut a 4-inch piece of dialysis tubing.

2 Securely tie off one end of the tubing with string and open the other end of the tubing by wetting the end of the tube and rubbing it between your fingers.

3 Fill the dialysis tubing about half-full with either animal blood or simulated blood, and securely tie off the open end of the tube with string.

4 Place the tied-off tube in a 200 mL beaker containing about 150 mL of deionized water.

5 Leave the tubing in the water for approximately 25 minutes.

6 After 25 minutes have passed, remove the tubing from the water.

7 Dip a URS-10 in the water and remove it quickly. Turn the bottle on its side and compare the colors of the pads for glucose, blood, leukocytes, and protein. You will notice that on the side of the bottle, time frames are listed for each substance that is tested. Wait these listed amounts of time to watch for reactions; otherwise, you could obtain false negative results. If you wait too long to read the results, though, the colors will tend to darken and may blend with adjacent colors.

8 Record your results below.

   a Glucose: _____

   b Erythrocytes (blood): _____

   c Leukocytes: _____

   d Protein: _____

9 Which substances were filtered out? Which substances stayed in the tubing? Explain your results.

_____

_____

_____

_____

Exercise **15-3**

Renal Physiology: Urinalysis

MATERIALS
- ☐ Samples of simulated urine
- ☐ Graduated cylinder or test tube
- ☐ URS-10 and bottle with key

For hundreds of years, healthcare providers have recognized the utility of urinalysis as a diagnostic tool. Historically, the urine was evaluated for color, translucency, odor, and taste (yes, taste!). Today, while these characteristics are still examined (except, thankfully, taste), we also utilize urinalysis test strips, as we did in Exercise 15-2 (p. 419), to test for the presence of various chemicals in the urine. Normal and abnormal readings both can give a healthcare provider a wealth of information about a patient's renal function and overall health.

In this exercise, you will analyze different urine samples using a URS-10, as in Exercise 15-2. For each urine sample, one or more results will be read as abnormal. In the second part of the exercise, you will research potential causes of the abnormalities that you detected.

Procedure **1** Urinalysis Part One: Identifying Abnormalities

Test a minimum of five samples of simulated urine using your URS-10 test strips. Use a new strip for each sample, and take care to clean your graduated cylinder thoroughly between each sample that you test. Please note that this is _simulated_ urine rather than real urine, although it contains the same chemicals as real urine and should be handled with equal caution.

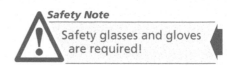

Safety Note
Safety glasses and gloves are required!

1 Randomly choose one sample of simulated urine, and pour approximately 3 mL of the sample into a test tube or graduated cylinder.

2 Submerge one URS-10 in the urine, and quickly remove it.

3 Compare the resulting colors and patterns on the test strip to those on the key on the bottle. Be sure to wait the appropriate amount of time to read the results.

4 Record your results in Table 15.3, and note any anomalous results.

5 Repeat this procedure for the remaining samples.

15

TABLE **15.3** Urinalysis Results

Reading	Sample 1	Sample 2	Sample 3	Sample 4	Sample 5
pH					
Leukocytes					
Nitrite					
Urobilinogen					
Protein					
Blood					
Specific gravity					
Ketones					
Bilirubin					
Glucose					

 Procedure 2 Urinalysis Part Two: Researching Possible Pathologies

Each sample that you tested should have had at least one abnormality. Use your textbook, the internet, or both, as a resource to determine which disease state(s) could lead to the abnormality you detected in each sample. Record your results in Table 15.4.

TABLE **15.4** Urinalysis Abnormalities and Disease States

Sample	Primary Abnormality	Potential Causes
1		
2		
3		
4		
5		

15

Exercise 15-4

Time to Trace!

MATERIALS
☐ Anatomical models of the kidney and nephron

L et's now trace the pathway of different substances through the vasculature and the microanatomy of the kidney. You will examine the path taken by (1) an erythrocyte (a red blood cell), (2) a molecule of glucose, and (3) a molecule of urea. The path that each takes differs in key ways—and it's your job to figure out how so.

Following are some hints to help you:

■ Remember the basic rules of blood flow: You must pass through a capillary bed in order to get from an artery to a vein. Also recall that there are two capillary beds in the microcirculation of the kidney: the glomerulus and the peritubular capillaries.

■ Table 15.1 in Pre-Lab Exercise 15-4 (p. 408) can be a big help, as it tells you three key pieces of information: (1) whether the item is filtered at the glomerulus, (2) whether a filtered item is reabsorbed, and (3) where reabsorption takes place.

■ Don't overthink things. If an item isn't filtered at the glomerulus, then that means it must stay in the blood, right? Similarly, if an item is filtered at the glomerulus, but is never reabsorbed, it must go through the whole nephron and collecting system to end up in the urine.

■ Refer to Figure 15.6 (p. 411) for help with the pathway of blood flow through the kidney and the kidney's microanatomy.

 Procedure 1 Tracing Substances through the Kidney

Complete the following tracing procedures, using Figure 15.6 (p. 411) for reference.

Part 1 Erythrocyte

Trace an erythrocyte from the renal artery to the renal vein (remember that glucose is filtered at the glomerulus, but is reabsorbed into the blood).

_____ → _____
_____ → _____
_____ → _____
_____ → _____
_____ → _____
_____ → _____ .

Part 2 Glucose

Trace a molecule of glucose from the renal artery to the renal vein (remember that glucose is filtered at the glomerulus, but is reabsorbed into the blood).

_____ → _____
_____ → _____
_____ → _____
_____ → _____
_____ → _____
_____ → _____ .

15

Part 3 Urea

Trace a molecule of urea from the renal artery to its final destination outside the body of a female.

_____ → _____

→ _____

→ _____

→ _____

→ _____

→ _____

→ _____

→ _____

→ _____

→ _____

_____ → exits the body.

1 Label the following parts of the kidney on **Figure 15.17.**

❑ Major calyx ❑ Renal column ❑ Renal pelvis
❑ Minor calyx ❑ Renal cortex ❑ Renal pyramid
❑ Renal artery ❑ Renal medulla ❑ Renal vein

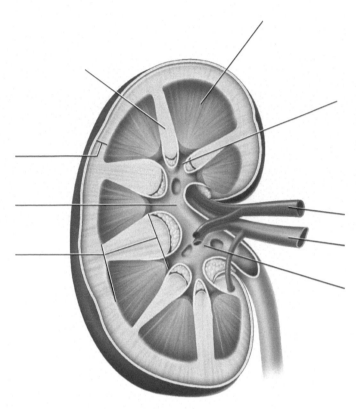

FIGURE **15.17** Right kidney, frontal section.

2 The blood flow through the kidney is special because
a. its first capillary beds drain into arterioles.
b. its second capillary beds drain into arterioles.
c. it is supplied by three renal arteries.
d. it contains no capillary beds.

15

3 Label the following parts of the nephron on **Figure 15.18**.

- ❏ Afferent arteriole
- ❏ Cortical collecting duct
- ❏ Distal tubule
- ❏ Efferent arteriole
- ❏ Glomerular capsule
- ❏ Glomerulus
- ❏ Nephron loop
- ❏ Proximal tubule

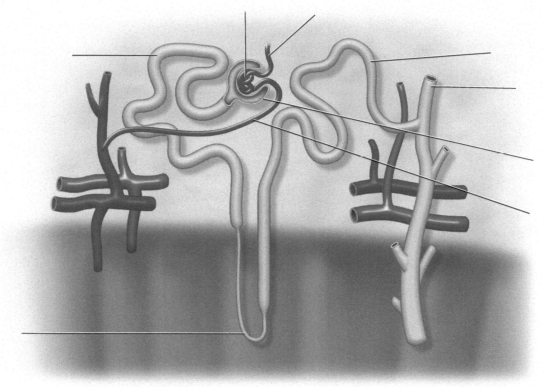

FIGURE **15.18** Structure of the nephron and collecting system.

4 Label the following parts of the urinary system on **Figure 15.19**.

- ❏ External urethral orifice
- ❏ Membranous urethra
- ❏ Pelvic (urogenital) diaphragm
- ❏ Prostatic urethra
- ❏ Spongy urethra
- ❏ Ureter
- ❏ Urinary bladder

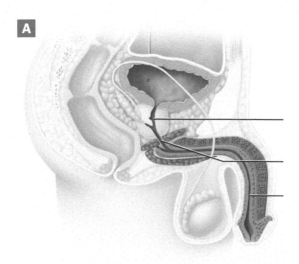 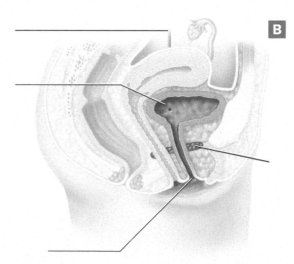

FIGURE **15.19** Organs of the male and female urinary systems, sagittal sections: **(A)** male; **(B)** female.

5 Number the following from the point the filtrate is first formed (with a number 1) to the point it drains into the renal pelvis (with a number 10).

_____ Major calyx

_____ Minor calyx

_____ Proximal tubule

_____ Cortical collecting duct

_____ Capsular space

_____ Nephron loop— descending limb

_____ Medullary collecting duct

_____ Nephron loop— ascending limb

_____ Papillary duct

_____ Distal tubule

6 Urine drains from the urinary bladder via the
a. renal pelvis.
b. urethra.
c. ureters.
d. papillary calyces.

7 Urine is expelled from the body by a process called
a. micturition.
b. parturition.
c. defecation.
d. procrastination.

8 Your lab partner wonders how the cells of the renal tubules and ducts of the collecting system obtain oxygen and nutrients when their capillary bed is located only in the renal corpuscle. What is your lab partner misunderstanding about the blood flow in the kidney? Explain his mistake to him.

_____

_____

_____

_____

9 Female patients suffer *urinary tract infections*, or bacterial infections of the urethra and urinary bladder, more frequently than do male patients. Why do you think this is so, considering the anatomy of the female and male urinary tracts?

_____

_____

_____

_____

15

10 *True/False:* Mark the following statements as true (T) or false (F). If the statement is false, correct it to make it a true statement.

_____ a. The fluid and solutes in the filtrate have been removed from the blood and are located in the renal tubule.

_____ b. The third and finest filter in the filtration membrane is created by the podocytes and their filtration slits.

_____ c. Nephron tubules can reabsorb large proteins and cells that enter the filtrate.

_____ d. Substances such as glucose, proteins, and erythrocytes are secreted into the filtrate.

11 Which of the following is not a component of the filtration membrane?
a. Glomerular endothelial cells.
b. Podocytes.
c. Smooth muscle.
d. Basal lamina.

12 Urinalysis can tell you
a. whether there is an infection in the kidney or urinary tract.
b. whether blood sugar is elevated.
c. whether a person is dehydrated.
d. whether a person's kidneys function normally.
e. All of the above.

13 In Exercise 15-4 (p. 424), you traced items that were filtered at the glomerulus. Now consider a hydrogen ion that is not filtered but is instead secreted from the peritubular capillaries into the filtrate at the distal tubule. Trace the pathway this hydrogen ion would take from the afferent arteriole to the point at which it exits the body of a male in the urine.

Start: Afferent arteriole ➜ _____ ➜ _____ ➜

_____ ➜ _____ ➜ _____ ➜

_____ ➜ _____ ➜ _____ ➜

_____ ➜ _____ ➜ _____ ➜

_____ ➜ _____ ➜ _____ ➜

_____ ➜ _____ ➜ exits the body in the urine.

14 Calculate the net filtration pressure if the glomerular hydrostatic pressure measures 46 mmHg, the colloid osmotic pressure 34 mmHg, and the capsular hydrostatic pressure 10 mmHg. Does this differ from the normal value? If so, how? What effect, if any, would this have on the GFR?

_____

_____

_____

_____

_____

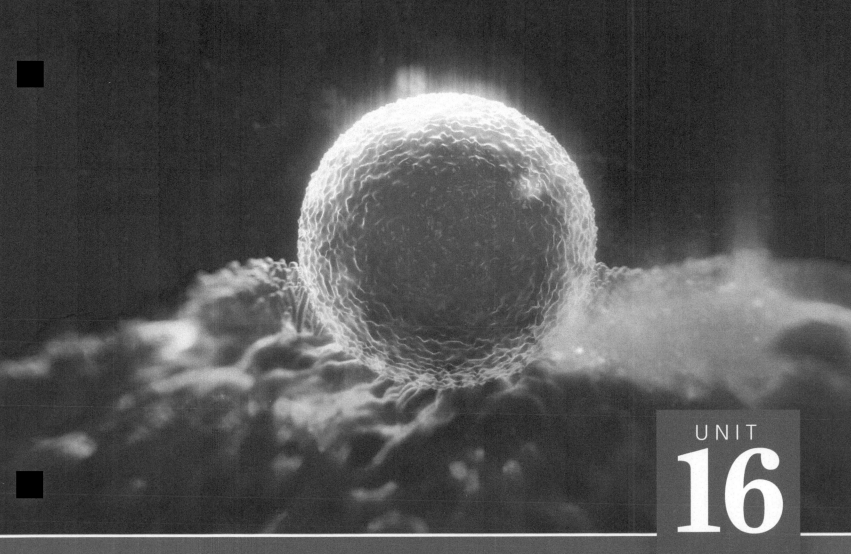

Reproductive Systems and Development

When you have completed this unit, you should be able to:

1 Describe and identify structures of the male and female reproductive systems.

2 Describe and identify the stages of meiosis, and describe how meiosis differs from mitosis.

3 Trace the pathway a developing gamete takes from its development to the time of fertilization.

4 Trace a developing zygote (or conceptus) from fertilization to implantation.

5 Describe and identify structures and membranes associated with the fetus.

6 Describe and identify the stages of development and structures of the fetal circulation.

PRE-LAB EXERCISES

Complete the following exercises prior to coming to lab, using your lab manual and textbook for reference.

Pre-Lab Exercise 16-1

✎ Key Terms

You should be familiar with the following terms before coming to lab.

Term	Definition

Structures of the Male Reproductive System

Testes _____

_____

Seminiferous tubules _____

_____

Epididymis _____

_____

Ductus deferens _____

_____

Spermatic cord _____

_____

Prostate gland _____

_____

Structures of the Female Reproductive System

Ovaries _____

_____

Uterine tube _____

_____

Uterus _____

_____

Vagina _____

_____

16

Vulva _____

_____

Mammary glands _____

_____

Gametogenesis Terms

Meiosis _____

_____

Haploid _____

_____

Spermatogenesis _____

_____

Oogenesis _____

_____

Development

Fertilization _____

_____

Zygote _____

_____

Embryo _____

_____

Fetus _____

_____

Fetal Structures

Chorion _____

_____

Amnion _____

_____

Placenta _____

16

_____

Pre-Lab Exercise **16-2**

Male Reproductive Anatomy

Color the structures of the male reproductive system in Figure 16.1, and label them with the following terms from Exercise 16-1 (p. 438). Use Exercise 16-1 in this unit and your text for reference.

- ❑ Scrotum
- ❑ Testis
- ❑ Epididymis
- ❑ Ductus deferens
- ❑ Ejaculatory duct
- ❑ Urethra
  - ☐ Prostatic urethra
  - ☐ Membranous urethra
  - ☐ Spongy urethra

- ❑ Glands
  - ☐ Seminal vesicle
  - ☐ Prostate gland
- ❑ Penis
  - ☐ Corpora cavernosa
  - ☐ Corpus spongiosum
    - ▪ Glans penis

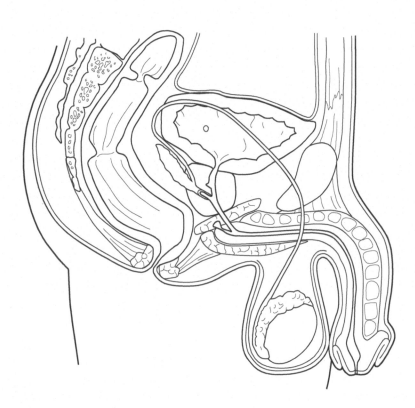

FIGURE **16.1** Midsagittal section of the male pelvis.

16

Pre-Lab Exercise **16-3**

Female Reproductive Anatomy

Color the structures of the female reproductive system in Figure 16.2, and label them with the following terms from Exercise 16-2 (p. 442). Use Exercise 16-2 in this unit and your text for reference.

❑ Ovary
❑ Uterine tube
❑ Uterus
    ☐ Fundus
    ☐ Body
    ☐ Cervix

❑ Vagina
❑ Vulva
    ☐ Labium major
    ☐ Labium minus
    ☐ Clitoris
    ☐ Urethral orifice

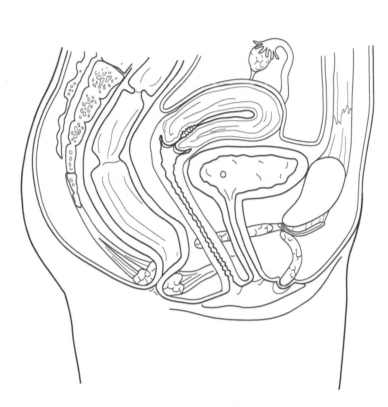

FIGURE **16.2** Midsagittal section of the female pelvis.

Pre-Lab Exercise **16-4**

Fetal Cardiovascular Anatomy

Color the structures of the fetal cardiovascular system in Figure 16.3, and label them with the following terms from Exercise 16-6 (p. 453). Use Exercise 16-6 in this unit and your text for reference.

- ❏ Placenta
- ❏ Umbilical cord
  - ☐ Umbilical arteries
  - ☐ Umbilical vein
- ❏ Foramen ovale
- ❏ Ductus venosus
- ❏ Ductus arteriosus

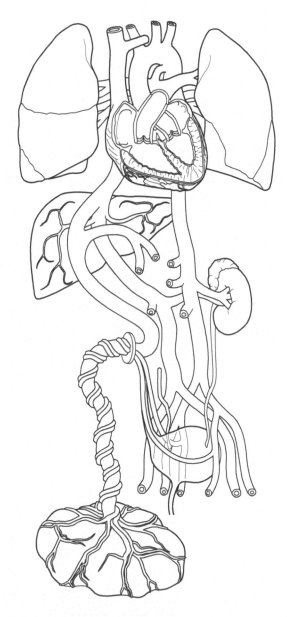

FIGURE **16.3** Fetal cardiovascular anatomy.

Ovum surrounded by cilia lining the uterine tubes.

EXERCISES

The other organ systems we have discussed all function in some manner to help maintain homeostasis of the body. The **reproductive system**, however, plays a lesser role in maintaining homeostasis and a greater role in perpetuating the species. The main organs of the reproductive system are the **gonads** (GOH-nadz)—the testes and the ovaries—which produce **gametes** (GAM-eetz), or sex cells, for reproduction.

We begin this unit with the anatomy of the male and female reproductive systems. Then we turn to the main functions of these organs: **gametogenesis** (gah-meet-oh-JEN-uh-sis), the formation of new gametes. Finally, we examine the process of development, and the unique anatomy and structures of the developing human.

Exercise 16-1

Male Reproductive Anatomy

MATERIALS
- ❑ Male reproductive models
- ❑ Human torso models

The **testes**, the gamete-producing organs of the male, are situated outside the body in a sac known as the **scrotum** (Fig. 16.4). They are located externally because sperm production requires a temperature of about 34°C (about 94°F), which is lower than body temperature. Along the midline, the scrotum has a connective tissue septum that divides it into two chambers, one for each testis. Smooth muscle tissue called the **cremaster muscle** (kreh-MASS-ter) surrounds each testis in the scrotum. This muscle tissue contracts to control the testes' distance from the body, which controls their temperature.

The layer of connective tissue deep to the scrotum extends up into the pelvic cavity to form a structure known as the **spermatic cord**. Notice in Figure 16.4 that the spermatic cord surrounds the cremaster muscle, the testicular artery, a group of veins called the **pampiniform venous plexus** that drain the testis, a duct that transports sperm known as the *ductus deferens*, and nerves that supply the testes. The spermatic cord passes into the pelvic cavity through the *external inguinal ring*, which is the opening to a small passageway called the *inguinal canal* (IN-gwin-uhl).

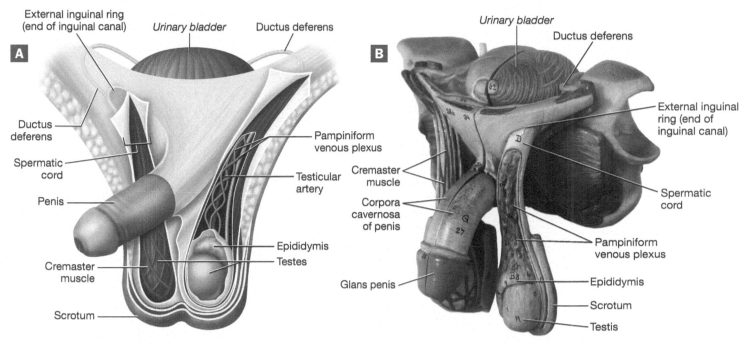

FIGURE **16.4** Structure of the scrotum and spermatic cord: **(A)** illustration; **(B)** anatomical model photo.

Deep to the cremaster muscle, we find connective tissue that divides the interior of each testis into lobules (Fig. 16.5). Each lobule contains a tightly coiled **seminiferous tubule** (sem-ih-NIF-er-us) where *spermatogenesis* (sper-mat-oh-JEN-ih-sis), or the formation of sperm cells, takes place.

The seminiferous tubules converge near the superior part of the testis to form a structure called the **rete testis** (REE-tee TES-tis). The rete testis exits the testis to join the first segment of the duct system of the male reproductive tract, the **epididymis** (ep-ih-DID-ih-miss; Figs. 16.5 and 16.6). Immature sperm produced by the seminiferous tubules migrate through the epididymis to finish their maturation, after which they move to a long tube called the **ductus** (or *vas*) **deferens** (DUK-tuss DEF-er-ahnz). Notice in Figure 16.6 that the ductus deferens ascends through the pelvic cavity and curls posteriorly and superiorly around the urinary bladder. On the posterior surface of the urinary bladder, the ductus deferens widens to form its terminal portion, the *ampulla*.

The duct leaving the ampulla merges with a duct leaving a gland called the **seminal vesicle**. Together, the merged ducts form a short passage known as the **ejaculatory duct**. This duct passes through a gland on the inferior surface of the urinary bladder—the **prostate gland** (PRAW-stayt—be sure not to call it the "prostrate"). Here its contents drain into the first segment of the urethra, the **prostatic urethra**. Recall from the urinary unit that the prostatic urethra becomes the **membranous urethra** as it exits the prostate, and then becomes the **spongy urethra** as it enters the corpus spongiosum of the penis. The urethra terminates at the tip of the penis at the external urethral orifice.

The male reproductive tract includes three exocrine glands: the previously mentioned prostate gland and seminal vesicles, and a set of two small glands located at the root of the penis called the **bulbourethral glands** (bul-boh-yoo-REETH-ruhl; Figs. 16.6 and 16.7). The seminal vesicles and the prostate gland together produce about 90 percent of the volume of **semen**, a fluid that contains chemicals to nourish and activate sperm. The bulbourethral glands produce an alkaline fluid secreted prior to the release of sperm during ejaculation.

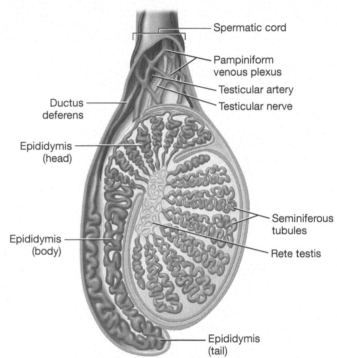

FIGURE **16.5** Midsagittal section through the testis.

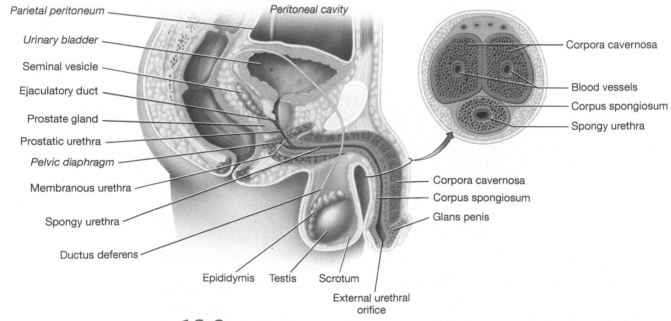

FIGURE **16.6** Midsagittal section through the male pelvis.

The **penis** is the male copulatory organ. It is composed of three erectile bodies: the single **corpus spongiosum** (KOHR-pus spun-jee-OH-sum) and the paired, dorsal **corpora cavernosa** (kohr-POHR-uh kah-ver-NOH-suh). The corpus spongiosum, which surrounds the spongy urethra, enlarges distally to form the **glans penis**. All three bodies consist of vascular spaces that fill with blood during an erection.

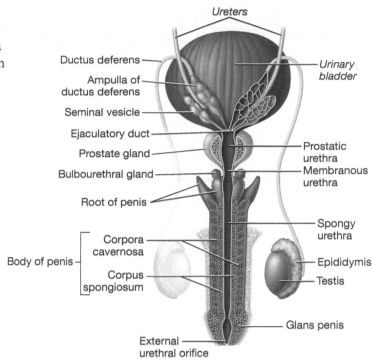

FIGURE **16.7** Posterior view of the male reproductive system with a frontal section of the penis.

Procedure **1** Model Inventory for the Male Reproductive System

Identify the following structures of the male reproductive system on models and diagrams, using this unit and your textbook for reference. As you examine the anatomical models and diagrams, record the name of the model and the structures you were able to identify on the model inventory in Table 16.1.

1. Scrotum
2. Cremaster muscle
3. Spermatic cord
   a. Testicular artery
   b. Pampiniform venous plexus
   c. Inguinal canal
4. Testes
   a. Seminiferous tubules
   b. Rete testis

5. Epididymis
6. Ductus deferens
7. Seminal vesicle
8. Ejaculatory duct
9. Prostate gland
10. Urethra
   a. Prostatic urethra
   b. Membranous urethra
   c. Spongy urethra
   d. External urethral orifice

11. Bulbourethral glands
12. Penis
   a. Corpus spongiosum
     (1) Glans penis
   b. Corpora cavernosa

16

TABLE **16.1** Model Inventory for the Male Reproductive System

Model/Diagram	Structures Identified

16

Exercise 16-2

Female Reproductive Anatomy

MATERIALS
- ❑ Female reproductive models
- ❑ Human torso models

The female reproductive organs are located in the pelvic cavity, with the exception of the almond-shaped **ovaries**, which are found in the peritoneal cavity (Fig. 16.8). The ovaries are the female gonads, which produce **oocytes** (OH-oh-syt′z) that travel through the reproductive tract to be fertilized. Oocytes are located within small sacs in the ovary called *follicles*.

Each ovary is held in place by several ligaments, including the **ovarian ligament**, which attaches the ovary to the uterus; a sheet of connective tissue called the **broad ligament**, which anchors it to the lateral pelvic wall; and the **suspensory ligament**, which attaches the ovary to the posterolateral pelvic wall. Notice in Figure 16.8 that the suspensory ligaments also carry with them the ovaries' blood supply.

The duct system of the female reproductive system begins with the **uterine tube**. The uterine tube is not directly connected to the ovary. For this reason, when an oocyte is released from the ovary, it is actually released into the pelvic cavity and the uterine tube must "catch" it and bring it into the tube. This is accomplished by fingerlike extensions of the tube called **fimbriae** (FIM-bree-ay) that wrap around the ovary. Like the ovaries, the uterine tubes are anchored by the broad ligament.

The uterine tubes join the superolateral portion of the **uterus** (YOO-ter-us), which is the organ in which a fertilized ovum implants and in which a conceptus develops. The uterus has three regions: the dome-shaped **fundus**, the central **body**, and the narrow **cervix**. The opening of the cervix is known as the **cervical os**. The uterine wall is quite thick, and its thickness changes as it progresses through the 28-day uterine cycle. There are three layers to the uterine wall: the inner epithelial and connective tissue lining called the **endometrium** (en-doh-MEE-tree-um), in which a fertilized ovum implants; the middle, muscular **myometrium** (my-oh-MEE-tree-um), composed of smooth muscle tissue; and the outermost **perimetrium** (pehr-ee-MEE-tree-um), which is composed of a serous membrane on the posterior uterus and connective tissue on the anterior uterus.

As you can see in Figure 16.9, the uterus is situated posterior to the urinary bladder and anterior to the rectum. It is held in place by several ligaments, such as the broad ligament, which anchors it to the anterior and lateral pelvis, and the **round ligaments,** which anchor it to the anterior abdominal wall (visible in Figure 16.9).

The **vagina** is a passage about 8–10 cm (3.1–3.9 in.) long that extends inferiorly from the cervical os and terminates at the **vaginal orifice.** The superior end of the vagina forms a recess around the cervical os called the **fornix**. Note that the vagina is entirely internal, although it is frequently and incorrectly referred to as the female's external genitalia.

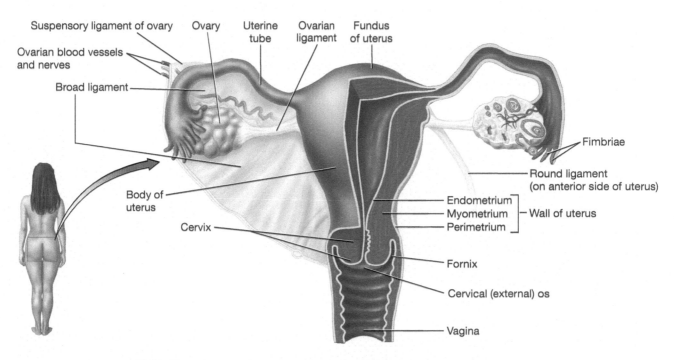

FIGURE **16.8** Posterior view of the female reproductive organs in a frontal section.

The external genitalia of the female, shown in Figure 16.10, are collectively called the **vulva**. It begins with the **mons pubis** (MAHNS PYOO-biss), the rounded area over the pubic symphysis that is covered in pubic hair after puberty. Posterior to the mons pubis are the **labia majora** and **labia minora** (LAY-bee-ah; singular: labium major and minus, respectively), folds that enclose an area called the **vestibule**. Within the vestibule we find the urethral and vaginal orifices. Anterior to the urethral orifice is the **clitoris** (KLIT-uhr-iss), which is composed of erectile tissue.

The **mammary glands** are not true reproductive organs—indeed, they are modified sweat glands and part of the integumentary system—but do have an associated reproductive function in milk production (Fig. 16.11; note that this figure shows a lactating mammary gland). Mammary glands are present in both males and females, but their anatomy is most appropriately discussed with female anatomy. Internally, mammary glands consist of 15–25 **lobes**, each of which has smaller **lobules** that contain milk-producing **alveoli**. Milk leaves the alveoli through **lactiferous ducts**, which join to form storage areas called **lactiferous sinuses**. Milk leaves through the **nipple**, which is surrounded by the darkly pigmented **areola** (aehr-ee-OH-lah).

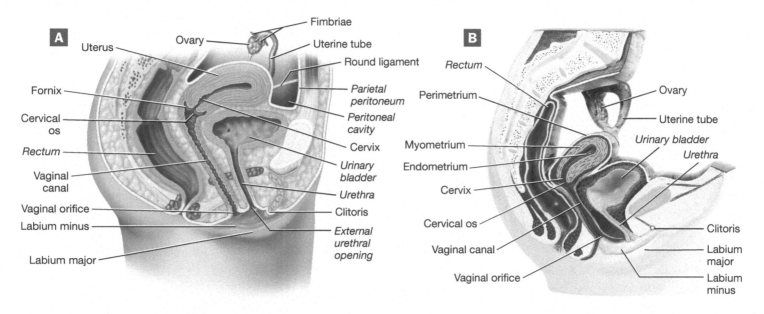

FIGURE **16.9** Midsagittal section of the female pelvis: **(A)** illustration; **(B)** anatomical model photo.

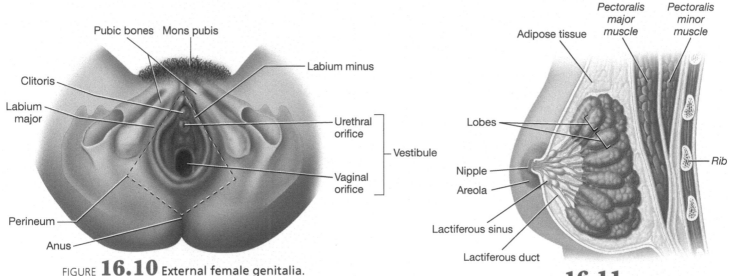

FIGURE **16.10** External female genitalia.

FIGURE **16.11** Structure of a lactating mammary gland.

Procedure 1 Model Inventory for the Female Reproductive System

Identify the following structures of the female reproductive system on models and diagrams, using this unit and your textbook for reference. As you examine the anatomical models and diagrams, record the name of the model and the structures you were able to identify on the model inventory in Table 16.2.

1. Ovary
   a. Ovarian ligament
   b. Broad ligament
   c. Suspensory ligament
2. Uterine tube
   a. Fimbriae
3. Uterus
   a. Round ligaments
   b. Fundus
   c. Body
   d. Cervix
      (1) Cervical os
   e. Layers
      (1) Endometrium
      (2) Myometrium
      (3) Perimetrium

4. Vagina
   a. Fornix
5. Vulva
   a. Mons pubis
   b. Labia majora
   c. Labia minora
      (1) Vestibule
      (2) Urethral orifice
      (3) Vaginal orifice
   d. Clitoris
6. Mammary glands
   a. Lobe
   b. Nipple
   c. Areola

TABLE **16.2** Model Inventory for the Female Reproductive System

Model/Diagram	Structures Identified

16

Exercise 16-3

Meiosis

MATERIALS
☐ Meiosis models
☐ Mitosis models
☐ Pop-bead chromosomes

As you may recall from Unit 3 (p. 43), somatic cells divide by a process called **mitosis**. During mitosis, a cell replicates its 23 pairs of homologous chromosomes and divides its DNA and organelles into two identical daughter cells. Each new diploid (DIH-ployd) cell is identical to the original cell. Gametogenesis (oogenesis and spermatogenesis), however, must proceed in a different way for two reasons:

- If each gamete were to have the same genetic material, we all would be genetically identical to our siblings.

- If each gamete were to have two pairs of chromosomes, our offspring would have four sets of chromosomes, and we would be tetraploid.

For these reasons, gametes undergo a process known as meiosis (MY-oh-sis) rather than mitosis.

In **meiosis**, gametes proceed through two rounds of cell division, and each gamete ends up with only one set of chromosomes, or a haploid cell (HAP-loyd). As you can see in Figure 16.12, meiosis begins in a manner similar to mitosis: The homologous chromosomes have replicated during the S phase of the cell cycle to yield pairs of sister chromatids. The cell then begins **meiosis** I. During middle to late prophase I, homologous chromosomes line up very tightly next to each other—a phenomenon called **synapsis**. Each synapse has two pairs of sister chromatids, and, therefore, the entire structure is called a **tetrad**. Within a tetrad, the homologous chromosomes overlap at points called **chiasmata** (ky-az-MAH-tah), or crossover. Near the end of prophase I, pieces of sister chromatids break and DNA is exchanged between homologous chromosomes at the points of crossover.

As meiosis I progresses, the homologous chromosomes line up along the equator of the cell, and then spindle fibers pull the homologous chromosomes to opposite poles of the cell. When telophase I and cytokinesis are complete, the resulting cells have 23 pairs of sister chromatids, and the amount of genetic material is halved because homologous chromosomes have been separated. The cells then proceed through the second round of division, **meiosis** II, during which the sister chromatids separate. At the end of telophase II, the result is four haploid gametes with only 23 individual homologous chromosomes.

Procedure 1 Model Meiosis

In the following procedure, you will model meiosis with either a set of cell meiosis models or with pop-bead chromosomes.

1 Obtain a set of meiosis and mitosis models.

2 Arrange the mitosis models in the correct order.

3 Now arrange the meiosis models in the correct order.

4 Compare the models of the two processes. List all the differences you can see between meiosis and mitosis in the space provided.

_____

_____

_____

_____

16

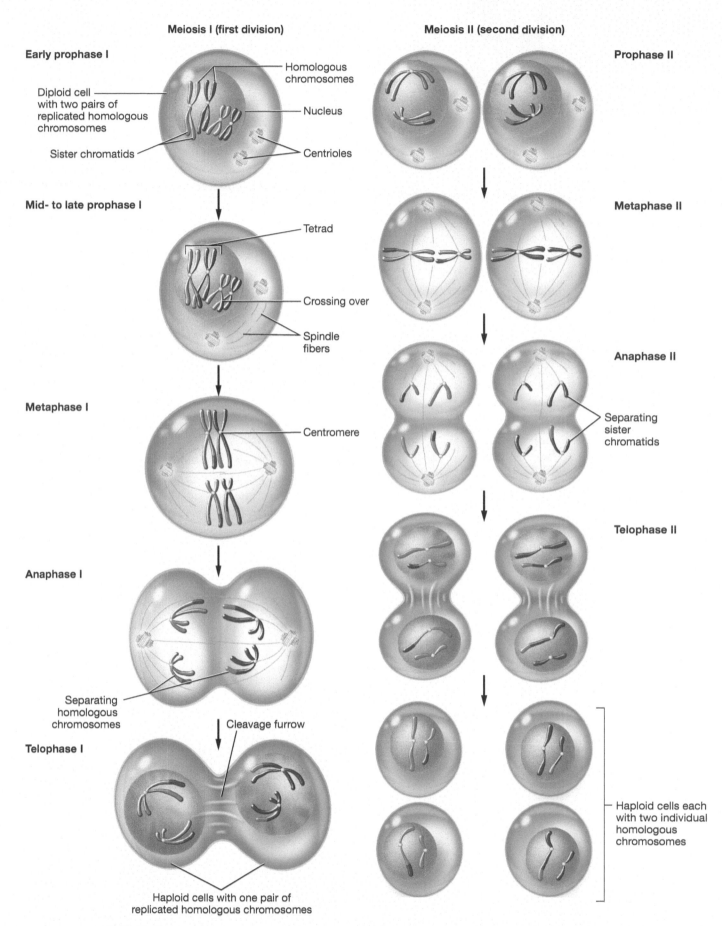

Meiosis I (first division)

Meiosis II (second division)

Early prophase I

Diploid cell with two pairs of replicated homologous chromosomes

Homologous chromosomes

Nucleus

Sister chromatids

Centrioles

Prophase II

Mid- to late prophase I

Tetrad

Crossing over

Spindle fibers

Metaphase II

Metaphase I

Centromere

Anaphase II

Separating sister chromatids

Anaphase I

Separating homologous chromosomes

Cleavage furrow

Telophase II

Telophase I

Haploid cells with one pair of replicated homologous chromosomes

Haploid cells each with two individual homologous chromosomes

FIGURE **16.12** Meiosis. For simplicity, a cell with only two pairs of homologous chromosomes is shown.

Exercise 16-4

Spermatogenesis and Oogenesis

Gametogenesis is the process of producing sperm cells by the testes (spermatogenesis) and oocytes by the ovaries (oogenesis). **Spermatogenesis**, shown in Figure 16.13, begins with **spermatogonia** (sper-mat-oh-GOH-nee-ah), stem cells located at the outer edge of the seminiferous tubules along the basement membrane. Before puberty, these cells undergo repeated rounds of mitosis to increase their numbers. As puberty begins, some spermatogonia divide into two different cells— one cell that remains in the basement membrane as a spermatogonium, and another cell that becomes a diploid **primary spermatocyte**. The primary spermatocyte then undergoes meiosis I and divides into two haploid **secondary spermatocytes** that migrate closer to the lumen of the tubule. The two secondary spermatocytes undergo meiosis II and give rise to four haploid **spermatids**.

Spermatids then move to the epididymis to mature into functional gametes by a process known as **spermiogenesis**. By the time the cells reach the end of the epididymis, they are mature sperm cells (or spermatozoa) that contain three parts: the **head**, in which the DNA resides; the **midpiece**, which contains an axoneme (the central strand of the flagellum) and mitochondria; and the **flagellum** (Fig. 16.14).

Like spermatogenesis, **oogenesis** (oh-oh-JEN-ih-sis) proceeds through meiosis to yield a haploid gamete, the **ovum** (Fig. 16.15). But the two processes differ in some notable ways:

■ *The number of oocytes is determined before birth.* During the fetal period, stem cells called **oogonia** (oh-oh-GOHN-ee-uh) undergo mitosis, increasing their numbers to between five and seven million in each ovary. This is the total number of oogonia a woman will ever produce. Oogonia and, later, primary oocytes degenerate during the fetal period and childhood, and by the time a girl reaches puberty, between 80,000 and 400,000 primary oocytes remain in each ovary. This is in sharp contrast to spermatogenesis, which begins at puberty and continues throughout a male's lifetime.

■ *Meiosis I begins during the fetal period but is arrested.* Still during the fetal period, the oogonia become encased in a ring of cells known as a **primordial follicle**, then enlarge and become **primary oocytes** (see the ovary in Figure 16.16). The primary oocytes begin prophase I but are arrested at this stage. Meiosis I does not resume until puberty.

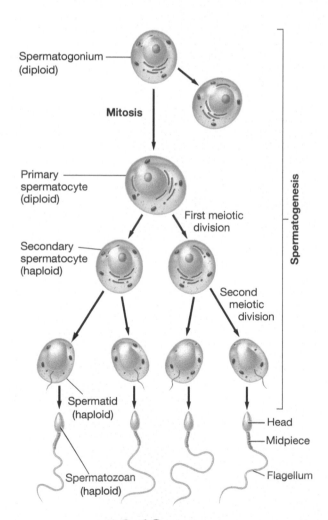

FIGURE **16.13** Spermatogenesis.

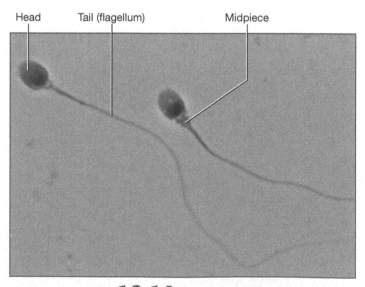

FIGURE **16.14** Mature sperm cells.

■ *The first meiotic division results in one secondary oocyte and one polar body*. During childhood, some primordial follicles enlarge and begin developing into **primary follicles**. At puberty, rising levels of hormones, such as **estrogens**, trigger the beginning of the ovarian cycle, during which some primary follicles enlarge to become **secondary follicles**. As the cycle progresses, estrogens and follicle-stimulating hormone (FSH) usually trigger only one of the secondary follicles to continue growing. When the secondary follicle develops a large, fluid-filled space called the **antrum**, it is considered to be a **vesicular follicle**. Under the influence of luteinizing hormone (LH), the primary oocyte then completes meiosis I to produce a **secondary oocyte** and a small bundle of nuclear material called a **polar body**. The formation of a polar body allows the oocyte to conserve cytoplasm, which will have to sustain the newly formed conceptus for several days if fertilization occurs.

■ *The secondary oocyte undergoes meiosis II only if fertilization takes place*. Rising levels of LH, called the *LH surge*, stimulate the secondary oocyte to begin meiosis II, but the process is arrested in metaphase II. The LH surge also stimulates rupture of the vesicular follicle and release of the secondary oocyte during ovulation. But the secondary oocyte only completes meiosis to form an ovum and a second polar body if fertilization occurs. If fertilization does not occur, the secondary oocyte degenerates.

Notice in Figure 16.16 that the ruptured vesicular follicle forms a structure called a **corpus luteum** (KOHR-pus LOO-tee-um), which is a temporary endocrine organ. The corpus luteum secretes hormones, mostly **progesterone** as well as some estrogen, that maintain the endometrial lining in case fertilization takes place. When the corpus luteum stops secreting hormones, it is degraded by macrophages, leaving behind a white scar known as the **corpus albicans**.

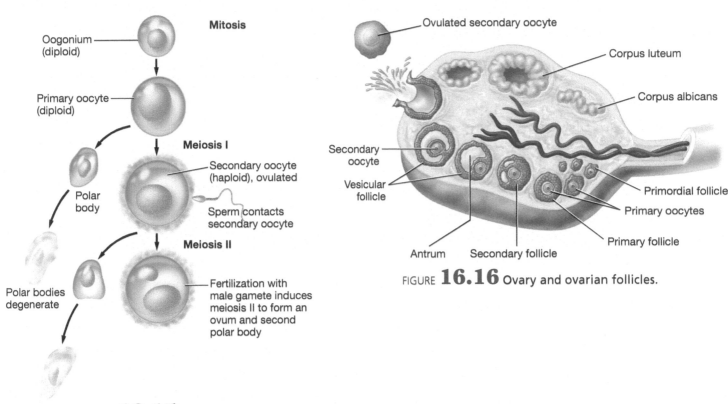

FIGURE **16.15** Oogenesis.

FIGURE **16.16** Ovary and ovarian follicles.

Procedure 1 Time to Trace!

In this procedure, you will be tracing the gametes from their production in the male and female gonads to the point at which they meet in the uterine tube, where fertilization takes place. This tracing involves the anatomy of the male and female reproductive systems as well as the steps of gametogenesis.

Step 1: Male Gamete

Trace the male gamete from its earliest stage—the spermatogonium in the seminiferous tubule—through the stages of spermatogenesis to a mature sperm. Then trace the mature sperm cell through the male reproductive tract until it exits the body and enters the uterine tube of the female reproductive tract. Trace the pathway using Figure 16.17, and also fill in the space provided.

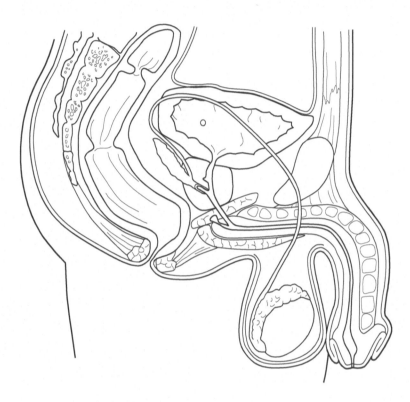

FIGURE **16.17** Male reproductive tract.

Start: Spermatogonium

_____

_____

_____

_____

_____

_____

_____ **End:** Uterine tube

Step 2: Female Gamete

Trace the female gamete from its earliest stage, the oogonium in a primordial follicle, to the time at which it is ovulated and enters the uterine tube. Trace the pathway using Figure 16.18, and also fill in the space provided.

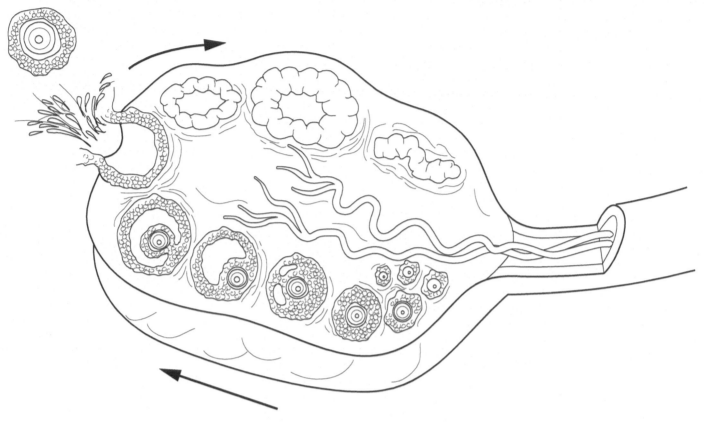

FIGURE **16.18** Ovary and ovarian follicles.

Start: Oogonium

_____

_____

_____

_____

_____

_____ **End:** Uterine tube

16

The first event in human development is **fertilization**, which generally takes place in the uterine tube when sperm cells encounter a secondary oocyte. Fertilization begins with a process called the **acrosomal reaction** (ak-roh-ZOH-muhl), in which the acrosomes in the heads of the sperm release digestive enzymes by exocytosis. The acrosomal enzymes digest the secondary oocyte's outer coverings.

When the oocyte's outer coverings have been cleared, the head and midpiece of a single sperm enter the oocyte. Note that although mitochondria from the sperm cell (in the midpiece) enter the secondary oocyte, they are destroyed. For this reason, offspring have only maternal mitochondria.

As the sperm cell enters, the secondary oocyte completes meiosis II, yielding an ovum and a second polar body. The sperm and ovum nuclei swell to become **pronuclei,** and their nuclear membranes rupture. The two sets of exposed chromosomes then fuse to form a single-celled structure called a **zygote** (ZY-goh't; Fig. 16.19).

The zygote, also called a **conceptus** (kun-SEPT-uhs), undergoes multiple changes during the period of time known as **gestation,** the 40-week period extending from the mother's last menstrual period to birth. The changes that the conceptus undergoes include an increase in cell number, cellular differentiation, and the development of organ systems.

The increase in cell number begins about 30 hours after fertilization, as the zygote travels down the uterine tube. At this time, a process called **cleavage** occurs, in which the cell undergoes a series of mitotic divisions to produce two cells, then four, and so on. As cleavage begins, the conceptus starts to migrate from the uterine tube to the uterus. At about day 3, the conceptus reaches the uterus as a 16-cell ball called a **morula** (MOHR-yoo-luh, which means "little mulberry"). The morula floats around the uterus for another 2 to 3 days and continues to divide until it becomes a hollow sphere called a **blastocyst** (BLAST-oh-sist). Notice in Figure 16.19 that the blastocyst has two populations of cells: the rounded **inner cell mass** (also called the *embryoblast*) and an outer layer of cells called the **trophoblast** (TROHF-oh-blast).

Around day 6, the blastocyst adheres to the endometrium and begins the process of **implantation.** The **implanting blastocyst** secretes digestive enzymes that eat away the endometrial lining. The endometrium reacts to the injury by growing over and enclosing the blastocyst. The process of implantation is generally complete by the second week after fertilization, about the time a woman's menstrual period would begin if fertilization had not taken place.

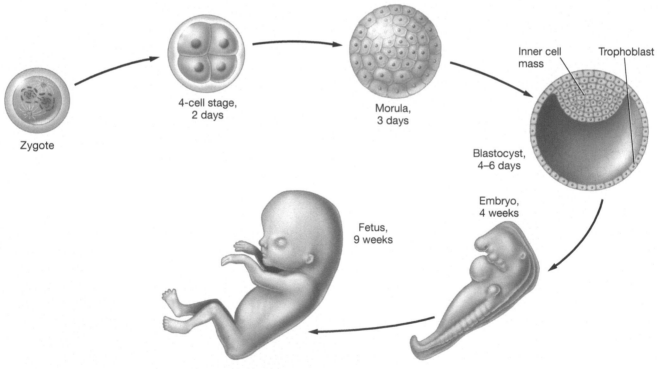

Zygote

4-cell stage, 2 days

Morula, 3 days

Inner cell mass   Trophoblast

Blastocyst, 4–6 days

Embryo, 4 weeks

Fetus, 9 weeks

FIGURE **16.19** Stages of development.

Procedure 1 Time to Trace!

Trace the events of development from fertilization to implantation. In your tracing, you will do two things: (1) describe each of the major changes that the conceptus undergoes from the time that fertilization takes place until the time that it implants; and (2) trace the conceptus' location as these events occur. Trace the pathway using Figure 16.20, and also fill in the space provided.

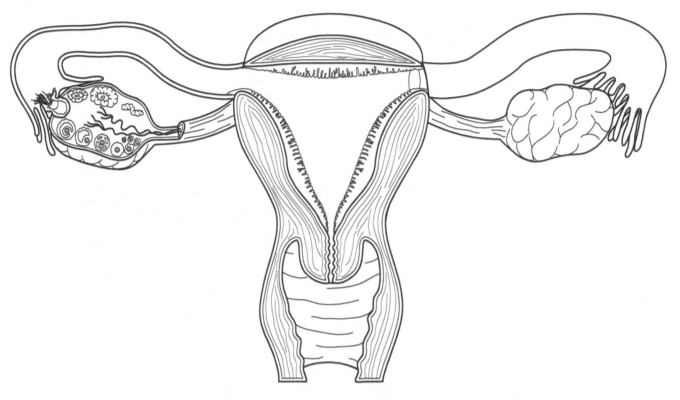

FIGURE **16.20** Female reproductive tract.

Start: Sperm and ovum meet in uterine tube

_____

_____

_____

_____

_____

_____

_____ **End:** Implanted blastocyst

Exercise 16-6

Embryogenesis, Fetal Development, and Fetal Cardiovascular Anatomy

MATERIALS

- ❑ Sequence of embryonic development model
- ❑ Female reproductive system with fetus model
- ❑ Fetal circulation model

As the blastocyst implants, it begins undergoing a process known as **embryogenesis** (em-bree-oh-JEN-ih-sis) during which cellular differentiation takes place. During embryogenesis, the inner cell mass differentiates into three primary tissue layers known as **germ layers**, from which all other tissues arise. The outermost germ layer is called **ectoderm**, the middle germ layer is called **mesoderm**, and the innermost germ layer is called **endoderm**. These germ layers are formed by the end of the second week after fertilization, at which point the conceptus is considered an **embryo**.

The next six weeks of the embryonic period are marked by differentiation of the germ layers into rudimentary organ systems, formation of extraembryonic membranes, and the development of a structure called the *placenta* (Fig. 16.21). The first extraembryonic membrane to develop is the **yolk sac**, which serves as a source of cells for the digestive tract, the first blood cells and blood vessels, and *germ cells* (which are the precursors to gametes). Next, the innermost membrane, called the **amnion** (AM-nee-ahn), develops. The amnion completely surrounds the conceptus and suspends it in **amniotic fluid** within the **amniotic cavity**.

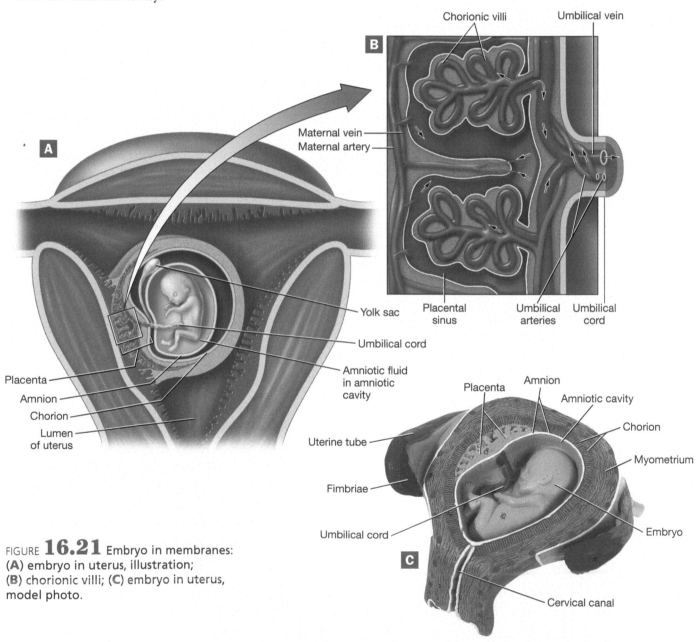

FIGURE **16.21** Embryo in membranes: (**A**) embryo in uterus, illustration; (**B**) chorionic villi; (**C**) embryo in uterus, model photo.

16

As the placenta and extraembryonic membranes form, the primitive organ systems develop. This process completes by the end of the eighth week after fertilization, at which point the conceptus is considered a **fetus**. For the duration of gestation, the organ systems become progressively more specialized and the fetus continues to grow and develop.

The **placenta** (plah-SIN-tuh) is the structure derived from embryonic and maternal tissues that provides oxygen and nutrients to the conceptus. It begins to form at day 11, when the trophoblast forms the outermost extraembryonic membrane, the **chorion** (KOHR-ee-ahn), which is the fetal portion of the placenta. As you can see in Figure 16.21B, the chorion develops elaborate projections called the **chorionic villi** (KOHR-ee-ahn-ik VILL-aye), which eat into uterine blood vessels to create the **placental sinus**, a space filled with maternal blood. Nutrients and oxygen delivered by *maternal arteries* diffuse from the placental sinus to the chorionic villi and are delivered to the conceptus by the large **umbilical vein**. Wastes are drained from the conceptus via the paired **umbilical arteries**, which drain into the chorionic villi, diffuse into the placental sinuses, and finally back into the *maternal veins*. The umbilical arteries and vein travel to and from the conceptus through the **umbilical cord**.

The fetus' cardiovascular system differs significantly from that of the neonate (Fig. 16.22). Within the fetal cardiovascular system are three **shunts** that bypass the relatively inactive liver and lungs and reroute the blood to other, more metabolically active organs. The three shunts include the following:

1. **Ductus venosus.** The **ductus venosus** (veh-NOH-suss) is a shunt that bypasses the liver. The umbilical vein delivers a small amount of blood to the liver but sends the majority of its oxygenated blood through the ductus venosus to the fetal inferior vena cava. This shunt closes about the time of birth and leaves behind a remnant called the *ligamentum venosum*.

2. **Foramen ovale.** The **foramen ovale** (oh-VAL-ay) is a hole in the interatrial septum that shunts blood from the right atrium to the left atrium. It allows blood to bypass the right ventricle and pulmonary circuit and go straight to the left heart and systemic circuit. The foramen ovale closes shortly after birth and leaves behind a dent in the interatrial septum called the *fossa ovalis*.

3. **Ductus arteriosus.** The **ductus arteriosus** (ahr-TEER-ee-oh-suss) is a vascular bridge between the pulmonary trunk and the aorta. It is a second shunt that allows blood to bypass the pulmonary circuit. When the ductus arteriosus closes after birth, it leaves behind a remnant called the *ligamentum arteriosum*.

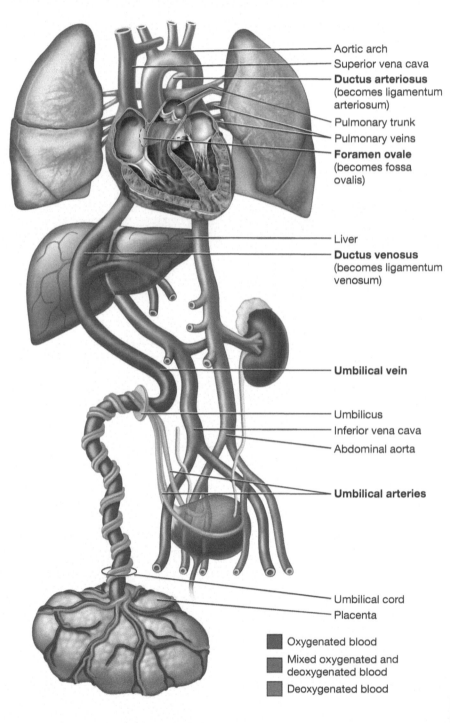

- Aortic arch
- Superior vena cava
- **Ductus arteriosus** (becomes ligamentum arteriosum)
- Pulmonary trunk
- Pulmonary veins
- **Foramen ovale** (becomes fossa ovalis)
- Liver
- **Ductus venosus** (becomes ligamentum venosum)
- **Umbilical vein**
- Umbilicus
- Inferior vena cava
- Abdominal aorta
- **Umbilical arteries**
- Umbilical cord
- Placenta

■ Oxygenated blood
■ Mixed oxygenated and deoxygenated blood
■ Deoxygenated blood

FIGURE **16.22** Fetal cardiovascular anatomy.

16

Notice how the pattern of oxygenation differs in the conceptus versus in the neonate. The umbilical vein delivers oxygenated blood to the inferior vena cava, but the blood doesn't remain fully oxygenated for long: It soon begins to mix with deoxygenated blood coming in from the inferior vena cava, and it mixes with more deoxygenated blood from the superior vena cava when it enters the right atrium. For this reason, the blood traveling through the systemic arterial circuit of the conceptus is never fully oxygenated and is represented as purple in Figure 16.22 instead of red.

Procedure 1 Model Inventory for Fetal Development

Identify the following structures of development on models and diagrams, using this unit and your textbook for reference. As you examine the anatomical models and diagrams, record the name of the model and the structures you were able to identify on the model inventory in Table 16.3.

1. Stages of development
   a. Zygote
   b. Morula
   c. Blastocyst
   d. Implanted blastocyst
   e. Embryo
   f. Fetus

2. Fetal membranes
   a. Yolk sac
   b. Amnion
      (1) Amniotic cavity
      (2) Amniotic fluid
   c. Chorion
      (1) Chorionic villi

3. Vascular structures
   a. Placenta
   b. Umbilical cord
      (1) Umbilical vein
      (2) Umbilical arteries
   c. Ductus venosus
   d. Foramen ovale
   e. Ductus arteriosus

TABLE **16.3** Model Inventory for the Stages of Development and Fetal Structures

Model/Diagram	Structures Identified

1 Label the following structures on **Figure 16.23.**

❑ Corpora cavernosa   ❑ Glans penis

❑ Corpus spongiosum   ❑ Prostate gland

❑ Ejaculatory duct   ❑ Scrotum

❑ Epididymis   ❑ Ductus deferens

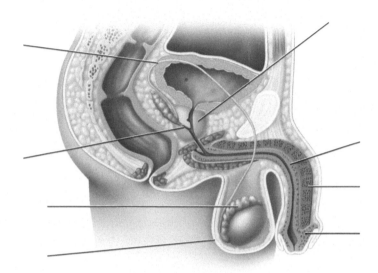

FIGURE **16.23** Midsagittal section through the male pelvis.

2 Label the following structures on **Figure 16.24.**

❑ Cervical os

❑ Labium major

❑ Labium minus

❑ Ovary

❑ Uterine tube

❑ Uterus

❑ Vaginal canal

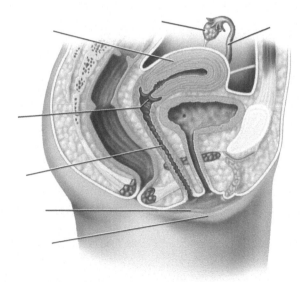

FIGURE **16.24** Midsagittal section through the female pelvis.

16

3 Label the following structures on **Figure 16.25**.

- ❏ Body of uterus
- ❏ Cervical canal
- ❏ Endometrium
- ❏ Fimbriae
- ❏ Fundus of uterus
- ❏ Myometrium
- ❏ Ovary
- ❏ Perimetrium
- ❏ Uterine tube

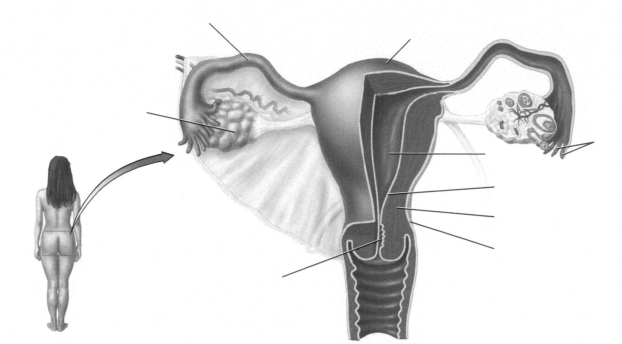

FIGURE **16.25** Posterior view of the female reproductive organs in a frontal section.

4 *True/False:* Mark the following statements as true (T) or false (F). If the statement is false, correct it to make it a true statement.

_____ a. Gametes result from two rounds of cell division.

_____ b. Meiosis results in diploid cells.

_____ c. Homologous chromosomes overlap at points called chiasmata or crossover.

_____ d. The DNA replicates prior to the start of meiosis II.

5 Number the following events of spermatogenesis in the proper order, with number 1 next to the first event and number 5 next to the final event.

_____ Primary spermatocytes undergo meiosis I to yield two secondary spermatocytes.

_____ Spermatids migrate to the epididymis to undergo spermiogenesis.

_____ The two secondary spermatocytes undergo meiosis II to produce four haploid spermatids.

_____ Spermatogonia divide into more spermatogonia and primary spermatocytes.

_____ Spermatogonia undergo repeated rounds of mitosis.

6 Describe the ways in which oogenesis differs from spermatogenesis.

_____

_____

_____

_____

7 The condition *benign prostatic hypertrophy*, in which the prostate gland is enlarged, often results in urinary retention—the inability to completely empty the bladder. Considering the anatomy of the male genitourinary tract, explain this symptom.

_____

_____

_____

_____

8 Circe has been unable to get pregnant for two years in spite of fertility treatments, and so she is undergoing an exploratory surgical procedure to look for possible anatomical causes of her infertility. During the procedure, Circe's physician notes that her fimbriae are extremely short and face laterally rather than surrounding the ovaries. How does this explain her infertility?

_____

_____

_____

_____

9 Which of the following is *not* one of the three basic changes a conceptus undergoes during development?
   a. An increase in cell number.
   b. Cellular differentiation.
   c. Development of organ systems.
   d. Neoplasia.

10 Explain why diseases caused by mutations in mitochondrial DNA are inherited exclusively from one's mother.

_____

_____

_____

16

11 *True/False:* Mark the following statements as true (T) or false (F). If the statement is false, correct it to make it a true statement.

_____ a. The umbilical arteries carry oxygenated blood to the conceptus.

_____ b. The ductus arteriosus is a vascular shunt from the pulmonary trunk to the aorta.

_____ c. The ductus arteriosus and foramen ovale bypass the fetal lungs.

_____ d. The ductus venosus is a vascular bridge between the umbilical artery and the inferior vena cava.

12 Label **Figure 16.26** with the terms below.
- ❑ Ductus arteriosus
- ❑ Ductus venosus
- ❑ Foramen ovale
- ❑ Placenta
- ❑ Umbilical arteries
- ❑ Umbilical vein

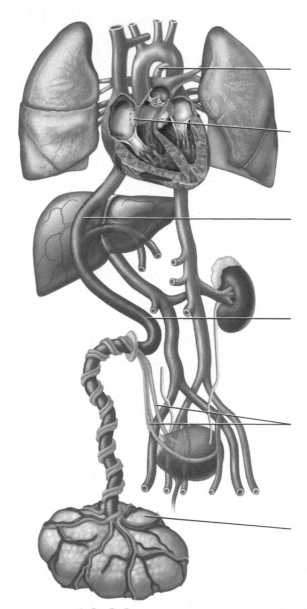

FIGURE **16.26** Fetal cardiovascular anatomy.

13 *Placental abruption* is a condition in which the placenta is pulled away and separated from the wall of the uterus. Why could this prove life-threatening for both the mother and the fetus?

_____

_____

16

_____

Congratulations!

*You've done what you may have thought was impossible—
completed this manual! I truly hope it has been fun
(at least a little!) and that you have enjoyed it as much
as I have. Best wishes with your future plans,
down whichever path they may take you.*

Photo Credits

Morton Publishing expresses thanks to the following sources for allowing us to use their photos:

Unit 1

Opener: Gustoimages/Science Source

Fig. 1.6D: CC Studio/Science Source

Fig. 1.9, Fig. 1.10: Kent M. Van De Graaff and John Crawley, *A Photographic Atlas for the Anatomy and Physiology Laboratory,* 6e ©Morton Publishing

Fig. 1.13A: Du Cane Medical Imaging/Science Source

Fig. 1.13B–C: Living Art Enterprises, LLC/Science Source

Unit 2

Opener: Laguna Design/Science Source

Fig. 2.2: Gusto/Science Source

Fig. 2.4: Justin Moore

Fig. 2.5: GIPhotoStock/Science Source

Unit 3

Opener: Rafe Swan/Science Source

Fig. 3.2: Courtesy of Olympus America

Fig. 3.3: Dr. Gopal Murti/Science Source

Fig. 3.5B: Dr. David Furness, Keele University/Science Source

Fig. 3.8: M. I. Walker/Science Source

Fig. 3.10A–C: David M. Phillips/Science Source

Fig. 3.12: Michael Abbey/Science Source

Fig. 3.13: Stem Jems/Science Source

Unit 4

Opener: CMEABG-UCBL/Science Source

Fig. 4.3A: Biophoto Assoc./Science Source

Fig. 4.3B: Biophoto Assoc./Science Source

Fig. 4.3C: Science Stock Photography/Science Source

Fig. 4.3D: Biophoto Assoc./Science Source and Eye of Science/Science Source

Fig. 4.4A: David M. Phillips/Science Source

Fig. 4.4B, 4.13D: Biophoto Assoc./Science Source

Fig. 4.4C: Steve Gschmeissner/Science Source

Fig. 4.5A: Biophoto Assoc./Science Source

Fig. 4.5B: Sci Stock Photography/Science Source

Fig. 4.5C, 4.13C: Alvin Telser/Science Source and Steve Gschmeissner/Science Source

Fig. 4.6A: Biophoto Assoc./Science Source and Steve Gschmeissner/Science Source

Fig. 4.6B: Biophoto Assoc./Science Source and David M. Phillips/Science Source

Fig. 4.6C: Biophoto Assoc./Science Source

Fig. 4.7A: Erin Amerman

Fig. 4.7B: M. I. Walker/Science Source

Fig. 4.7C: Chuck Brown/Science Source

Fig 4.8: Steve Gschmeissner/Science Source

Fig. 4.9: Biophoto Assoc./Science Source

Fig. 4.10A, 4.13A: Eric Grave/Science Source

Fig. 4.10B: Manfred Kage/Science Source

Fig. 4.10C: Biophoto Assoc./Science Source

Fig. 4.10D: Getty Images

Fig. 4.11: M. I. Walker/Science Source

Fig. 4.12: Biophoto Assoc./Science Source; Southern Illinois University/Science Source

Unit 5

Opener: Steve Gschmeissner/Science Source

Fig. 5.2: Biophoto Assoc./Science Source

Fig. 5.3B: BSIP/Science Source

Fig. 5.5A: Garry Delong/Science Source

Fig. 5.5B: Andrew Syred/Science Source

Unit 6

Opener: Professor Pietro M. Motta/Science Source

Fig. 6.9: Anatomical Travelogue/Science Source

Fig. 6.10A: Living Art Enterprises, LLC/Science Source

Fig. 6.10B: Biophoto Assoc./Science Source

Fig. 6.11B: VideoSurgery/Science Source

Fig. 6.12B: David Bassett/Science Source

Fig. 6.13B: Getty Images

Fig. 6.14B: Paul Rapson/Science Source

Fig. 6.15B: VideoSurgery/Science Source

Fig. 6.120A, B: Martin Shields/Science Source

Fig. 6.309B: VideoSurgery/Science Source

Fig. 6.31B: VideoSurgery/Science Source

Fig. 6.35B: Kent M. Van De Graaff and John Crawley, *A Photographic Atlas for the Anatomy and Physiology Laboratory*, 6e ©Morton Publishing; Video Surgery/Science Source

Fig. 6.36B: Kent M. Van De Graaff and John Crawley, *A Photographic Atlas for the Zoology Laboratory*, 6e ©Morton Publishing; VideoSurgery/Science Source

Fig. 6.40A: VideoSurgery/Science Source

Fig. 6.47: Dr. P. Marazzi/Science Source

Unit 7

Opener: SPL/Science Source

Fig. 7.5B: VideoSurgery/Science Source

Fig. 7.7A: Eric Grave/Science Source

Fig. 7.7B: Michael Leboffe

Fig. 7.9B: Geoff Tompkinson/Science Source

Fig. 7.15: CNRI/Science Source

Unit 8

Opener: Steve Gschmeissner/Science Source

Fig. 8.7: M. I. Walker/Science Source

Fig. 8.13: Biophoto Assoc./Science Source

Fig. 8.14B: Dr. Colin Chumbley/Science Source

Fig. 8.14C: BSIP/Science Source

Fig. 8.17, Fig. 8.18, Fig. 8.19: Kent M. Van De Graaff and John Crawley, *A Photographic Atlas for the Anatomy and Physiology Laboratory*, 6e ©Morton Publishing

Fig. 8.21B: Martin M. Rotker/Science Source

Fig. 8.28: BSIP/Science Source

Fig. 8.33: Ph. Gerbier/Science Source

Unit 9

Opener: Science Source

Fig. 9.5, Fig. 9.6, Fig. 9.7: Kent M. Van De Graaff and John Crawley, *A Photographic Atlas for the Anatomy and Physiology Laboratory*, 6e ©Morton Publishing

Unit 10

Opener: Biophoto Assoc./Science Source

Unit 11

Opener: Science Stock Photography/Science Source

Fig. 11.9, Fig. 11.10, Fig. 11.11: Justin Moore

Fig. 11.26, Fig. 11.27, Fig. 11.28: BSIP/Science Source

Fig. 11.35: Andy Levin/Science Source

Unit 12

Opener: SPL/Science Source

Fig. 12.2: Biophoto Assoc./Science Source

Fig. 12.3: Biophoto Assoc./Science Source and Michael Ross/Science Source

Fig. 12.5: Justin Moore

Fig. 12.6: BSIP/Science Source

Fig. 12.11B: BSIP/Science Source

Fig. 12.13: Biophoto Assoc./Science Source

Unit 13

Opener: Eye of Science/Science Source

Fig. 13.2B, 13.4B: BSIP/Science Source

Fig. 13.4C, Fig. 13.5B: VideoSurgery/Science Source

Fig. 13.6B: BSIP/Science Source

Fig. 13.9: Martyn F. Chillmaid/Science Source

Fig. 13.10A, B: Justin Moore

Unit 14

Opener: SPL/Science Source

Fig. 14.3B: BSIP/Science Source

Fig. 14.3C: L. Bassett/Visuals Unlimited/Getty Images

Fig. 14.5B: BSIP/Science Source

Fig. 14.5C: L. Bassett/Visuals Unlimited/Getty Images

Fig. 14.9A: BSIP/Science Source

Fig. 14.15, 14.16: Michael Leboffe

Unit 15

Opener: SPL/Science Source

Fig. 15.6C: BSIP/Science Source

Fig. 15.9C: BSIP/Science Source

Fig. 15.12A, B: VideoSurgery/Science Source

Fig. 15.16: Scott Camazine/Science Source

Unit 16

Opener: Science Picture Co/Science Source

Fig. 16.4B: Julie Dermansky/Science Source

Fig. 16.9B: Gregory Davies/Science Source

Fig. 16.14: Michael Abbey/Science Source

Fig. 16.21C: BSIP/Science Source

Index